Remote Sensing of Vegetation: Mapping, Trend Analysis, and Drivers of Change

Remote Sensing of Vegetation: Mapping, Trend Analysis, and Drivers of Change

Guest Editors

Sadegh Jamali
Torbern Tagesson
Feng Tian
Meisam Amani
Per-Ola Olsson
Arsalan Ghorbanian

MDPI

Basel • Beijing • Wuhan • Barcelona • Belgrade • Novi Sad • Cluj • Manchester

Guest Editors

Sadegh Jamali
Department of Technology and Society
Lund University
Lund
Sweden

Torbern Tagesson
Department of Physical Geography and Ecosystem Sciences
Lund University
Lund
Sweden

Feng Tian
School of Remote Sensing and Information Engineering
Wuhan University
Wuhan
China

Meisam Amani
Wood Environment & Infrastructure Solutions
Ottawa
Canada

Per-Ola Olsson
Department of Physical Geography and Ecosystem Sciences
Lund University
Lund
Sweden

Arsalan Ghorbanian
Department of Photogrammetry and Remote Sensing
K. N. Toosi University of Technology
Tehran
Iran

Editorial Office

MDPI AG
Grosspeteranlage 5
4052 Basel, Switzerland

This is a reprint of the Special Issue, published open access by the journal *Remote Sensing* (ISSN 2072-4292), freely accessible at: https://www.mdpi.com/journal/remotesensing/special_issues/vegetation_mapping_trend.

For citation purposes, cite each article independently as indicated on the article page online and as indicated below:

Lastname, A.A.; Lastname, B.B. Article Title. *Journal Name* **Year**, *Volume Number*, Page Range.

ISBN 978-3-7258-5969-6 (Hbk)
ISBN 978-3-7258-5970-2 (PDF)
https://doi.org/10.3390/books978-3-7258-5970-2

Contents

About the Editors

Sadegh Jamali

Sadegh Jamali received his Ph.D. from Lund University in 2014, where he specialized in analyzing vegetation trends using data from Earth observation (EO) satellites. He is currently a senior lecturer at Lund University. His research focuses on land surface change analysis, with particular emphasis on developing robust methods to detect trends and breakpoints in satellite data time series. He has led and contributed to several research projects that utilize EO data for drought monitoring, forest damage assessment, and high-resolution land use and land cover mapping. Additionally, he has a strong interest in applying machine learning and deep learning techniques to enhance land surface change detection using sensor data.

Torbern Tagesson

Torbern Tagesson was awarded his Ph.D. in 2011 from the Department of Physical Geography and Ecosystem Sciences of Lund University, where he currently holds a position as a Researcher. His main research interests of lie in the interaction of ecosystems with environmental change and include global change studies, climate change, estimates of vegetation productivity, carbon stocks and land–atmosphere exchange processes.

Feng Tian

Feng Tian is a professor of remote sensing at Wuhan University. His research uses Earth observation data for vegetation studies, including mapping and estimating vegetation functioning (e.g., phenology, productivity and carbon stock), revealing vegetation spatial and temporal dynamics, and exploring interactions between vegetation and the climate. He uses time series of optical (e.g., AVHRR, MODIS, Landsat, Sentinel-2) and passive microwave (e.g., VOD from SMOS and AMSR) data at global and local scales, with a particular focus on Africa and the Tibetan Plateau.

Meisam Amani

Meisam Amani received a B.Eng. degree in Geomatics Engineering from the University of Tabriz, Tabriz, Iran, in 2012, an M.Eng. degree in Remote Sensing Engineering from the K.N. Toosi University of Technology, Tehran, Iran, in 2014, and a Ph.D. in Electrical Engineering from Memorial University of Newfoundland, St. John's, NL, Canada, in 2018. He is currently an Earth Observation Research Scientist at Natural Resources Canada. Dr. Amani was an Associate Editor of *IEEE JSTARS* and a Guest Editor for three Special Issues in the journal *Remote Sensing*. He also serves as a regular Reviewer in about 15 international remote sensing journals. He was the recipient of the prestigious Professional Engineers and Geoscientists Newfoundland and Labrador Environmental Award in 2020 due to his contribution to wetland mapping in Canada using advanced machine learning and Big Data processing algorithms. His research works include over 120 peer-reviewed journal and conference papers.

Per-Ola Olsson

Per-Ola Olsson is a researcher at the Department of Physical Geography and Ecosystem Science, Lund University in Sweden. He holds a Master's degree in Geomatics, and during his PhD, he developed methods for the detection of insect disturbances in forests with satellite-based remote sensing, including methods for near-real-time monitoring and modelling the impact of carbon uptake from insect outbreaks. Dr. Olsson has a broad range of research interests, including UAV and satellite-based remote sensing, with a focus on time-series analyses, disturbance detection and machine learning as well as more GIS related 3D city modelling and standardization projects. Since his PhD, he has worked with UAV data, with a focus on radiometric correction of images, and satellite data for agriculture and forestry applications. He has been leading a project about remote sensing and risk mapping of European spruce bark beetle outbreaks, developing methods for the detection of bark-beetle-attacked trees with Sentinel-2 data, as well as risk mapping of outbreaks. In 3D city modelling, he is involved in projects where a national specification for 3D city models in Sweden has been developed. Recently he has also been involved in health-related studies, where he is working with urban heat islands and satellite-derived land surface temperature.

Arsalan Ghorbanian

Arsalan Ghorbanian received his B.Eng. degree in Geodesy and Geomatics Engineering (2016), M.Eng. degree in Remote Sensing (2018), and Ph.D. degree in Remote Sensing (2023) from K. N. Toosi University of Technology, Tehran, Iran. In 2022, he was a Visiting Ph.D. Researcher at Lund University, Sweden, supported by the Erasmus+ International Credit Mobility (ICM) scholarship. His research interests focus on satellite image and video processing using statistical, machine, and deep learning approaches, with applications in land use/land cover (LULC) mapping, drought monitoring, geo–big data analytics, wetland and mangrove ecosystem studies, time-series analysis, and, recently, interdisciplinary research at the intersection of remote sensing and public health. He has worked with diverse remote sensing data, including multispectral, hyperspectral, synthetic aperture radar, and thermal infrared imagery, and has authored and co-authored over 40 peer-reviewed publications in these fields. He has served as a Guest Editor for two Special Issues of the journal *Remote Sensing*, and since 2022, he has been a member of the Advisory Board of the journal *Ecological Indicators*.

Preface

Vegetation forms the living foundation of the Earth's ecological systems, serving as the critical nexus between the biosphere, atmosphere, hydrosphere, and lithosphere. As the primary engine of biogeochemical cycles, it regulates global climate patterns through carbon sequestration, generates life-sustaining oxygen, and maintains soil integrity while modulating surface energy budgets. These indispensable ecosystem services make vegetation monitoring not just an academic pursuit but a vital necessity for achieving the United Nations Sustainable Development Goals (SDGs). In our current era of accelerating climate change and expanding human influence, understanding vegetation dynamics through long-term observation has become particularly crucial. Such monitoring reveals how plant communities adapt to—or suffer from—the combined stresses of anthropogenic activities and natural environmental shifts, providing essential baseline data for conservation planning and climate adaptation strategies.

Recent advances in remote sensing have revolutionized vegetation monitoring, enabling precise mapping of vegetation structure, composition, and physiological status across scales, while robust time-series analysis distinguishes gradual environmental changes from sudden disturbances and identifies their underlying drivers. This Reprint showcases how these technological breakthroughs are being applied to address pressing environmental challenges, featuring cutting-edge research in vegetation mapping, change detection, and driver attribution. By presenting these innovative approaches, we aim to bridge the gap between technological potential and practical solutions, fostering interdisciplinary collaboration among researchers, policymakers, and practitioners to protect and sustainably manage the Earth's vital vegetation resources

Sadegh Jamali, Torbern Tagesson, Feng Tian, Meisam Amani, Per-Ola Olsson, and Arsalan Ghorbanian
Guest Editors

remote sensing

MDPI

Article

Enhanced Detection of Artisanal Small-Scale Mining with Spectral and Textural Segmentation of Landsat Time Series

Alejandro Fonseca [1], Michael Thomas Marshall [2,*] and Suhyb Salama [3]

[1] IABG Industrieanlagen-Betriebsgesellschaft, Hermann-Reicheltelt-Straße 3, 01109 Dresden, Germany
[2] Department of Natural Resources, Faculty of Geo-Information Science and Earth Observation, University of Twente, P.O. Box 217, 7500 AE Enschede, The Netherlands
[3] Department of Water Resources, Faculty of Geo-Information Science and Earth Observation, University of Twente, P.O. Box 217, 7500 AE Enschede, The Netherlands
* Correspondence: m.t.marshall@utwente.nl

Abstract: Artisanal small-scale mines (ASMs) in the Amazon Rainforest are an important cause of deforestation, forest degradation, biodiversity loss, sedimentation in rivers, and mercury emissions. Satellite image data are widely used in environmental decision-making to monitor changes in the land surface, but ASMs are difficult to map from space. ASMs are small, irregularly shaped, unevenly distributed, and confused (spectrally) with other land clearance types. To address this issue, we developed a reliable and efficient ASM detection method for the Tapajós River Basin of Brazil—an important gold mining region of the Amazon Rainforest. We enhanced detection in three key ways. First, we used the time-series segmentation (LandTrendr) Google Earth Engine (GEE) Application Programming Interface to map the pixel-wise trajectory of natural vegetation disturbance and recovery on an annual basis with a 2000 to 2019 Landsat image time series. Second, we segmented 26 textural features in addition to 5 spectral features to account for the high spatial heterogeneity in ASM pixels. Third, we trained and tested a Random Forest model to detect ASMs after eliminating irrelevant and redundant features with the Variable Selection Using Random Forests "ensemble of ensembles" technique. The out-of-bag error and overall accuracy of the final Random Forest was 3.73 and 92.6%, which are comparable to studies mapping large industrial mines with the normalized difference vegetation index (NDVI) and LandTrendr. The most important feature in our study was NDVI, followed by textural features in the near and shortwave infrared. Our work paves the way for future ASM regulation through large area monitoring from space with free and open-source GEE and operational satellites. Studies with sufficient computational resources can improve ASM monitoring with advanced sensors consisting of spectral narrow bands (Sentinel-2, Environmental Mapping and Analysis Program, PRecursore IperSpettrale della Missione Applicativa) and deep learning.

Keywords: deforestation; forest degradation; Amazon; remote sensing; machine learning; LandTrendr; Google Earth Engine

1. Introduction

Global deforestation has slowed in the past few decades because of widespread afforestation campaigns in high-income countries outside the tropics [1]. Forest loss in low-income tropical countries offsets the gains. Countries in South America have experienced the largest declines despite improved forest management and the expansion of protected areas [2]. Gold mining near protected areas caused much of this loss because of a surge in gold prices leading up to the global economic crisis of 2008 [3]. The analysis of satellite image data plays an integral role in monitoring industrial mines because it captures surface conditions consistently and frequently over large areas through time [4].

The detection of artisanal small-scale mines (ASMs) with satellite image data is challenging for three main reasons [5]. First, pixels are mixed at the ground sampling distance

(GSD) of moderate-resolution (~30 m) monitoring satellites. Second, ASMs are irregularly shaped and unevenly distributed. Third, the spectral response is like that of other land clearance types.

Studies involving the pixel-wise detection of ASMs with satellite image data typically consist of two main phases: (i) the transformation of spectral information into meaningful environmental indicators and (ii) classification or probabilistic mapping derived from these indicators with simple decision trees or more advanced data mining techniques. Image transformation compensates for mixed pixel effects and the irregular shape of the ASMs. The most common transformations are vegetation indices (VIs) such as the normalized difference vegetation index (NDVI) [6]. Other transformations include orthogonal rotation with principal components analysis [7] or spectral mixture analysis [3,8]. Isidro et al. [5] fused higher spatial resolution panchromatic imagery with coarser resolution multispectral broadband imagery (i.e., pan-sharpening) to improve ASM detection. Forkuor et al. [9] developed simple thresholds of ASM occurrence from summary statistics of Sentinel-1 backscatter metrics. Spatial image segmentation has also been widely used to extract ASM boundaries from spectral information [10]. More advanced data mining techniques determine ASM occurrence without or in concert with spatial image segmentation. These include support vector machines [11,12], classification and regression trees [13], canonical correlation analysis [14], and deep convolutional neural networks [15]. Distance to river networks, soil properties, and topographic features are sometimes included as covariates because of the close proximity of ASMs to water bodies [16].

The footprints of ASMs extend beyond a single pixel and consist of vegetation, pits, mounds of deposited soil, and standing pools of water, mercury, and amalgamated gold [3]. This makes them spatially heterogeneous. The pixel-wise detection of ASMs with spectral information could therefore be improved by considering spatial dependencies in neighboring pixels [17] through texture analysis. Textural metrics have been used alongside spectral metrics to segment single-date images to identify ASMs [5]. They have not been used for ASM detection in satellite image time series (SITS).

The pixel-wise segmentation of time series (i.e., time-series segmentation) is increasingly used to detect abrupt and gradual changes in vegetation [18]. This is due in large part to the availability of dense archives of free and open-access Landsat SITS and cloud-computing platforms such as Google Earth Engine (GEE) [19]. Time-series segmentation has not been applied to ASM detection, but results are promising from mapping industrial mining and mining impacts. Time-series segmentation of spectral information yielded accuracies above 80% for the detection of large open-cast (coal, mineral) mines and soil moisture declines caused by mining [20–22]. Temporally segmented spectral metrics derived from SITS better capture changes in forest cover through time when textural metrics are included as covariates [23]. The integration of temporally segmented spectral and textural features could therefore enhance ASM detection, but such an evaluation has not been previously reported.

We present a reliable and efficient ASM mapping method with the time-series segmentation of spectral and textural features derived from Landsat SITS. The method was tested in the most important gold mining region of the Brazilian portion of the Amazon Rainforest. The method involved building a Random Forest (RF) with Landsat-based detection of Trends in Disturbance and Recovery (LandTrendr) [18] segmented spectral and textural metrics, as well as proximity and topographic features. Our study utilized the Google Earth Engine (GEE) LandTrendr Application Programming Interface (API), which we modified for texture analysis. The study identified the most important spectral, textural, proximity, and topographic features and their functional relationships with the probability of ASM occurrence. We evaluated the model with ASM polygons derived from ground-based surveys and digitized manually from high spatial resolution imagery in the study area.

2. Study Area

The study area was in the Amazon Rainforest in the southeast corner of the Pará state of Brazil (Figure 1). The area spans 370 km^2 and falls within Landsat scene path 228, row 64. It is part of the Tapajós-Xingu moist forest ecoregion and includes the downstream contributing area of the Tapajós River, which is among the largest tributaries of the Amazon River. The Tapajós River Basin is predominantly covered by dense tropical rainforest, which serves as a major global carbon sink and critical habitat for numerous plant and animal species. The elevation varies widely, ranging from low-lying floodplains to mountainous terrain. The basin encompasses parts of the Amazon rainforest and the Brazilian Shield, with elevations ranging from sea level to over 2000 m. The climate in the Tapajós River Basin is typical tropical rainforest, which is characterized by high temperatures and humidity throughout the year. Precipitation is abundant, with a distinct wet season from December to May. The area undergoes a major flood pulse from February to May. The dry season occurs from June to September.

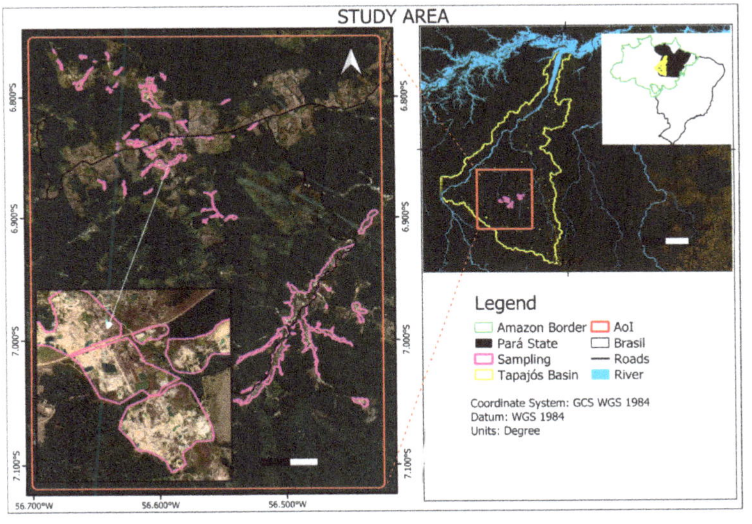

Figure 1. The Pará state and Tapajos River basin. The background of the area of interest (AOI) is a true color Landsat 8 image composite captured on 27 October 2020. The inset of the AOI b highlights ASM polygons in one of the three tiles sampled for model training and validation.

ASM activities in Pará state have significantly altered the landscape over the past 60+ years [24]. The first major gold rush began in 1958 and quickly made Pará state the principal gold producer of Brazil [25]. A new gold rush began in the early 1980s, which led the Brazilian government to create a gold mining reserve to support local miners and slow the impact of mining on the environment [26]. Gold mining not only causes forest loss. It is traditionally performed by removing topsoil near rivers or dragging out the sediments from river bottoms using suction and separating the gold by gravity [27]. Both techniques discharge sediments composed chiefly of fine organic particles and trace amounts of mercury used to amalgamate gold [28]. Mercury is also carried by air and deposited in waters further downstream. In water, mercury is methylated by microorganisms, which bioaccumulates in fish, humans, and other animals. It is an endocrine disruptor that interferes with genetic and enzyme systems and damages the nervous system and developing embryos.

Creporizinho and Creporizão are the two main settlements in the study area. Both are located close to the Transgarimpeira road (06°50′14.1″S–56°35′00.0″W). The establishment of Creporizão is directly linked to the construction of the Transgarimpeira in the late 1980s. It is the local center of commerce because it connects the road to a landing on the river. Creporizinho, on the other hand, is a typical gold-mining village. Creporizinho emerged in the early 1960s with the first artisanal extractions in the gold-mining district and increased in the 1980s when miners started to work with some machinery for processing alluvial–colluvial terraces.

3. Material and Methods

The technical workflow consisted of five main steps: (i) data acquisition from GEE and other online sources; (ii) pre-processing and transformation of spectral, textural, proximity, and topographic features; (iii) temporal segmentation with LandTrendr; (iv) feature selection with the Variable Selection Using Random Forests (VSURF) algorithm [29]; and (v) RF classification [30] of occurrence and non-occurrence (Figure 2).

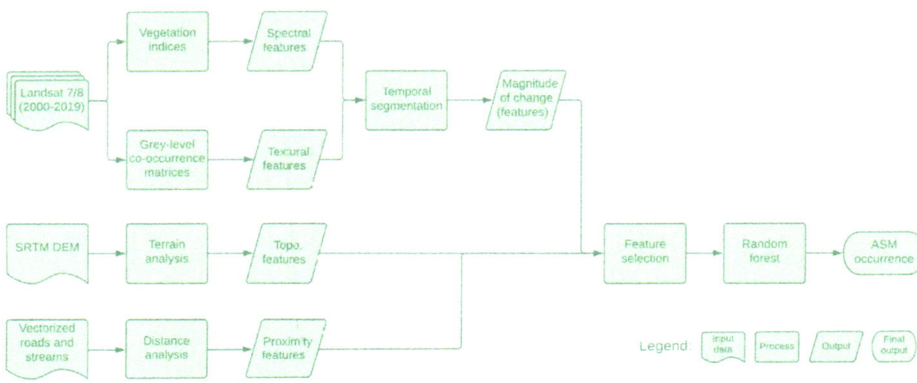

Figure 2. Technical workflow.

3.1. Geospatial Data Acquisition and Processing

We accessed Landsat 7 Enhanced Thematic Mapper (ETM+) and Landsat 8 Operational Land Imager (OLI) surface reflectance (Level-2) images, WGS-84 in GEE, from 2000 to 2019 to develop annual spectral and textural features for temporal segmentation. Landsat has a 16-day return frequency and a GSD of 30 m. Compositing was performed for the low-water period corresponding to the dry season from June to September when ASM alluvium is more easily detected remotely [16]. Six Landsat bands (Table 1) were considered for the analysis. All pre-processing steps were performed in GEE. This included Landsat 7 and 8 image harmonization; the removal of pixels with cloud, cloud shadow, and other poor-quality pixels; and compositing. Coefficients taken from Roy et al. [31] linearly transformed Landsat 7 spectral bands to minimize radiometric inconsistencies with Landsat 8. Cloud and cloud shadow were flagged and masked with the CFmask algorithm [32]. The unmasked pixels were composited on an annual basis using the medoid method [33]. The medoid method is commonly used to generate time series for LandTrendr [34]. The composites were filtered for pixels containing more than 50% cloud or cloud shadow.

Table 1. Description of Landsat 7 and 8 spectral bands.

Sensor	Designation	Spectral Range (nm)
Landsat 7 (8)	BLUE	450–520 (450–510)
	GREEN	520–600 (530–590)
	RED	630–690 (640–670)
	Near-infrared/NIR	770–900 (850–880)
	Shortwave infrared 1/SWIR 1	1550–1750 (1570–1650)
	Shortwave infrared 2/SWIR 2	2090–2350 (2110–2290)

We downloaded the National Aeronautics and Space Administration (NASA) Jet Propulsion Laboratory Shuttle Radar Topography Mission V3 digital elevation model (DEM) at a resolution of 1 arcsecond (approximately 30 m on the equator) in WGS-84 [35]. This product includes derived products from the DEM—aspect, elevation, hill shade, and slope.

The input data for the distances to roads and streams were derived from the 2019 annual Landsat image composite. The Brazilian geo web service at the Institute of Geography and Statistics Monitroamento da Cobertura e Uso da Terra do Brasil [36] roads and water bodies datasets aided the interpretation of the composite. Euclidean distance was used for proximity, which was defined as the nearest orthogonal distance from each ASM site to either roads or streams.

3.2. Spectral Transformations

We used eight spectral vegetation indices (VIs) for model building (Table 2). VIs are more robust than raw spectral bands for changing atmospheric and surface conditions [37]. Three of the indices (NBR, NDMI, NDVI) are standard functions in the LandTrendr API. We also segmented EVI, LSWI, MNDWI, NDPI, and NDWI because of their sensitivity to changes in vegetation growth and development. EVI uses coefficients and the blue band to reduce the effects of varying soil backgrounds and atmospheric constituents that adversely affect NDVI. LSWI includes a SWIR band, which is sensitive to the moisture content of canopies and soil background. The NIR band in LSWI increases as the abundance and complexity of vegetation increases. The NIR band also acts to normalize the sensitivity of the SWIR band to noise. The NDWI and MNDWI indices both enhance water features using the visible green band. Water reflects more strongly in the visible green compared to longer wavelengths. MNDWI normalizes the green band with an SWIR band. NDPI employs red, NIR, and SWIR bands. Chlorophyll in vegetation absorbs strongly in the red [38]. The red band, however, is sensitive to soil moisture, so the index includes an SWIR band to correct this effect. NDPI is scalable and able to spectrally separate water bodies from other surfaces.

Table 2. Spectral vegetation indices for temporal segmentation.

Spectral Vegetation Index	Formula	Source
Enhanced Vegetation Index	$EVI = 2.5 * \frac{(NIR-RED)}{(NIR+6*RED-7.5*BLUE+1)}$	[37]
Land Surface Water Index	$LSWI = \frac{(SWIR\ 1-NIR)}{(SWIR1+NIR)}$	[39]
Modified Normalized Difference Water Index	$MNDWI = \frac{(GREEN-SWIR\ 1)}{(GREEN+SWIR\ 1)}$	[40]
Normalized Burn Ratio	$NBR = \frac{(NIR-SWIR\ 2)}{(NIR+SWIR\ 2)}$	[41]
Normalized Difference Moisture Index	$NDMI = \frac{(NIR-SWIR\ 1)}{(NIR+SWIR\ 1)}$	[42]
Normalized Difference Phenology Index	$NDPI = \frac{(NIR-SWIR\ 1)}{(NIR+SWIR\ 1)}$	[43]
Normalized Difference Vegetation Index	$NDVI = \frac{(NIR-RED)}{(NIR+RED)}$	[6]
Normalized Difference Water Index	$NDWI = \frac{(GREEN-NIR)}{(GREEN+NIR)}$	[44]

3.3. Textural Transformations

We used twelve grey-level co-occurrence matrix (GLCM) textural metrics [45,46] for model building (Table 3). The metrics were calculated for each Landsat band with GLCM, yielding a total of 72 textural features. GLCMs establish the frequency of occurrence of angular relationships and spatial dependence between image pixel pairs. The features are essentially statistical inferences of the frequency distributions. The features were calculated in GEE with a directional averaging kernel of 3×3 (90×90 m^2). Kernel size was based on the smallest patch size of an ASM in the training dataset (~1 ha) and the GSD of the Landsat imagery. Angular second momentum is an indication of textural uniformity in the grey levels [47]. Contrast shows local differences in high and low pixel values. The higher the contrast, the higher the difference in these values. Correlation measures the linear strength of the relationship between a given pixel with a point in the image. Variance measures the overall deviation of gray levels from the mean. High variance tends to correspond to high contrast in grey levels. Cluster prominence is a measure of asymmetry. Inertia is indicative of the contrast in intensity between a pixel and its neighbors. Cluster shade accounts for grayscale matrix skewness and uniformity [48].

Table 3. Textural metrics for temporal segmentation.

Texture Metric	Designation	Description
Angular Second Moment	Asm	Measures the number of repeated pairs
Contrast	Con	Measures the local variability of an image
Correlation	Corr	Measures the linear dependency between pixel pairs
Variance	Var	Measures the spread of the grey-level distribution
Inverse Difference Moment	IDM	Measures the homogeneity
Entropy	Ent	Measures the randomness of the gray-level distribution
Difference Variance	Dvar	Measures the variance of the gray-level distribution
Difference Entropy	DEN	Measures the difference in the randomness of the gray-level distribution
Cluster Prominence	Prom	Measures clusters by the gray-level occurrence
Dissimilarity	Diss	Measures the variation between pairs of pixels
Inertia	Iner	Measures the intensity between a pixel and its neighborhood
Shade	Shade	Measures the cluster shade of gray-level distribution

3.4. Temporal Segmentation with LandTrendr

We used the LandTrendr version and procedure described by Kennedy et al. [49] to segment the spectral and textural annual Landsat time series on a pixel basis. LandTrendr detects abrupt or gradual pixel-wise changes according to a set of breakpoints and line segments defined by the time series. The curve shown in Figure 3 represents the trajectory and LandTrendr segmentation of a 20-year NDVI time series representing a pixel that likely experienced ASM encroachment in 2008. The figure illustrates LandTrendr's ability to detect interannual noise as well as the timing and magnitude of vegetation disturbance and recovery. LandTrendr employs piecewise linear regression to determine the best-fitting trajectory along breakpoints or vertices in the time series. Like other time series analytical techniques, model tuning is required, which can lead to under- or over-fitting. Eight control parameters are required for LandTrendr to define the breakpoints and for line segment fitting. These include the maximum number of segments, spike threshold, vertex count overshoot, prevent one-year recovery (true/false), recovery threshold, p-value threshold, best model proportion, and minimum observations needed. LandTrendr uses root mean square error (RMSE) to identify the "best" model fit. Multiple combinations and different values for the eight control parameters were tested, and the set of eight parameters with the lowest RMSE was chosen for the time series analysis.

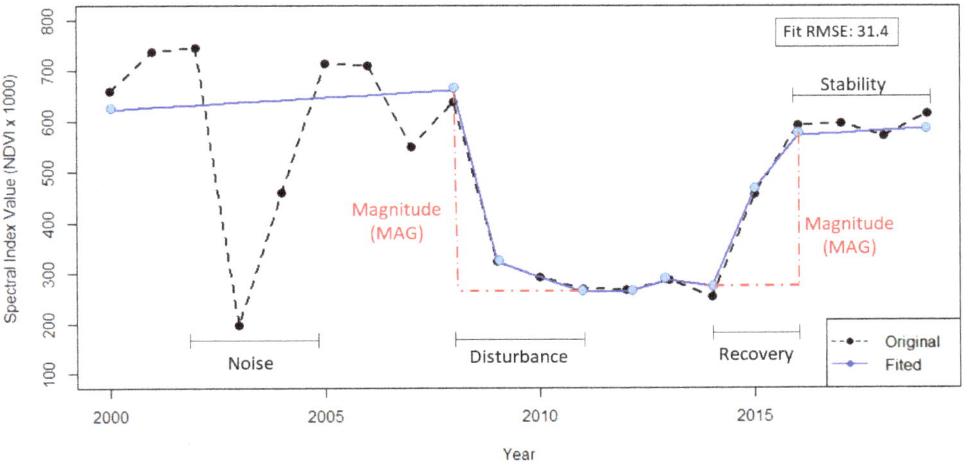

Figure 3. Temporal segmentation of a 20-year NDVI time series from an ASM disturbed pixel (6°50′35.2″S 56°36′40.7″W) using LandTrendr. The dotted dashed red lines represent the delta by loss drop in magnitude and disturbance duration.

Only the magnitude (MAG) of spectral and textural features was considered for Random Forest. MAG in LandTrendr is calculated as the greatest delta by loss drop in magnitude and disturbance duration, representing the largest rate of change over the time series. Ultimately, the final output included 80 MAG features (12 textures per 6 raw spectral bands plus 8 vegetation indices).

3.5. Ancillary Geospatial Information

Six ancillary data predictors were considered for model building to boost the predictive power of the spectral and textural features. These included the Euclidean distance to roads and streams, aspect, elevation, slope, and the topographic wetness index (TWI) [50]. TWI is indicative of the upstream contributing area. A high (low) TWI corresponds to stream beds (hillslopes) where water accumulation is high (low).

3.6. Feature Selection with VSURF and Random Forest Classification

RF is an ensemble machine learning technique widely used in satellite image classification [30]. RF takes the ensemble mean of a number of "weak" bootstrapped independent decision tree classifications, which tends to yield higher accuracies for land cover change detection than other classification methods [51]. RF is prone to over-fitting, however, when the number of features is high [52]. We therefore removed redundant and irrelevant features with the VSURF package in the R software version 4 environment [29] before model building. VSURF is essentially an "ensemble of ensembles" technique. It consists of three main steps: (i) thresholding; (ii) interpretation; and (iii) prediction. The first stage uses 50 iterations of RF to determine a cut-off threshold for features based on the standard deviation of feature importance. Highly irrelevant features are eliminated because they have small standard deviations. In the second stage, 25 iterations of RF are performed to identify important explanatory variables that yield a parsimonious model with low error. Some redundancy in features can remain after the interpretation stage, so the prediction stage removes any remaining redundancies in a stepwise fashion with 25 iterations of RF. Each RF consisted of 500 trees and the number of variables per split was set to the square root of the input bands as recommended by [53]. Partial dependence plots (PDPs: [54]) were extracted from the Random Forest with the PDP package in the R software version

4 environment [55]. PDPs express the relationship between ASM occurrence and a target feature, irrespective of the other features.

3.7. Random Forest Model Training and Testing

We acquired the ground reference data for RF modeling from the Rede Amazônica de Informação Socioambiental (RAISG) [56]. RAISG consists of civil societies in countries of the Amazon concerned with the health and sustainability of the Amazon rainforest. It produces vectorized data on the geolocation of mining in the Amazon basin. The dataset is a compilation of more than two decades of multidisciplinary fieldwork. We subset the dataset for mine type (gold and tin) from 2000 to 2019 to coincide with the SITS. The subset consisted of 826 "point" samples and 182 polygons. We utilized very high-resolution imagery from Google Earth to identify and digitize mine polygons based on the point data. The combination of hand-digitized polygons and RAISG polygons resulted in more than 1000 polygons in total for the analysis. We employed a pixel-stratified sampling framework to ensure spatial representativeness and prevent autocorrelation. The spatial disjoining involved dividing the entire sample set into three main tiles according to their cardinal direction in the area of interest (east, west, south). Two of the tiles were used for training and one was used for validation. Considering the nature of the problem and limitations with ground truth data, we addressed class imbalance and overfitting by employing a dual strategy [57]. All the occurrence pixels from mine polygons were used for training, which was balanced by an under-sampling of the non-occurrence class.

We used the out-of-bag (OOB) error and confusion matrix to assess model performance. RF includes an internal validation. At each iteration, RF uses a portion of the data for calibration. The unused portion is the out-of-bag sample. The OBB is the prediction error of approximately 1/3 of the samples not used in each iteration to build a Random Forest. The OBB is averaged over all iterations to give an unbiased estimate of model performance [58]. The confusion matrix expresses the frequency of correctly and misclassified samples. The diagonal of the matrix indicates the number of correctly calculated samples. Overall accuracy is calculated by summing the diagonal and dividing it by the total number of samples classified. The off-diagonals report the error of omission (Type 1 error) and error of commission (Type II error).

We converted ASM occurrence to the probability of occurrence to aid decision-makers who may need to focus resources in "hot spots" (i.e., areas of high probability of ASM occurrence). In RF, the probability of occurrence is calculated with logistic regression, probability machines, or vote counting [59]. We selected vote counting. The probability is interpreted as the proportion of votes in favor of the majority class (i.e., non-occurrence), which tends to reduce noise and increases recognition accuracy for binary classification [60].

4. Results

4.1. Temporal Segmentation with LandTrendr

LandTrendr produced maps that integrated the full temporal range of ASM disturbances over the study period (2001–2019). This is illustrated for NDVI MAG and occurrence year (YoD) in Figure 4. The occurrence year represented the year in which the temporal segment showed the first significant drop in NDVI. For example, we see near the urban areas of Creporizao and Creporizinho the expansion of the ASMs, evident by large declines in NDVI. Areas representing agricultural expansion and other forms of land clearance (e.g., logging, topsoil removal, bushfires) experienced much smaller and/or earlier (pre-2000) declines in NDVI. Other forested areas remained largely intact as they showed no change from 2000 to 2019. The occurrence of ASM expansion varied but mainly followed the global economic crisis of 2008, as can be seen in Figure 5. The figure depicts the temporal segmentation according to a red–green–blue (RGB) color composite to summarize spatial patterns. In this representation, red indicates changes occurring in the year 2000, green represents changes in 2010, and blue signifies changes in 2019. Consequently, pixels appearing yellow denote changes within the initial decade of the timeframe and purple denotes the last ten years. The inset at the bottom of the figure depicts the temporal segmentation of an ASM pixel. It was generated from NDVI in LandTrendr to improve the interpretation of changes when they occurred according to the RGB composite. Thirty-five percent of the ASM pixels occurred before 2008. From 2008 to 2010, detected ASM pixels dropped to 25%. This figure increased to 40% after 2010.

We evaluated the performance of 86 proximity, spectral, textural, and topographic features with VSURF. Thresholding (phase 1) and interpretation (phase 2) eliminated more than half of the predictors (49 out of 86). The model at the end of these phases had an average OBB error and overall accuracy of 3.76% and 92.6%. The prediction phase reduced the number of features to 33. The overall accuracy increased by 3.33% and the OBB error decreased by 0.03%, even though an additional 16 features were removed from the model.

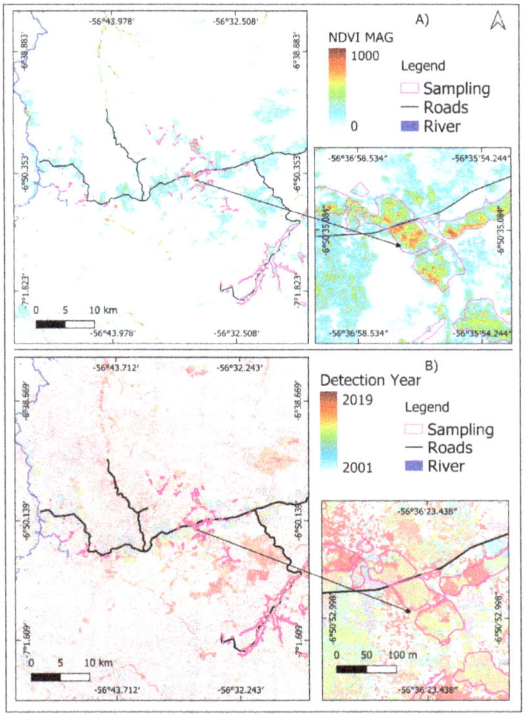

Figure 4. NDVI MAG (**A**) and YoD (**B**) calculated with LandTrendr. NDVI was expressed as an integer (×1000) in GEE LandTrendr to reduce computational demand. The insets illustrate ASM activity around the towns of Creporizão and Creporizinho.

Figure 6 shows the final subset of features we used for model building. We display only the 18 most important features for visualization purposes. NDVI ranked first. Removing it from the model led to an increase in OBB MSE of 0.086. Textural features were the next most important. If they were removed from the model, textural features led to an increase in OBB MSE of 0.066. The features included textural properties of the NIR and SWIR bands (B7 Var, B5 Prom, B5 Shade, B6 Shade, B6 contrast).

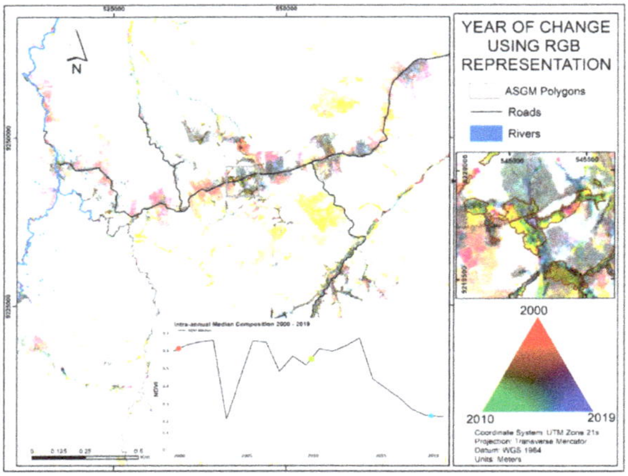

Figure 5. NDVI year of occurrence using a red–green–blue (RGB) 2000–2010–2019 composite. The inset shows the trajectory of a pixel experiencing change due to ASM activity with respect to the RGB coloring scheme. The disturbance occurred between 2010 and 2019, so the pixel appears purple in the map.

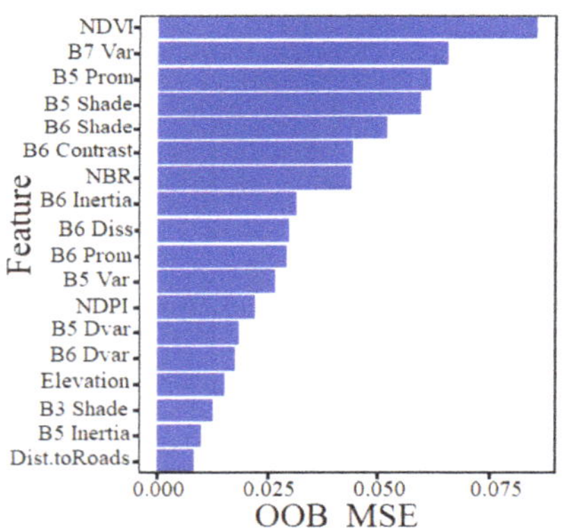

Figure 6. Features in descending order of importance versus the change in MSE that occurred if the feature was eliminated from the model.

4.2. Random Forest Classification

The overall accuracy of the Random Forest for the hold-out sample set was 85.2% for the binary classification (Table 4). Type I errors were higher in occurrence (6.4%) than non-occurrence (2.0%). Type II errors, on the other hand, were higher for non-occurrence (5.2%) than occurrence (2.4%). The distribution of the categories was approximately balanced as occurrence and non-occurrence represented 40.6% and 59.3% of the hold-out sample set, respectively. Type I errors for occurrence were relatively high because ASMs exhibited similar spectral and textural characteristics of other land clearance types. The binary classification was able to distinguish ASMs mainly along streams and roads that connect informal settlements. ASMs cluster along streams because it is relatively easy to develop mining ponds for mineral treatment and extraction. ASMs are clustered along roads to facilitate storage and transportation in a strategy locally known as "garimperios."

Table 4. Confusion matrix (classified versus reference samples) reported for the hold-out sample set.

	Reference			
	Class	**Occurrence**	**Non-occurrence**	**Total**
Classification	Occurrence	76%	24%	6402
	Non-occurrence	5%	95%	6367
	Total	5194	7575	12,769

4.3. Probabilistic Classification

The probability of occurrence provides a more nuanced interpretation of the binary classification (Figure 7). In the figure insets, it appears only the core areas of the ASM clusters especially along stream networks are likely mines (probability $\geq$ 70%). Probabilities tend to drop off sharply away from the core areas. Forest disturbances and other land clearance types unrelated to ASM activity correspond to probabilities below 50%. Some of these areas were misclassified as ASMs in the binary classification.

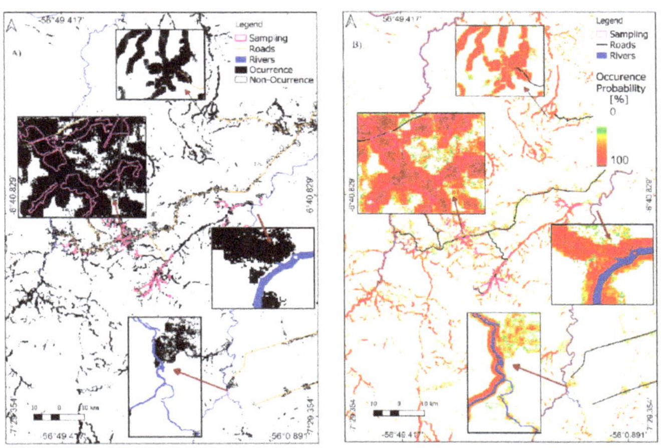

Figure 7. ASM binary (**A**) and probabilistic (**B**) classification. The three insets illustrate areas with a high frequency of ASM occurrence along streams.

4.4. Partial Dependence of Important Features

The relationships between the six most important features and the probability of ASM occurrence are shown in Figure 8. We only show six because the PDPs became less meaningful as the contribution of a feature to the predictive power of the RF declined. The probability of occurrence steadily rose as NDVI increased from 0.10 to 0.45. Values below 0.1 represent bare soils, burned areas, or some other type of non-vegetative surface. The probability gradually declined as pixels became more vegetated and NDVI increased beyond 0.5. Similarly, the likelihood of an ASM increased as differences in pixel grey levels in the SWIR increased. At extreme differences in grey levels (>500), however, ASM likelihood declined. ASMs tended to occur where NIR and SWIR shade were low (100–200 MAG). Shade is a skewness indicator; lower values represent a lack of spatial symmetry, while higher values represent spatial uniformity. On the other hand, high cluster prominence (B5 Prom) indicates high variation (contrast) in grayscale levels according to the peak of the mean in the GCLM matrix. Local (i.e., within the processing window) variations in the SWIR increased the likelihood of an ASM until around 100. Beyond 400, the probability of an ASM declined.

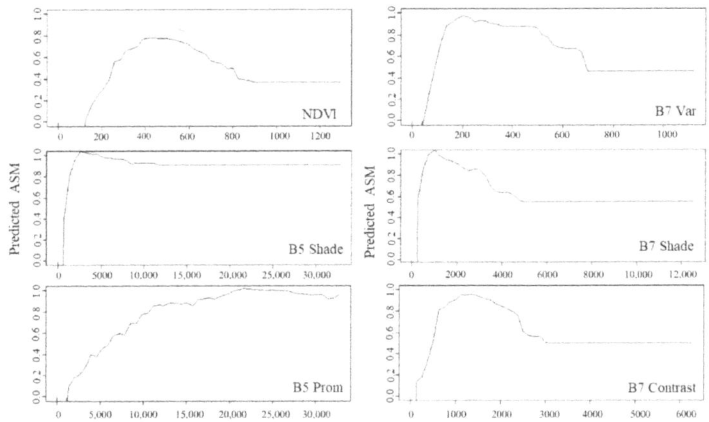

Figure 8. Partial dependence plots. Features are expressed as an integer (×1000).

5. Discussion

The results showed that the ASMs can be mapped with high accuracy at 30 m resolution with temporally segmented Landsat spectral and textural features using free and open source GEE. The analysis made three important observations. First, temporal segmentation confirmed large and widespread increases in ASM-driven natural vegetation disturbance in the study area following the global financial crisis of 2008. Second, NDVI, the traditional LandTrendr input, was the most important feature in ASM detection. However, textural features in the NIR and SWIR boosted model performance. Other spectral, topographic, and proximity features, which are commonly used to map various land cover types, were not as important. Lastly, feature selection with VSURF and balanced sampling increased the robustness of the RF model.

The performance of our RF model classification was quite high considering the small footprint of ASMs and the mixed pixel effect. Studies in inner Mongolia, China [21] and KwaZulu-Natal, South Africa [20], employed LandTrendr and Landsat SITS to interrogate disturbance caused by large open cast (coal, mineral) industrial mines. Both studies used NDVI as a model input only for RF. The models were assessed with high spatial resolution image interpretation in Google Earth. The former yielded an overall accuracy of 86.5%, while the latter yielded an overall accuracy of 99.0%. NDVI is effective at separating, spectrally, vegetative from non-vegetative surfaces. It is therefore widely used to monitor land degradation [61] caused by droughts [62], soil erosion [63], deforestation [64], and prescribed burning [65]. In this study, the PDPs revealed a proportional response of NDVI to ASM occurrence until 0.45. NDVI beyond this threshold was indicative of dense vegetation such as forest. ASMs, unlike other forms of land degradation, however, do not have a consistent spatial distribution or well-defined shape. Textural information, which characterizes the spatial structure of a landscape, can enhance the predictive power of spectral information. Meng et al. [23] used textural metrics derived from NBR to map the restoration of forests in China from orchards and other land cover types. NBR, like NDVI, is commonly used to monitor forest disturbance and recovery. The accuracy of the model improved from 20% to 63% when the temporally segmented GLCM-based texture metrics were included. We analyzed the raw spectral bands only. NBR is a ratio-based index with NIR and SWIR as input. NIR and SWIR bands were important texture metrics in our study. NIR is sensitive to changes in canopy structure, while SWIR is sensitive to changes in soil/canopy moisture. The high variance and cluster prominence together with low shade corresponded to increased ASM occurrence. This is logical since increases in these metrics are indicative of high spatial heterogeneity in terms of vegetation and moisture. Other spectral, topographic, and proximity features were likely not as critical to

ASM detection because they exhibited strong collinearity with NDVI and textural NIR and SWIR metrics. These variables were either removed during rigorous feature selection or only accounted for a small portion of the feature space after more important variables were added to the model.

Cluster Prominence, Cluster Shade, Variance, and Contrast were the most important textural features for ASM detection. They have large separations and small standard deviations compared to other metrics [48]. Cluster shade and prominence are indicative of local spectral asymmetry/heterogeneity. As both increase, the likelihood that pixels represent multiple land uses over small areas increases [66]. The second most important feature was variance in the SWIR-2 band (B7 Var). Var is a measure of the deviation of the values around the mean, indicating how spread out the sum of the gray level of a pixel pair is. Var accounts for the local gray-level variations (smoothness) of an image. Smoother images yield lower Var. Variance measures the variability by the sum of the squares of the differences between the intensity of the central pixel and its neighbors. Areas with high variance are prone to be high in contrast and thus are direct indicators of irregular shapes and image heterogeneity. Cluster Prominence texture (B5 Prom) was the third most important feature. This texture is an asymmetry indicator. SWIR-1 shade (B5 Shade) and SWIR-2 shade (B6 Shade) were the fourth and fifth most important features. Contrast (ranked sixth most important) identified local variations while calculating the differences between a pixel and neighboring pixels over an entire image.

The spatial distribution of the ASMs in the study area was dynamic and multifaceted, influenced by various factors that reflected the interplay of environmental, socioeconomic, and regulatory factors shaping where and when mining activities occur. ASMs tend to be concentrated in areas where geological formations indicate the presence of gold-bearing deposits. In our study, miners clearly targeted riverbeds, floodplains, and areas with visible signs of mineralization, such as quartz veins or altered rocks. The accessibility of mining sites plays a crucial role since miners may prefer areas with relatively easy access, such as along rivers or existing road networks. To a much lesser extent, ASM operations were in remote and difficult-to-reach areas. Miners face logistical challenges in these areas but also benefit from reduced competition and regulatory oversight. Many current mining sites are located near historical gold rushes, reflecting the continuity of mining traditions and the persistence of gold-bearing deposits.

Commission errors in the classification could be reduced by following two approaches: (i) applying strict data sampling criteria so that only the core ($\geq$70% probability) mining pixels are used for training the RF classifier and (ii) eliminating uncertain samples based on local probabilities of class membership (i.e., proximity). A similar approach using TM/ETM+ images for mapping burned areas is described in Bastarrika et al. [67].

The integration of other sensors and analytical approaches into the technical workflow could improve the accuracy of ASM detection but would incur a loss of efficiency. We used Landsat 5 TM and 8 OLI whenever possible, but some Landsat 7 ETM+ image data were used to fill gaps in the time series. The SLC-off (i.e., null data) issue with Landsat 7 ETM+ was still noticeable in the final LandTrendr outputs even though LandTrendr applies an interpolation routine to correct for SLC-off. Sentinel-2 through the Harmonized Landsat and Sentinel-2 project [68] or new experimental (ENMAP, PRISMA) and planned (ESA CHIME, NASA SBG) hyperspectral missions could be added to overcome this and other spatial, spectral, and temporal limitations of Landsat. The Sentinel-2 constellation (a+b) has been acquiring images since 2015 with a higher spatial resolution (20 m) and return frequency (5 days) than Landsat [69]. It includes four additional narrow (<=20 nm) bands in the red edge and NIR. These bands are particularly sensitive to changes in vegetation growth and development. SWIR was an important region for ASM detection. Hyperspectral missions retrieve spectral information at $\leq$10 nm intervals in the SWIR, which is much higher than Landsat or Sentinel-2. The tropics experience persistent cloud cover, which limits the application of optical sensors such as these, particularly during the growing season. Polarimetric indices derived from cloud-penetrating Sentinel-1 or other synthetic aperture radar overcome this challenge. This could further improve ASM detection by providing additional spectral information during the growing season when spectral differences between vegetation and other land cover types are most striking [70]. The analytical approach could be re-oriented toward semantic segmentation with convolutional neural networks (i.e., deep learning), given the importance of both spectral and textural features as well as the abundance of reference data. These approaches were recently demonstrated for ASM detection in Ghana [15]. The temporal dimension was not addressed in the Ghanaian study, possibly due to the computational demand of deep learning. A deep learning approach that more efficiently integrates time series, such as long short-term memory networks, could be a powerful new direction for monitoring ASMs.

The accurate and efficient detection of ASMs via a monitoring system could help government agencies, non-governmental organizations, and other decision-making bodies to intervene to achieve the following positive outcomes:

- Environmental Protection: ASMs often involve harmful practices such as deforestation, mercury pollution, and habitat destruction.
- Regulatory Enforcement: ASMs typically do not comply with environmental and land use laws, which leads to the illegal exploitation of natural resources.
- Community Health and Safety: ASMs have adverse effects on nearby communities, including health risks from exposure to toxic substances such as mercury as well as safety hazards from unstable mining structures.
- Human Rights: ASMs are often associated with human rights abuses such as forced labor, child labor, and exploitation.
- Conflict Prevention: In some regions, ASMs fuel conflicts and contribute to instability.

6. Conclusions

Mapping ASMs is challenging because they are small, spatially heterogeneous, and difficult to separate spectrally from other land cover/use types. We demonstrate a reliable and efficient method for ASM binary and probabilistic classification over a 20-year period (2000–2019) in the Pará state of Brazil—an important gold mining region of the Amazon River Basin. Previous studies mostly provide image snapshots of large industrial mines using the spatial segmentation of spectral and textural features with or without machine learning. More recent studies used time-series segmentation of NDVI derived from Landsat SITS to detect large industrial mines. Our study integrated the time-series segmentation of several Landsat spectral and textural features into an RF to detect ASMs. More than 1000 polygons were vectorized from a series of RAISG field surveys collected in the study area from 2000 to 2019. It employed rigorous feature selection with VSURF—an ensemble of ensembles technique—to eliminate all but 33 (38%) of the most important features. Feature selection increased the overall accuracy by 3.33% and decreased the OBB error by 0.03%. The final model produced an overall accuracy of 92.6%, which was 6.1% higher than a study mapping large industrial mines with the time-series segmentation of NDVI. Like other mining studies, NDVI was the most important feature in our study. However, textural features (variance, prominence, shade, contrast) in the NIR and SWIR regions of the electromagnetic spectrum enhanced model performance. The PDPs of the RF showed that as vegetation abundance (NDVI) decreased and the spatial heterogeneity in NIR and SWIR bands within ASMs and between ASMs and other land use types increased, the likelihood of a ASM activity increased.

Author Contributions: Conceptualization, A.F., M.T.M. and S.S.; Methodology, A.F., M.T.M. and S.S.; Software, A.F.; Validation, A.F.; Formal analysis, A.F.; Investigation, A.F.; Resources, A.F.; Writing—original draft, A.F., M.T.M. and S.S.; Writing—review & editing, A.F., M.T.M. and S.S.; Visualization, A.F.; Supervision, M.T.M. and S.S. All authors have read and agreed to the published version of the manuscript.

Funding: This research received no external funding.

Data Availability Statement: The original contributions presented in the study are included in the article, further inquiries can be directed to the corresponding author.

Acknowledgments: This publication was largely based on the MSc thesis of the lead author: Detecting Artisanal Small-Scale Gold mines with LandTrendr multispectral and textural features at the Tapajós river basin, Brazil. We would like to thank Roshanak Darvish and Harald van der Werff at ITC—University of Twente for their contributions to the MSc thesis.

Conflicts of Interest: The authors declare no conflict of interest.

References

1. Sloan, S.; Sayer, J.A. Forest Resources Assessment of 2015 shows positive global trends but forest loss and degradation persist in poor tropical countries. *For. Ecol. Manag.* **2015**, *352*, 134–145. [CrossRef]
2. Keenan, R.J.; Reams, G.A.; Achard, F.; de Freitas, J.V.; Grainger, A.; Lindquist, E. Dynamics of global forest area: Results from the FAO Global Forest Resources Assessment 2015. *For. Ecol. Manag.* **2015**, *352*, 9–20. [CrossRef]
3. Asner, G.P.; Llactayo, W.; Tupayachi, R.; Luna, E.R. Elevated rates of gold mining in the Amazon revealed through high-resolution monitoring. *Proc. Natl. Acad. Sci. USA* **2013**, *110*, 18454–18459. [CrossRef] [PubMed]
4. Werner, T.T.; Mudd, G.M.; Schipper, A.M.; Huijbregts, M.A.J.; Taneja, L.; Northey, S.A. Global-scale remote sensing of mine areas and analysis of factors explaining their extent. *Glob. Environ. Chang.* **2020**, *60*, 102007. [CrossRef]
5. Isidro, C.M.; McIntyre, N.; Lechner, A.M.; Callow, I. Applicability of Earth Observation for Identifying Small-Scale Mining Footprints in a Wet Tropical Region. *Remote Sens.* **2017**, *9*, 945. [CrossRef]
6. Rouse, J.W.; Haas, R.H.; Schell, J.A.; Deering, D.W. *Monitoring Vegetation Systems in the Great Plains with ERTS*; Scientific and Technical Information Office, National Aeronautics and Space Administration: Washington, DC, USA, 1974; Volume 351, p. 309.
7. Almeida-Filho, R.; Shimabukuro, Y.E. Digital processing of a Landsat-TM time series for mapping and monitoring degraded areas caused by independent gold miners, Roraima State, Brazilian Amazon. *Remote Sens. Environ.* **2002**, *79*, 42–50. [CrossRef]
8. Caballero Espejo, J.; Messinger, M.; Román-Dañobeytia, F.; Ascorra, C.; Fernandez, L.E.; Silman, M. Deforestation and Forest Degradation Due to Gold Mining in the Peruvian Amazon: A 34-Year Perspective. *Remote Sens.* **2018**, *10*, 1903. [CrossRef]
9. Forkuor, G.; Ullmann, T.; Griesbeck, M. Mapping and Monitoring Small-Scale Mining Activities in Ghana using Sentinel-1 Time Series (2015–2019). *Remote Sens.* **2020**, *12*, 911. [CrossRef]
10. Simionato, J.; Bertani, G.; Osako, L.S. Identification of artisanal mining sites in the Amazon Rainforest using Geographic Object-Based Image Analysis (GEOBIA) and Data Mining techniques. *Remote Sens. Appl. Soc. Environ.* **2021**, *24*, 100633. [CrossRef]
11. Ibrahim, E.; Lema, L.; Barnabe, P.; Lacroix, P.; Pirard, E. Small-scale surface mining of gold placers: Detection, mapping, and temporal analysis through the use of free satellite imagery. *Int. J. Appl. Earth Obs. Geoinf.* **2020**, *93*, 102194. [CrossRef]
12. Ngom, N.M.; Mbaye, M.; Baratoux, D.; Baratoux, L.; Ahoussi, K.E.; Kouame, J.K.; Faye, G.; Sow, E.H. Recent expansion of artisanal gold mining along the Bandama River (Côte d'Ivoire). *Int. J. Appl. Earth Obs. Geoinf.* **2020**, *112*, 102873. [CrossRef]
13. Lobo, F.D.L.; Souza-Filho, P.W.M.; de Moreas Novo, E.M.L.; Carlos, F.M.; Barbosa, C.C.F. Mapping Mining Areas in the Brazilian Amazon Using MSI/Sentinel-2 Imagery (2017). *Remote Sens.* **2018**, *10*, 1178. [CrossRef]
14. Snapir, B.; Simms, D.M.; Waine, T.W. Mapping the expansion of galamsey gold mines in the cocoa growing area of Ghana using optical remote sensing. *Int. J. Appl. Earth Obs. Geoinf.* **2017**, *58*, 225–233. [CrossRef]
15. Gallwey, J.; Robiati, C.; Coggan, J.; Vogt, D.; Eyre, M. A Sentinel-2 based multispectral convolutional neural network for detecting artisanal small-scale mining in Ghana: Applying deep learning to shallow mining. *Remote Sens. Environ.* **2020**, *248*, 111970. [CrossRef]
16. Lobo, F.; Costa, M.; Novo, E.; Telmer, K. Distribution of Artisanal and Small-Scale Gold Mining in the Tapajós River Basin (Brazilian Amazon) over the Past 40 Years and Relationship with Water Siltation. *Remote Sens.* **2016**, *8*, 579. [CrossRef]
17. Li, M.; Zang, S.; Zhang, B.; Li, S.; Wu, C. A Review of Remote Sensing Image Classification Techniques: The Role of Spatio-contextual Information. *Eur. J. Remote Sens.* **2014**, *47*, 2279–7254. [CrossRef]
18. Kennedy, R.E.; Yang, Z.; Cohen, W.B. Detecting trends in forest disturbance and recovery using yearly Landsat time series: 1. LandTrendr—Temporal segmentation algorithms. *Remote Sens. Environ.* **2010**, *114*, 2897–2910. [CrossRef]
19. Vogelmann, J.E.; Gallant, A.L.; Shi, H.; Zhu, Z. Perspectives on monitoring gradual change across the continuity of Landsat sensors using time-series data. *Remote Sens. Environ.* **2016**, *185*, 258–270. [CrossRef]
20. Dlamini, L.Z.D.; Xulu, S. Monitoring Mining Disturbance and Restoration over RBM Site in South Africa Using LandTrendr Algorithm and Landsat Data. *Sustainability* **2019**, *11*, 6916. [CrossRef]
21. Xiao, W.; Deng, X.; He, T.; Chen, W. Mapping Annual Land Disturbance and Reclamation in a Surface Coal Mining Region Using Google Earth Engine and the LandTrendr Algorithm: A Case Study of the Shengli Coalfield in Inner Mongolia, China. *Remote Sens.* **2020**, *12*, 1612. [CrossRef]
22. Yi, Z.; Liu, M.; Liu, X.; Wang, Y.; Wu, L.; Wang, Z.; Zhu, L. Long-term Landsat monitoring of mining subsidence based on spatiotemporal variations in soil moisture: A case study of Shanxi Province, China. *Int. J. Appl. Earth Obs. Geoinf.* **2021**, *102*, 102447. [CrossRef]
23. Meng, Y.; Liu, X.; Wang, Z.; Ding, C.; Zhu, L. How can spatial structural metrics improve the accuracy of forest disturbance and recovery detection using dense Landsat time series? *Ecol. Indic.* **2021**, *132*, 108336. [CrossRef]

24. Berzas Nevado, J.J.; Rodríguez Martín-Doimeadios, R.C.; Guzmán Bernardo, F.J.; Jiménez Moreno, M.; Herculano, A.M.; do Nascimento, J.L.M.; Crespo-López, M.E. Mercury in the Tapajós River basin, Brazilian Amazon: A review. *Environ. Int.* **2010**, *36*, 593–608. [CrossRef] [PubMed]
25. Villas Bôas, R.C.; Beinhoff, C.; da Silva, A.R.B. *Mercury in the Tapajos Basin*; CYTED: Rio de Jeniro, Brazil, 2001; 198p.
26. Roulet, M.; Lucotte, M.; Canuel, R.; Farella, N.; De Freitos Goch, Y.G.; Pacheco Peleja, J.R.; Guimarães, J.-R.D.; Mergler, D.; Amorim, M. Spatio-temporal geochemistry of mercury in waters of the Tapajós and Amazon rivers, Brazil. *Limnol. Oceanogr.* **2001**, *46*, 1141–1157. [CrossRef]
27. Lobo, F.L.; Costa, M.P.F.; Novo, E.M.L.M. Time-series analysis of Landsat-MSS/TM/OLI images over Amazonian waters impacted by gold mining activities. *Remote Sens. Environ.* **2015**, *157*, 170–184. [CrossRef]
28. Castello, L.; Macedo, M.N. Large-scale degradation of Amazonian freshwater ecosystems. *Glob. Chang. Biol.* **2016**, *22*, 990–1007. [CrossRef] [PubMed]
29. Genuer, R.; Poggi, J.-M.; Tuleau-Malot, C. VSURF: An R Package for Variable Selection Using Random Forests. *R J.* **2015**, *7*, 19–33. [CrossRef]
30. Breiman, L. Random Forests. *Mach. Learn.* **2001**, *45*, 5–32. [CrossRef]
31. Roy, D.P.; Kovalskyy, V.; Zhang, H.K.; Vermote, E.F.; Yan, L.; Kumar, S.S.; Egorov, A. Characterization of Landsat-7 to Landsat-8 reflective wavelength and normalized difference vegetation index continuity. *Remote Sens. Environ.* **2016**, *185*, 57–70. [CrossRef]
32. Zhu, Z.; Wang, S.; Woodcock, C.E. Improvement and expansion of the Fmask algorithm: Cloud, cloud shadow, and snow detection for Landsats 4-7, 8, and Sentinel 2 images. *Remote Sens. Environ.* **2015**, *159*, 269–277. [CrossRef]
33. Flood, N. Seasonal Composite Landsat TM/ETM+ Images Using the Medoid (a Multi-Dimensional Median). *Remote Sens.* **2013**, *5*, 6481–6500. [CrossRef]
34. Cohen, W.B.; Yang, Z.; Healey, S.P.; Kennedy, R.E.; Gorelick, N. A LandTrendr multispectral ensemble for forest disturbance detection. *Remote Sens. Environ.* **2018**, *205*, 131–140. [CrossRef]
35. Farr, T.G.; Rosen, P.A.; Caro, E.; Crippen, R.; Duren, R.; Hensley, S.; Kobrick, M.; Paller, M.; Rodriguez, E.; Roth, L.; et al. The shuttle radar topography mission. *Rev. Geophys.* **2007**, *45*, 2004. [CrossRef]
36. Instituto Brasileiro de Geografia e Estatística (IBGE). Cartas e Mapas de Bases Cartograficas Continuas. 2020. Available online: http://geoftp.ibge.gov.br/ (accessed on 4 May 2021).
37. Huete, A.; Didan, K.; Miura, T.; Rodriguez, E.P.; Gao, X.; Ferreira, L.G. Overview of the radiometric and biophysical performance of the MODIS vegetation indices. *Remote Sens. Environ.* **2002**, *83*, 195–213. [CrossRef]
38. Xu, D.; Wang, C.; Chen, J.; Shen, M.; Shen, B.; Yan, R.; Li, Z.; Karnieli, A.; Chen, J.; Yan, Y.; et al. The superiority of the normalized difference phenology index (NDPI) for estimating grassland aboveground fresh biomass. *Remote Sens. Environ.* **2021**, *264*, 112578. [CrossRef]
39. Xiao, X.; Boles, S.; Frolking, S.; Salas, W.; Moore Iii, B.; Li, C.; He, L.; Zhao, R. Observation of flooding and rice transplanting of paddy rice fields at the site to landscape scales in China using VEGETATION sensor data. *Int. J. Remote Sens.* **2002**, *23*, 3009–3022. [CrossRef]
40. Xu, H. Modification of normalised difference water index (NDWI) to enhance open water features in remotely sensed imagery. *Int. J. Remote Sens.* **2006**, *27*, 3025–3033. [CrossRef]
41. Keeley, J.E. Fire intensity, fire severity and burn severity: A brief review and suggested usage. *Int. J. Wildland Fire* **2009**, *18*, 116–126. [CrossRef]
42. Kaptué, A.T.; Hanan, N.P.; Prihodko, L. Characterization of the spatial and temporal variability of surface water in the Soudan-Sahel region of Africa. *J. Geophys. Res.* **2013**, *118*, 1472–1483. [CrossRef]
43. Wang, C.; Chen, J.; Wu, J.; Tang, Y.; Shi, P.; Black, T.A.; Zhu, K. A snow-free vegetation index for improved monitoring of vegetation spring green-up date in deciduous ecosystems. *Remote Sens. Environ.* **2017**, *196*, 1–12. [CrossRef]
44. McFeeters, S.K. The use of the Normalized Difference Water Index (NDWI) in the delineation of open water features. *Int. J. Remote Sens.* **1996**, *17*, 1425–1432. [CrossRef]
45. Haralick, R.M.; Dinstein, I.; Shanmugam, K. Textural Features for Image Classification. *IEEE Trans. Syst. Man Cybern.* **1973**, *SMC-3*, 610–621.
46. Conners, R.W.; Trivedi, M.M.; Harlow, C.A. Segmentation of a high-resolution urban scene using texture operators (Sunnyvale, California). *Comput. Vis. Graph. Image Process.* **1984**, *25*, 273–310. [CrossRef]
47. Ramola, A.; Shakya, A.K.; Van Pham, D. Study of statistical methods for texture analysis and their modern evolutions. *Eng. Rep.* **2020**, *2*, e12149. [CrossRef]
48. Yang, X.; Tridandapani, S.; Beitler, J.J.; Yu, D.S.; Yoshida, E.J.; Curran, W.J.; Liu, T. Ultrasound GLCM texture analysis of radiation-induced parotid-gland injury in head-and-neck cancer radiotherapy: An in vivo study of late toxicity. *Med. Phys.* **2012**, *39*, 5732–5739. [CrossRef] [PubMed]
49. Kennedy, R.E.; Yang, Z.; Gorelick, N.; Braaten, J.; Cavalcante, L.; Cohen, W.B.; Healey, S. 5 LT-GEE Outputs | LT-GEE Guide [WWW Document]. Kennedy RE Yang Z Gorelick N Braaten J Cavalcante Cohen WB Heal. 2018 Implement. LandTrendr Algorithm Google Earth Engine. *Remote Sens.* **2018**, *10*, 691. [CrossRef]
50. Beven, K.J.; Kirkby, M.J. A physically based, variable contributing area model of basin hydrology/Un modèle à base physique de zone d'appel variable de l'hydrologie du bassin versant. *Hydrol. Sci. J.* **1979**, *24*, 43–69. [CrossRef]

51. Talukdar, S.; Singha, P.; Mahato, S.; Shahfahad Pal, S.; Liou, Y.A.; Rahman, A. Land-use land-cover classification by machine learning classifiers for satellite observations-A review. *Remote Sens.* **2020**, *12*, 1135. [CrossRef]
52. Chavent, M.; Genuer, R.; Saracco, J. Combining clustering of variables and feature selection using random forests. *Commun. Stat. Simul. Comput.* **2021**, *50*, 426–445. [CrossRef]
53. Belgiu, M.; Drăguţ, L. Random forest in remote sensing: A review of applications and future directions. *ISPRS J. Photogramm. Remote Sens.* **2016**, *114*, 24–31. [CrossRef]
54. Friedman, J.H. Greedy function approximation: A gradient boosting machine. *Ann. Stat.* **2001**, *29*, 1189–1232. [CrossRef]
55. Greenwell, B.M. pdp: An R Package for Constructing Partial Dependence Plots. *R J.* **2022**, *9*, 421–436. [CrossRef]
56. RAISG. Amazon Geo-Referenced Socio-Environmental Information Network [WWW Document]. 2020. Available online: https://mineria.amazoniasocioambiental.org/ (accessed on 6 April 2021).
57. Singh, P.S.; Singh, V.P.; Pandey, M.K.; Karthikeyan, S. Enhanced classification of hyperspectral images using improvised oversampling and undersampling techniques. *Int. J. Inf. Tecnol.* **2022**, *14*, 389–396. [CrossRef]
58. Matthew, W. Bias of the random forest out-of-bag (OOB) error for certain input parameters. *Open J. Stat.* **2011**, *1*, 205–211.
59. Svetnik, V.; Liaw, A.; Tong, C.; Wang, T. Application of Breiman's random forest to modeling structure-activity relationships of pharmaceutical molecules. In Proceedings of the Multiple Classifier Systems: 5th International Workshop, MCS 2004, Cagliari, Italy, 9–11 June 2004; Springer: Berlin/Heidelberg, Germany, 2004; pp. 334–343. [CrossRef]
60. Joutsijoki, H.; Juhola, M. Kernel selection in multi-class support vector machines and its consequence to the number of ties in majority voting method. *Artif. Intell. Rev.* **2013**, *40*, 213–230. [CrossRef]
61. Jin, J.; Wang, Q. Assessing ecological vulnerability in western China based on Time-Integrated NDVI data. *J. Arid Land* **2016**, *8*, 533–545. [CrossRef]
62. Liu, S.; Wei, X.; Li, D.; Lu, D. Examining Forest Disturbance and Recovery in the Subtropical Forest Region of Zhejiang Province Using Landsat Time-Series Data. *Remote Sens.* **2017**, *9*, 479. [CrossRef]
63. Del Río-Mena, T.; Willemen, L.; Vrieling, A.; Nelson, A. Understanding Intra-Annual Dynamics of Ecosystem Services Using Satellite Image Time Series. *Remote Sens.* **2020**, *12*, 710. [CrossRef]
64. Asner, G.P. Automated mapping of tropical deforestation and forest degradation: CLASlite. *J. Appl. Remote Sens.* **2009**, *3*, 033543. [CrossRef]
65. Key, C.H.; Benson, N.C. Landscape Assessment (LA). Sampling and Analysis Methods. In USDA Forest Service—General Technical Report RMRS-GTR. 2006. Available online: https://gsp.humboldt.edu/OLM/Courses/GSP_216/labs/rmrs_gtr164_13_land_assess.pdf (accessed on 1 June 2021).
66. Caelli, T.; Julesz, B. Psychophysical Evidence for Global Feature Processing in Visual Texture Discrimination. *J. Opt. Soc. Am.* **1979**, *69*, 675–678. [CrossRef]
67. Bastarrika, A.; Chuvieco, E.; Martín, M.P. Mapping burned areas from Landsat TM/ETM+ data with a two-phase algorithm: Balancing omission and commission errors. *Remote Sens. Environ.* **2011**, *115*, 1003–1012. [CrossRef]
68. Claverie, M.; Ju, J.; Masek, J.G.; Dungan, J.L.; Vermote, E.F.; Roger, J.-C.; Skakun, S.V.; Justice, C. The Harmonized Landsat and Sentinel-2 surface reflectance data set. *Remote Sens. Environ.* **2018**, *219*, 145–161. [CrossRef]
69. Frampton, W.J.; Dash, J.; Watmough, G.; Milton, E.J. Evaluating the capabilities of Sentinel-2 for quantitative estimation of biophysical variables in vegetation. *ISPRS J. Photogramm.* **2013**, *82*, 83–92. [CrossRef]
70. Trivedi, M.; Marshall, M.; Estes, L.; de Bie, K.; Chang, L.; Nelson, A. Cropland Mapping in Tropical Smallholder Systems with Seasonally Stratified Sentinel-1 and Sentinel-2 Spectral and Textural Features. *Remote Sens.* **2023**, *15*, 3014. [CrossRef]

remote sensing

MDPI

Article

Detection and Monitoring of Woody Vegetation Landscape Features Using Periodic Aerial Photography

Damjan Strnad [1,*], Štefan Horvat [1], Domen Mongus [1], Danijel Ivajnšič [2,3] and Štefan Kohek [1]

[1] Faculty of Electrical Engineering and Computer Science, University of Maribor, 2000 Maribor, Slovenia
[2] Faculty of Natural Sciences and Mathematics, University of Maribor, 2000 Maribor, Slovenia
[3] Faculty of Arts, University of Maribor, 2000 Maribor, Slovenia
* Correspondence: damjan.strnad@um.si

Abstract: Woody vegetation landscape features, such as hedges, tree patches, and riparian vegetation, are important elements of landscape and biotic diversity. For the reason that biodiversity loss is one of the major ecological problems in the EU, it is necessary to establish efficient workflows for the registration and monitoring of woody vegetation landscape features. In the paper, we propose and evaluate a methodology for automated detection of changes in woody vegetation landscape features from a digital orthophoto (DOP). We demonstrate its ability to capture most of the actual changes in the field and thereby provide valuable support for more efficient maintenance of landscape feature layers, which is important for the shaping of future environmental policies. While the most reliable source for vegetation cover mapping is a combination of LiDAR and high-resolution imagery, it can be prohibitively expensive for continuous updates. The DOP from cyclic aerial photography presents an alternative source of up-to-date information for tracking woody vegetation landscape features in-between LiDAR recordings. The proposed methodology uses a segmentation neural network, which is trained with the latest DOP against the last known ground truth as the target. The output is a layer of detected changes, which are validated by the user before being used to update the woody vegetation landscape feature layer. The methodology was tested using the data of a typical traditional Central European cultural landscape, Goričko, in north-eastern Slovenia. The achieved F_1 of per-pixel segmentation was 83.5% and 77.1% for two- and five-year differences between the LiDAR-based reference and the DOP, respectively. The validation of the proposed changes at a minimum area threshold of 100 m^2 and a minimum area percentage threshold of 20% showed that the model achieved recall close to 90%.

Keywords: woody vegetation landscape features; change detection; segmentation neural network; cyclic aerial photography; digital orthophoto

1. Introduction

The loss of landscape and biotic diversity due to intensified land use for agricultural production and urbanization is one of the major problems in the European Union (EU), whose strategy for reversing the ecosystem degradation is a core part of the European Green Deal (https://commission.europa.eu/strategy-and-policy/priorities-2019-2024/european-green-deal_en, accessed on 22 May 2023). The European biodiversity strategy for 2030 (https://environment.ec.europa.eu/strategy/biodiversity-strategy-2030_en, accessed on 22 May 2023) includes specific actions for the preservation and restoration of important habitats. In Slovenia, the landscape diversity is threatened by both agricultural expansion in some areas, and its abandonment in others. The National Environmental Action Programme for 2020–2030 (https://www.gov.si/assets/ministrstva/MOP/Publikacije/okolje_en.pdf, accessed on 22 May 2023) specified the conservation of biodiversity and protection of valuable natural features in Slovenia as some of the main goals. The protection of specific high nature value habitat types (HT) is one of the guidelines, which is considered in the adoption of future agricultural development plans. Because the prevalent land cover type in the Slovenian rural environment is a mosaic (i.e., a mixture of several different HT), its characteristic components are small woody vegetation landscape features, such as hedges, tree lines, and riparian vegetation. These structures are crucial elements of ecosystem diversity and have many positive effects on the environment since they prevent erosion and offer shelter to (endangered) wildlife. The importance of their preservation and restoration has been recognized and incorporated into European ecological strategies and national policies. Landscape features are acknowledged in the Slovenian 2023-2027 Common Agricultural Policy Strategic plan (https://skp.si/en/cap-2023-2027, accessed on 22 May 2023) as the driving forces of biotic diversity and representatives of national identity with both cultural and aesthetic value.

Following the practice of European countries with established advanced systems for the registration and monitoring of landscape features in agroecosystems, a Land Parcel Identification System (LPIS) (https://eprostor.gov.si/imps/srv/api/records/8c8072f5-2075-49c3-b3e5-56ee58f8db8d, accessed on 22 May 2023) is used in Slovenia as a central registry of control layers related to areas of ecological importance. In recent years, several studies have been funded by the government to provide the necessary data layers for LPIS. The most elaborate product of this type so far, covering the whole national territory, was commissioned by the Institute of the Republic of Slovenia for Nature Conservation (https://iaps.zrc-sazu.si/en/programi-in-projekti/feasibility-study-and-mapping-of-vegetative-landscape-structures-important-for, accessed on 22 May 2023). It contains more than 3 million woody vegetation landscape features (outside woodland), which are classified into five classes [1]. The layer is derived from the LiDAR point clouds provided by the airborne laser scanning (ALS) of Slovenia, which was conducted for the last time in 2014 (http://gis.arso.gov.si/evode/profile.aspx?id=atlas_voda_Lidar@Arso, accessed on 22 May 2023). Due to this, the layer is already an inaccurate representation of the current situation in the field.

In order to support periodic updates of the woody vegetation landscape feature databases in agroecosystems, there is a desire to build a system for the automated detection of vegetation changes from remote sensing data. In the paper, we propose a methodology that uses DOP from periodic aerial surveys to generate the change layers. The methodology is envisioned as the basis for a future decision support tool where the user can set the significance level of detected changes and validate them before updating the database. The goal of the study is to evaluate the efficiency and suitability of the proposed approach for such purposes.

In terms of temporal resolution, an attractive source of remote sensing data for change detection is satellite imagery, such as Sentinel-2 multispectral data. In the past, a lot of research has been devoted to the detection of land cover changes on a large scale from satellite images [2–10], where the typical value of a minimum mapping unit (MMU) is on the order of 100 m^2. The main advantage of using Sentinel-2 data is its low cost. However, the spatial resolution of 10 m/pixel for the color and NIR channels is insufficient for reliable detection of small-scale vegetation changes driven mainly by man for agricultural purposes. The monitoring of such woody vegetation landscape features requires higher precision to capture changes in narrow linear structures and small patches of non-forest greenery, which can be only a few meters in width or diameter. Using very high spatial resolution (VHR) satellite images can provide a necessary solution [11,12], but is less attractive due to the higher costs involved in periodic updates. Satellite data were, therefore, not used in this study.

High-resolution images of the ground can also be produced by unmanned aerial vehicles (UAVs), which have been used on a limited local scale for precise vegetation mapping and classification from hyperspectral [13,14], multispectral [15,16], and RGB data [17–19], as well as for land cover change detection [20,21]. However, extending such an approach to a regional scale is, in the best case, impractical (e.g., time-consuming, incurring high costs, requiring high data capacity and computational power). Some of the recent approaches have, thus, started addressing the middle ground, where high-resolution vegetation monitoring on a regional or state level is based on periodic aerial surveys [22]. For instance, the cyclic aerial photography of Slovenia (CAS) has been performed in digital form regularly since 2006, with a period of approximately three years in which the whole area of the country is covered. The product of CAS is a color digital orthophoto (DOP) with a spatial resolution of 25 cm/pixel, which currently presents the best compromise in terms of spatial/temporal resolution and cost for use in landscape feature monitoring. In this paper, an automated procedure is proposed for periodic detection of changes in woody vegetation landscape features, which can be combined with manual validation for regular updating of vegetation layers.

Previous studies demonstrated that high-resolution imagery is required for the precise detection of small woody vegetation features. Several object-based approaches for this task, which typically involves image segmentation, feature extraction, and object classification, have been proposed in the past. The problem of mapping trees outside forests was addressed using object-based image analysis techniques [23] and random forest classification [24], where it was shown that combining UAV and multi-temporal satellite imagery can provide the best overall accuracy. In order to map linear woody vegetation, an object-based approach was proposed, using discriminant functions with spectral, textural, and shape features [25]. The problem of mapping riparian habitats was addressed using a random forest classifier in combination with airborne [26] and UAV imagery [27]. A similar approach was employed to perform wide-scale mapping of non-cropped habitats in high-resolution aerial photography [28], while classification trees were used for the detection of shrub encroachment in areas where agricultural management practices have been abandoned [29]. Integration of multiple classifier outputs was shown to improve overall classification accuracy [30].

In most cases, manual labeling of training samples is necessary to provide the ground truth or masks for vegetation mapping, but some studies rely on existing datasets, such as Small Woody Features (SWF) [31–33]. SWF is a product available through Copernicus Land Monitoring Services (https://land.copernicus.eu/pan-european/high-resolution-layers/small-woody-features, accessed on 22 May 2023), but only for the year 2015, while the SWF for 2018 is, at the time of this writing, still in production. The spatial resolution of the SWF raster layer is 5 m/pixel, while the vector layer classifies woody features into linear, patchy, and additional features. The latter are features that are, according to the rules, neither linear nor patchy but enhance the connectivity of other features or represent isolated features with an area larger than 1500 m^2.

In recent years, the field of semantic image segmentation has been dominated by deep learning (DL) models, which have achieved substantial performance improvements over the previous state of the art [34]. Many DL-based approaches to semantic image segmentation employ an end-to-end convolutional neural network (CNN), such as the fully convolutional network (FCN) [35], in order to classify individual pixels into predefined semantic classes. The architecture used by several popular segmentation neural networks is the encoder-decoder [36–38]. The encoder transforms the input image into a compressed latent representation using a backbone feature extractor, which is typically a CNN pre-trained on a large dataset such as ImageNet [39]. The decoder reconstructs from the latent representation a segmented image of original size through a series of up-sampling operations and a final pixel-wise classification. The encoder-decoder architecture has been employed for vegetation mapping from both UAV [40,41] and high-resolution satellite images [42]. Improvements over the baseline have later been achieved by introducing dilated separable convolutions [43] and the channel attention mechanism [44].

In the context of aerial and satellite image segmentation, recent advances comprise techniques like transfer learning from high-resolution satellite datasets [45], the use of segmentation neural network ensembles [46], hybrid architectures [47,48], adaptive CNNs that improve the classification of easily confused classes [49], and deformable convolutions that adjust the receptive field to geometric deformations of shape and size [50,51]. Some alternative approaches to image segmentation employ graph neural networks (GNN) [52–54], which operate on graph nodes constructed from an image in a preprocessing step. Another recent line of research in remote sensing image segmentation uses attention-based transformers to replace or supplement the backbone CNN [55–58].

In this paper, we propose a methodology for detecting woody vegetation landscape features in cyclic aerial photography and marking areas where significant changes with respect to a reference layer are indicated by the model. Besides the segmentation neural network-based detector, the methodology includes post-processing steps to match the detected features in consecutive DOPs and indicate the locations of potentially important changes for visual or field inspection. Such a methodology could be used primarily for semi-automated layer updates in the periods between consecutive ALSs. The proposed methodology is a step towards systematic registration and monitoring of woody vegetation landscape features, which is important for the shaping of future environmental and agricultural policies, including issuing Subsidy Control Acts to encourage establishing new landscape features and preventing their removal.

The main contributions of the paper are:

- machine learning based detection of small-scale woody vegetation landscape features in aerial photography,
- segmentation neural network training with LiDAR-based targets and weighted training loss, masked by combined land use, cadastre, and public infrastructure masks,
- a methodology for automated generation of change layers at user-specified levels of detail,
- experimental evaluation of system efficiency for a region in north-eastern Slovenia,
- discussion of limitations and analysis of conditions that impact model accuracy.

The rest of the paper is organized as follows. In Section 2, an overview of the proposed methodology is presented, and its individual components are described in detail. The experimental setup is described, and the results are reported in Section 3. This is followed by an analysis of the results and a discussion of limitations in Section 4, while Section 5 concludes the paper.

2. Materials and Methods

In this section, we first provide an overview of the proposed methodology, and then describe individual components in more detail. We also describe the way the methodology is evaluated from the perspective of its use as a decision-support tool.

2.1. Overview

The goal of the proposed methodology is to support the automated generation of a woody vegetation change layer from the input DOP. The changes are indicated with respect to a reference layer, which is typically the last known state of the vegetation. The significance of change is determined based on a user-specified threshold, which defines the minimum difference area or some other property of the mismatch between the old and the new layers. The core element of the methodology is a segmentation neural network, which is trained from the perspective that the extent of significant changes from the last cycle is small compared with the non-changes. This means that periodically updated targets remain largely representative of the situation in the field, which allows their use for re-training the network on a new DOP. The expectation is that, with a large enough training set, the segmentation neural network will be forced to disagree with the training targets in regions where there is a true change.

A high-level schematic of the proposed methodology for updating the landscape feature layer is shown in Figure 1. A segmentation neural network is first trained using the new DOP against the targets sampled from the last known ground truth (Figure 1 top), and the trained model is used to perform the segmentation of the whole region (Figure 1 bottom).

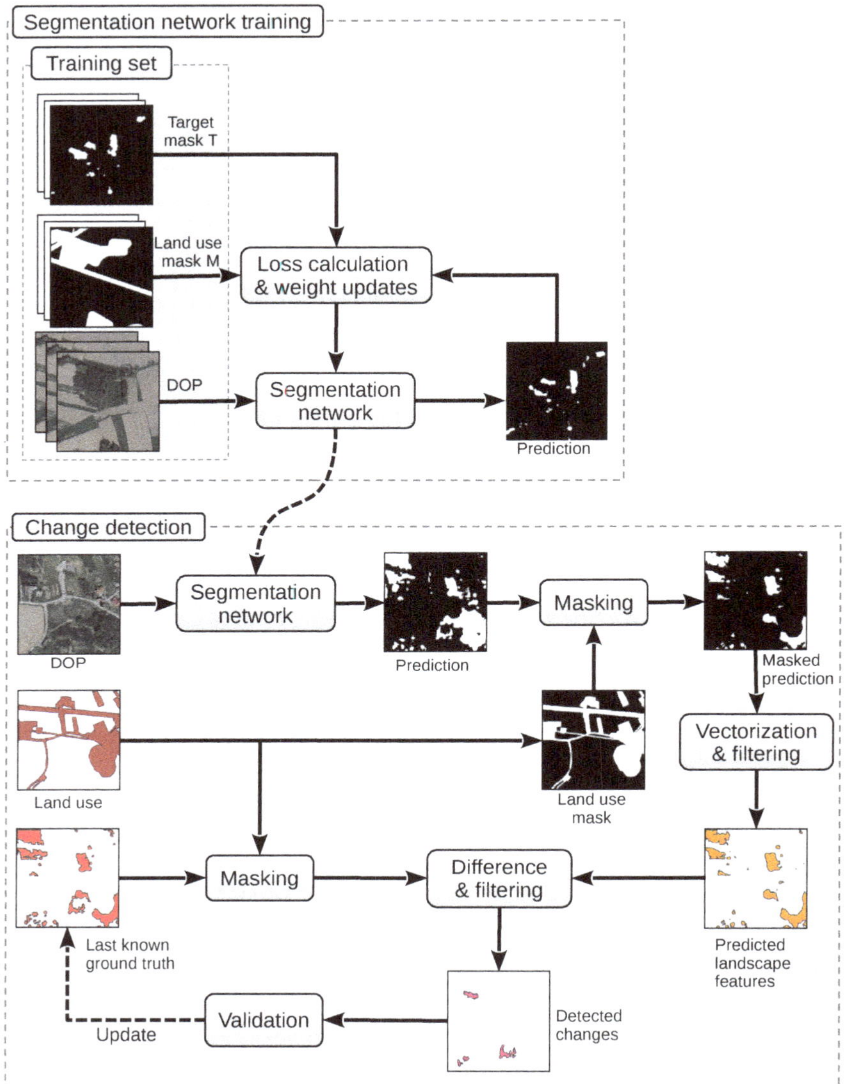

Figure 1. The structure of the proposed methodology for change detection in woody vegetation landscape features.

In both the training and final segmentation, the regions of specific land use, such as forests and intensive orchards, are masked out in order to exclude from model evaluation the vegetation that does not fall under the category of suitable woody vegetation landscape features. The obtained raster is vectorized, and features with areas smaller than 10 m^2 are removed following the methodology proposed in [1]. This eliminates potential leftover vegetation fragments from the filtering step, as well as individual bushes and small trees, which are not considered standalone woody vegetation landscape features. A vegetation change layer is produced next as a symmetric difference between the new layer and the last ground truth, and the area attributes are computed for the new features. The change layer is finally filtered with respect to user-specified threshold values for minimum area and percentage of change. This allows the user to focus on the most significant changes first or specify the minimum required extent of change to be registered. The changes are validated manually by visual inspection of the DOP or through field inspection before they are used to update the ground truth. In the case of CAS, such updates would be performed once per year after a new DOP is obtained for the target areas where woody vegetation landscape features are lacking owing to intensive agricultural practices.

2.2. Study Area

The selected study area is the region of Goričko, which lies between 46°61′–46°88′ N and 15°98′–16°43′ E, covering approximately 462 km^2 of hilly landscape in the northeasternmost part of Slovenia (Figure 2). It is the largest landscape park in the country (with the exception of the National parks). The region was selected because it is one of the highest priority areas for the preservation of existing and the formation of new woody vegetation landscape features in the country, and environmental actions are implemented here much more easily owing to the specific land ownership structure. The whole landscape park also falls under the Natura 2000 network of protected areas for biodiversity conservation, which makes Goričko an interesting target for continuous surveillance of agricultural practice implementations.

2.3. Data

The datasets used in this study are the LiDAR point clouds, the DOP, the building cadastre, the public infrastructure cadastre and the land use database for the selected study area of Goričko.

The LiDAR scanning of Goričko was performed in early 2014. The average density of first returns calculated over 10 m^2 squares was 5 points/m^2, while the root mean square error (RMSE) of absolute height measurements for the entire survey block was 8 cm (http://gis.arso.gov.si/related/lidar_porocila/b_24_izdelava_izdelkov.pdf, accessed on 22 May 2023). The geo-referenced and classified point clouds are available for download as a grid of 1 km^2 cells from the website of the Slovenian Environment Agency (http://gis.arso.gov.si/evode/profile.aspx?id=atlas_voda_Lidar@Arso&culture=en-US, accessed on 22 May 2023).

The DOPs of Goričko for the years 2014, 2016, and 2019 were used in this study. Aerial images was produced with a 25 cm ground sample distance. The orthorectification of images was performed using a 5 m photogrammetric digital terrain model (DTM) for the year 2014 and a LiDAR-based DTM with 1 m resolution for the years 2016 and 2019. The public digital orthophoto is accessible through the e-survey data portal maintained by the Surveying and Mapping Authority of the Republic of Slovenia (https://egp.gu.gov.si/egp/?lang=en, accessed on 22 May 2023).

The building cadastre is a vector layer containing approximately 1.18 million shapes that represent the ground plans of registered buildings for the whole country. The public infrastructure cadastre contains vector layers for different public infrastructure objects and networks, such as roads, railways, water supply networks, communication networks, and electric grids. In this study, the power transmission line layer was used, which contains the medial axes of transmission corridors. One of the important object attributes is the power line voltage level, which determines the required width of the corridor (ranging from 1.5 m to 40 m). The vegetation encroaching on the transmission corridors is not considered a landscape feature, as it is removed during periodic corridor maintenance. Both datasets can be obtained from the same data portal as DOP.

The last dataset is the national record of land use, which contains a vector layer with approximately 1.7 million objects. The objects are classified into 25 different types of land use based on photo-interpretation, field inspections, and third-party information. The updates to the layer correspond to the coverage of yearly CAS surveys, which include approximately one third of the national territory. The land use dataset is published on the website of the Ministry for Agriculture, Forestry and Food (https://rkg.gov.si/vstop/, accessed on 22 May 2023).

The main properties of the used datasets are summarized in Table 1.

Table 1. Main properties of datasets used in the study.

Dataset	Data Type	Date	Properties
LiDAR	Point cloud	4/2014	Density: 5 pts/m^2
DOP	RGB images	5/2014, 4/2016, 4/2019	Resolution: 25 cm/px
Buildings	Polygons	2020	#Features: 1.18 M
Transmission lines	Lines	2020	#Features: 319,455
Land use	Polygons	2014, 2017, 2020	#Features: 1.55 M–1.74 M

2.4. Data Processing

The original data were processed in order to produce the masks and layers needed in the later stages of the proposed pipeline. Individual pre-processing steps were performed using available open-source software tools and libraries for working with GIS data, which are described in the continuation.

The LiDAR point clouds were used to generate the reference layer of woody vegetation landscape features by roughly following the approach by Kokalj et al. [1]. The points already classified as ground were first used to generate a digital terrain model (DTM) at 1 m resolution, which matches the resolution used in the orthorectification of aerial images during the production of the DOP. The generated DTM was used to extract the raster canopy height model (CHM) using the pit-free method [59]. These steps were fully automated using the lidR package (https://github.com/r-lidar/lidR, accessed on 22 May 2023), which provides a set of tools and algorithms for the manipulation of airborne LiDAR data [60].

The vectorization of CHM was performed next, where only the pixels with a height above 2 m were considered. The result was subsequently filtered in order to remove all segments not corresponding to potential regions of interest (RoI). The filtering was performed by sequentially subtracting the geometry of individual masking layers from the vectorized CHM. In particular, the building cadastre, the public infrastructure layer, and the land use layer were used to respectively mask out urban areas, power line corridors, and registered special purpose areas, such as hop fields, greenhouses, vineyards, intensive orchards, forests, and water bodies. The power line corridors were derived from their axial representation by applying buffers of prescribed clearance widths depending on the voltage levels of the power lines.

In order to prevent detecting narrow strips of forest border and connected vegetation as separate landscape features due to the imprecision of the land use layer, a 10 m buffer was used to expand forest segments before filtering. In a similar way, a 2 m buffer was used for the building cadastre, and a 1 m buffer for the power lines. In contrast, the regular fields were not filtered out because this would remove most of the narrow linear vegetation along field edges, which are rarely delineated separately in the land use layer. Finally, the remaining segments with areas smaller than 10 m^2 were discarded, and the result was rasterized to produce the candidate target mask at a spatial resolution of 25 cm. All of the steps were performed using the QGIS (https://qgis.org/, accessed on 22 May 2023) software tool and automated with the GDAL (https://gdal.org/, accessed on 22 May 2023) library, which includes Python utilities for operations on raster and vector layers as well as conversions between different formats.

In order to reduce the fragmentation due to the many small gaps in the vegetation picked up by LiDAR, a morphological closing with a square structuring element of 3 pixels width was applied to the raster layer. The resulting target mask T was used as ground truth in the model training as well as to derive the potential vegetation changes based on the disagreement between the baseline and the model predictions. The joined mask M of areas that were filtered during the target mask preparation was also rasterized. It was used during the training to limit the calculation of segmentation loss to regions of interest as well as to perform the final filtering of the model's predictions.

The training dataset for the segmentation neural network was built using 32 sample tiles corresponding to rectangular patches of land selected from various parts of the Goričko region. Each tile was 4000 × 4000 pixels and covered an area of 1 km^2. Another 8 tiles were used for the validation dataset during training, while the final 8 tiles were used for the test set (Figure 2). The plots were chosen such that the number of pixels belonging to woody vegetation landscape features represented at least 3% of the non-masked area in the tile.

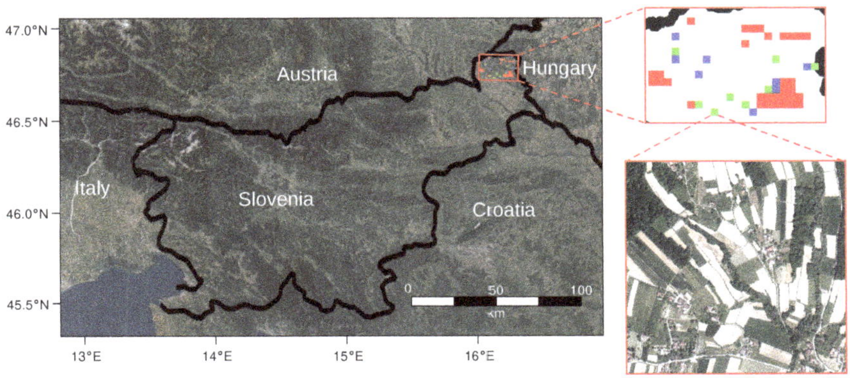

Figure 2. Location of the target study area (the Goričko region) within Slovenia, and the locations of the training tiles (red), validation tiles (blue), and test tiles (green) within Goričko.

The training, validation, and test samples consisted of spatially aligned parts of the DOP, land use mask M, and target mask T. The actual training samples were generated from the training tiles on the fly by cropping random subimages of size 1024 × 1024 pixels, which is the expected input dimension for the segmentation neural network. Training set augmentation was performed by flipping the extracted samples randomly in horizontal and vertical directions, each with a 50% probability. Unlike the training samples, the validation and test samples were obtained by uniform subdivision of the corresponding tiles into 4 × 4 cells with a 32 pixel overlap between neighboring images.

2.5. Segmentation Model

The core step of the proposed change detection methodology involved training a segmentation neural network, which uses the most recent DOP as an input and outputs a segmentation probability map. The last known state of woody vegetation landscape features was used as the target reference T for computing the training loss, while the combined land use mask M was used to constrain the loss computation to relevant pixels. The expectation was that training with fresh input data and the last ground truth as the target would force the model to misclassify primarily in areas of actual changes because adjusting for them would lead to a significantly higher overall training loss.

In our experiments, we used a pre-trained segmentation neural network based on the U-Net architecture (Figure 3) from the Python segmentation models package (https://github.com/qubvel/segmentation_models, accessed on 22 May 2023). The model allows the selection of several backbone neural network architectures, which have been pre-trained on the ImageNet dataset. We selected the EfficientNet backbone [61] because it provides better or comparable performance using a much smaller number of network parameters than other architectures, which allows network training on a regular desktop GPU. The fine-tuning of the U-Net model to a particular task of vegetation detection in DOP was finally performed. In order to speed up the training and reduce the space requirements, the encoding layers of U-Net can be frozen during fine-tuning, and only the decoding layers are adapted. However, we found that fine-tuning the complete stack was beneficial for performance, presumably because aerial imagery contains visual features and patterns not present in typical image datasets.

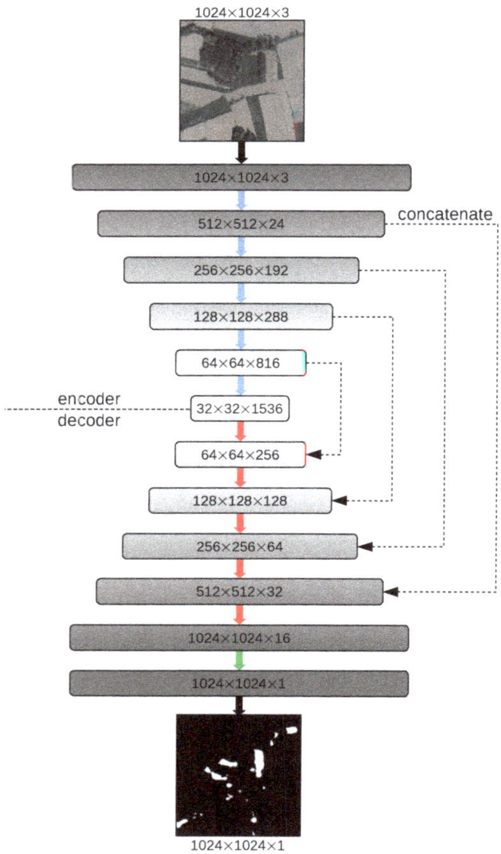

Figure 3. The U-Net architecture of the segmentation neural network. The blue arrows represent downscale operations, the red arrows are upscale operations, and the green arrow is a final per-pixel binary classification.

The training was implemented in Tensorflow using the Keras neural network library. The Adam optimizer with an initial learning rate $\eta = 10^{-4}$ and default values for other parameters was used for model optimization. The output prediction of the model was a 2D probability map $\hat{Y}$ with the same width and height as the input image. The loss function was a customized binary cross-entropy loss, which applied the mask of land use M to constrain loss contribution to non-masked pixels only. The errors made in pixels corresponding to regions of interest (i.e., woody vegetation landscape features), were weighted differently according to the following equation:

$$\mathcal{L} = -\frac{1}{|\hat{Y}|} \sum_i (1 - m_i)(w y_i \log(\hat{y}_i) + (1 - w)(1 - y_i) \log(1 - \hat{y}_i)).$$

Here, w is the weight for balancing false positive and false negative errors, and was, in our final experiments, set to 0.6 after extensive hyper-parameter tuning on a validation set. The mean loss was calculated over all pixels i in the output, where $m_i \in M$ is the corresponding mask value (1 for masked pixels), $y_i \in Y$ is the target value, and $\hat{y}_i \in \hat{Y}$ is the model prediction.

In the ideal case, re-training a model to a new DOP would not be required. However, there can be significant differences between photos from different surveys due to varying phenological stages of the target vegetation and illumination conditions. Some differences can be reduced by careful planning of flight dates, but this depends on many factors, including the weather. Alternatively, different training data augmentation techniques, such as brightness adjustment and adaptive histogram equalization, can be applied during the training of the original model in order to simulate varying environmental factors. The downside is that this may lower the model's performance on an otherwise homogeneous training set such as the CAS, where a selected target region is typically recorded in a short time span, during which the state of vegetation foliage and weather conditions are consistent. Additionally, not all differences can be adequately simulated by data augmentation, and the model used in the previous cycle is likely overfitted to conditions present at the time of the aerial imagery capture. Due to that, either a fine-tuning of the previous network or the training of a new model from scratch is necessary. We compared different approaches in the experiments. The results demonstrated that training a new model was more efficient than fine-tuning a trained model from the previous cycle. The most likely reason is that fine-tuning cannot escape the local optimum basin of a model that was adapted to the seasonal characteristics of aerial images from the previous training cycle, although training data augmentation can help in reducing the performance gap.

2.6. Post-Processing and Filtering

The post-processing of model predictions in the change detection phase was completely automated using the GDAL library. The value of each pixel in the output image represents the probability of vegetation's presence in the pixel. The image was first converted to binary using a probability threshold value of 0.5. The same mask M that was used during training was applied next in order to remove the areas of land use that were not compatible with the woody vegetation landscape feature definition. The images generated from subparts of the original tiles were concatenated back to form a single larger raster, and morphological closing with a 3-pixel window was applied to it. The result was vectorized, and the areas were computed for the so-obtained feature polygons. A symmetric difference was computed next, with the reference vector layer representing the last known state of woody vegetation landscape features. The land use mask was applied to the reference layer as well in order to eliminate areas whose land use designations may have changed since the last update (e.g., overgrown abandoned farmland can, after some time, be reclassified as forest). The resulting polygons corresponded to potential changes in nature, representing the locations of vegetation ingrowth or its removal. A final filtering step was performed in which the areas of the detected change polygons were computed first, as well as the ratio between the area of the change and that of the source polygon (reference or prediction). In the last step, all change polygons were removed whose area or percentage of change were below the user-specified thresholds.

2.7. Validation

The efficacy of the proposed approach was evaluated on an independent test set by human visual inspection. The test set included the DOPs and masks from the CAS years 2016 and 2019, which were used to extract separate woody vegetation change layers with respect to the reference LiDAR-based layer from 2014. The detected changes of woody vegetation landscape features were visualized and presented to the user in an overlay with the DOP 2014 and the DOP from the respective comparison year. The user was able to show or hide individual layers in order to validate the proposed changes. In this way, the user could label false positive (FP) and true positive (TP) cases of woody vegetation change by concentrating on the polygons in the generated layer. In order to expose the false negative (FN) cases, which represent the changes overlooked by the model, the user also had the ability to draw a polygon around the apparent change. Each such manually delineated polygon was clipped against the reference layer, and the area of the result was calculated. The percentage of area change for the polygon was finally calculated with respect to the polygon's total area. The resulting feature was counted as a false negative if it passed the threshold conditions.

3. Results

The aim of the experiments was to simulate a typical scenario, which is first to train the segmentation neural network using the DOP from the selected CAS year as input and the LiDAR-based ground truth as target. The trained model was then applied to a test set, and the results were used to generate the woody vegetation landscape feature change layer. Separate experiments were performed using the training data from the CAS years 2014, 2016, and 2019. The data from 2014 were used only to evaluate the extent of agreement between the DOP-based predictions and the LiDAR-based reference targets from the same year. The target layer was not updated between consecutive experiments because the goal was to assess the degradation of the landscape feature registry when no monitoring was performed for an extended period of time.

The experiments were conducted on a computer with an AMD Ryzen 9 5900X CPU and an NVIDIA Geforce RTX 3060 GPU with 12 GB of VRAM. The operating system was Manjaro Linux kernel 6.0. The efficientnetb3 backbone was used as a feature extractor in the segmentation neural network. The training data for the network were 32 tiles of the aligned DOP, land use mask, and target mask. In each epoch, a single random training sample of size 1024 × 1024 pixels was extracted from each tile. The network was trained for 200 epochs, which took approximately one hour to complete. A separate validation set, which contained 128 samples obtained by uniform subdivision of 8 larger tiles, was used during training to record the best performing model variant. After training, the stored best model was used to perform the segmentation of a test set with another 128 samples. Table 2 summarizes the results in terms of the achieved per-pixel precision (P), recall (R), and F_1 score on the training and test sets.

Table 2. Training and test precision (P), recall (R), and F_1 score of models trained on DOPs from different CAS years and the LiDAR-based ground truth

Year	Train			Test		
	P	R	F_1	P	R	F_1
2014	88.3%	88.3%	88.3%	88.5%	87.3%	87.9%
2016	81.4%	83.3%	82.3%	83.6%	83.3%	83.5%
2019	76.0%	73.9%	74.9%	78.7%	75.6%	77.1%

As explained before, the training of a new model for each CAS cycle is desirable for the best results, because not all seasonal differences in aerial photography can be adequately simulated by data augmentation. We tested this using a series of experiments. We first used the 2014 model directly to perform the segmentation of DOP 2016 and DOP 2019, which resulted in an F_1 equal to 71.7% and 67.2%, respectively, a substantially worse outcome than what we got with the dedicated model. As the next step, the 2014 training data were augmented using random combinations of brightness, contrast, hue/saturation/value, and histogram equalization. The F_1 scores of the resulting model improved to 75.7% and 69.8%. The last attempt involved additional fine-tuning of the 2014 model with new data for 50 epochs using a smaller initial learning rate $\eta = 10^{-5}$. This improved the F_1 score for the 2016 and 2019 datasets further to 81.2% and 74.9%, respectively, but the preference clearly remained for specialized models.

The two trained models for the years 2016 and 2019 were used to produce corresponding change layers for woody vegetation landscape features as described in Section 2.6. Two levels of thresholding were applied during the filtering stage in order to assess the usability of the approach at different granularities. For level 1, the more selective threshold values $T_1 = 100$ m^2 for the minimum area of change and $T_2 = 20\%$ for the minimum percentage of change were used in order to make the process of manual validation manageable. At level 2, the relaxed requirements $T_1 = 30$ m^2 and $T_2 = 10\%$ were used for comparison purposes only.

Figure 4 shows, for a selected region, the DOPs from different years, the detected landscape features, and the changes corresponding to the two filtering levels. Table 3 reports the number of detected changes for the case $T_1 = 100$ m^2 and $T_2 = 20\%$, as well as the counts of TP, FP, and FN determined by visual inspection of the results.

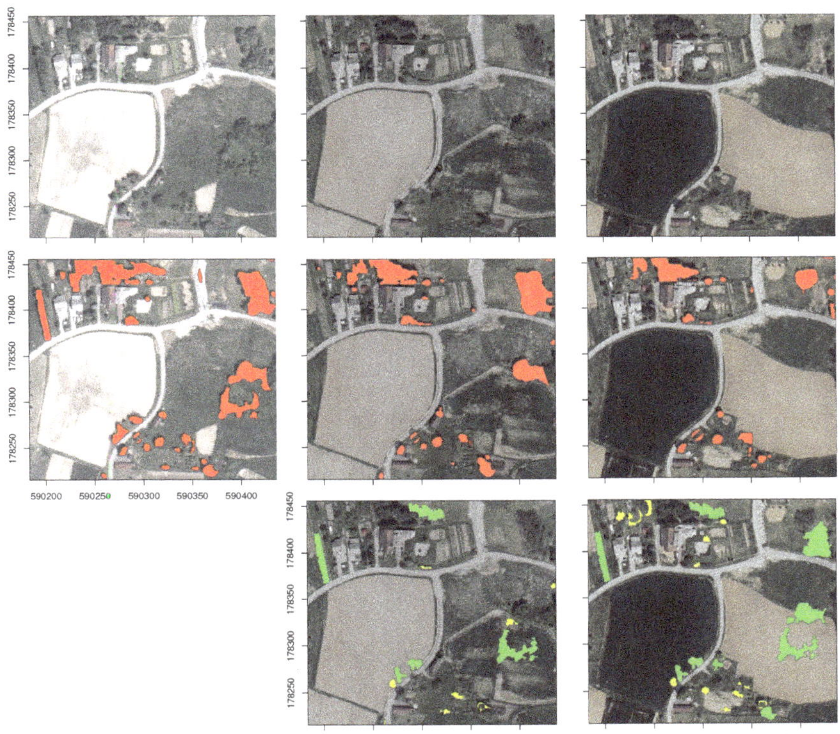

Figure 4. DOPs from the years 2014, 2016, and 2019 (top row, left to right), detected woody vegetation landscape features (red polygons, middle row), and detected changes at level 1 (green polygons) and level 2 (additional yellow polygons). The scale is in meters, and the coordinate reference system (CRS) is EPSG:3794.

Table 3. Validation results for the produced landscape feature change layer at the threshold values $T_1 = 100$ m^2 and $T_2 = 20\%$.

Year	*TP*	*FP*	*FN*	*P*	*R*	F_1
2016	84	30	10	0.737	0.894	0.808
2019	173	49	19	0.715	0.901	0.836

The high-detail change detection at level 2 resulted in 772 detected changes in 2016, and 1142 detected changes in 2019. It should be noted that, by reducing the threshold values, the increasing number of detected changes was due to the displacement of objects in orthorectified DOPs. There were also many cases where the gaps between previously sparse vegetation were overgrown. However, we considered this to be a valid case for joining previously fragmented woody vegetation landscape features into larger plots.

4. Discussion

The proposed approach to monitoring woody vegetation landscape features differs from related studies in that it addresses the complete workflow necessary to support semi-automated layer updates. A segmentation neural network is trained in order to detect woody vegetation landscape features in a DOP from periodic aerial surveys. The initial training target is generated from LiDAR point clouds, while the detected vegetation changes are presented for validation and used for subsequent database updates. Unlike the more general vegetation detection approaches, the proposed methodology uses additional layers for vegetation filtering in both the model training and the post-processing phases.

We compared the segmentation performance of our model with similar studies that used UAV [41] and high-resolution satellite images [42]. In ref. [41], a U-Net model was used for shrub segmentation, and achieved an F_1-score of 82% on images with a comparable spatial resolution of approximately 25 cm/pixel. This is below the 88% score of our model obtained with the contemporary DOP and reference layer. The authors reported a notable drop in performance on images with distinct seasonal characteristics, which is consistent with our findings about the use of the same model on images from different surveys. By using satellite images with a 1 m spatial resolution for mapping in woody vegetation, the approach in [42] achieved a precision and recall of around 88%, which is very close to our results. However, both reference approaches involved manual labeling of training data, while our method uses a LiDAR-based ground truth.

Based on the comparable training and test set results in Table 2, we can conclude that overfitting of the segmentation models to the training set did not occur. As demonstrated by the results, the best training performance can be achieved by training a separate model for each year of the aerial photography cycle. Besides assuring similar imaging conditions and vegetation states, an additional motivation for the use of specialized models in the case of CAS is that individual aerial surveys cover parts of the country in adjacent photogrammetric blocks where the landscape and vegetation are relatively homogeneous [62]. Indeed, in a country with highly diverse natural biotopes, an even finer granulation of models could be applied at the level of individual blocks at a reasonable additional cost of preparing separate training data. For instance, the eastern part of Slovenia is recorded in three photogrammetric blocks, which include both the Pohorje medium mountain range and the Pannonian plain. The survey of the western country part, similarly, encompasses four blocks of diverse alpine and karstic landscapes, which could benefit from sub-specialized models. Such an approach would be feasible given the low cost of model training.

Another finding of the study is that end-to-end fine-tuning of a pre-trained model is superior to training the decoder alone. This is consistent with the results of related studies, which have shown that fine-tuning of the encoder is beneficial for performance [63,64]. It was further established that transfer learning from a generic pre-trained model was more efficient than additional fine-tuning of a model from the previous cycle. Similar findings have been reported in the literature [65]. Although the surveys were made in the same period of the year, there were important year-to-year differences in the phenological stages of vegetation and capture conditions, and it was shown in [66] that such significant data distribution shifts cannot always be adequately simulated by synthetic data augmentation.

Regarding the validation of the detected changes, it can be observed from Table 3 that the number of false negatives was considerably lower than the number of false positives. A low false negative rate is desirable, since false positives are easier to discover by subsequent verification. There was, consequentially, a noticeable gap between the achieved precision and recall, owing to the model's preference for false positives. It can also be observed that the number of detected changes grew with the increasing temporal difference between the reference and the current time. From an ecological perspective, it was interesting to analyze the trend of the detected changes. According to the validated changes, the net area loss of woody vegetation landscape features in the test plots was approximately 11.5 ares in 2016 and 15 ares in 2019, which represents roughly 2% and 3% of the corresponding landscape feature areas in 2016 and 2019, respectively. Moreover, such information is of crucial importance for agricultural policy decision-makers as well as for parcel owners who can control and, with financial support, potentially improve the detected unfavorable ecological status of the environment.

Our results demonstrate that a reliable detection of changes in woody vegetation landscape features with a true positive rate of around 90% is possible. An important insight from the experiments is how the representativeness of the reference layer deteriorates without regular updates, since the number of validated changes in the test set more than doubled between consecutive surveys (from 94 in 2016 to 192 in 2019).

The analysis of the disagreement between the LiDAR-based ground truth and the woody vegetation landscape features detected in the DOP showed that there are two main sources of false differences. The first one is radial displacement of objects in orthorectified aerial photography, which results in bands of misclassified pixels along opposing edges in woody vegetation patches. This is illustrated in Figure 5, where the woody vegetation landscape feature is identified correctly by the model but is shifted slightly with respect to the LiDAR target, indicating a change. The majority of such cases are eliminated later during the filtering step because either the area of the change or its percentage within the feature polygon is below the threshold.

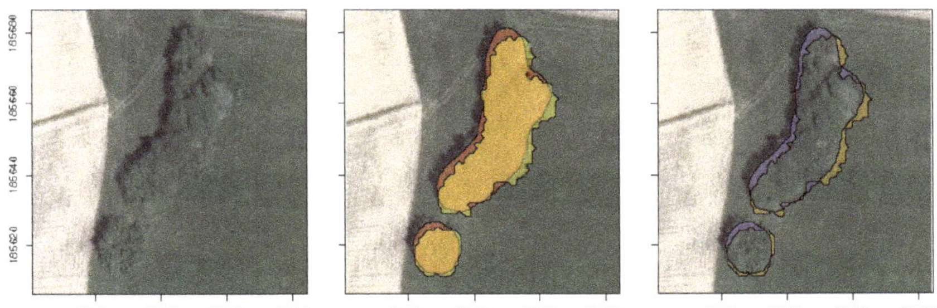

Figure 5. Radial displacement of objects in the DOP (**left** image) leads to a positional mismatch between the LiDAR-based target (the yellow polygon in the **middle** image) and segmented woody vegetation landscape feature (the red polygon in the **middle** image), which manifests as false detected changes with a recognizable pattern (**right** image). The scale is in meters, the CRS is EPSG:3794.

The second source of mismatches between the model predictions and LiDAR targets is the ability of LiDAR to detect low vegetation, smaller clearings, and gaps, which are not included in woody vegetation landscape feature delineation. It is also able to capture finer details in woody vegetation edges, which the up-sampling part of the segmentation neural network cannot reconstruct as precisely (Figure 6). There are two possible ways to address this in future work: by using a wider window for morphological closing of targets or by incorporating edge-preserving loss in model training [67]. However, most changes detected due to such differences are typically fragmented into many smaller patches, which can be filtered by a properly set minimum area threshold.

There are some limitations to the proposed DOP-based vegetation localization that affect the efficiency of change detection. Insufficient foliage in late-sprouting vegetation in particular can cause the segmentation model to underestimate the extent of the growth and indicate a change where there was none. In order to avoid an unnecessarily high number of such false positives, the phenological stage of woody vegetation would need to be considered in determining future surveys since their initial purpose was not specifically vegetation monitoring.

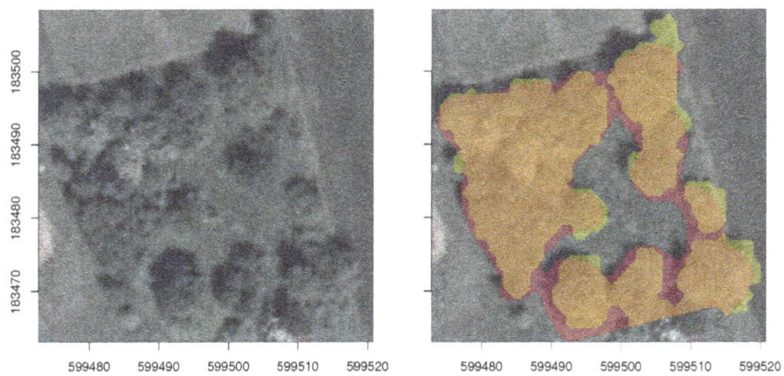

Figure 6. Vegetation polygon detected by the segmentation neural network (red polygon) merges narrow gaps and smooths vegetation edges in targets derived from LiDAR (yellow polygons). The scale is in meters, the CRS is EPSG:3794.

A more general limitation of monitoring woody vegetation landscape features is that, from an ecological perspective, not all plant species are equally important for biodiversity. The possibility of accurately determining the species composition of small woody features from single-shot aerial photography is severely constrained. In fact, the spreading of some fast-growing invasive species is becoming a serious problem globally [13,68], and such vegetation sites should not be considered valuable landscape features. The proposed framework could also be used to indicate the potential propagation of invasive species by detecting fast-expanding growth.

5. Conclusions

In the paper, we propose a methodology for automated detection of woody vegetation landscape features in cyclic aerial photography and the construction of change layers with respect to the reference ground truth. Systematic registration and monitoring of woody vegetation is essential for the efficient implementation of ecosystem and biodiversity preservation strategies. While LiDAR and high-resolution satellite imagery are the best sources for reliable vegetation detection, their high cost makes them less suitable for continuous monitoring. As this study demonstrated, a viable solution is to utilize the remote sensing data of periodic aerial surveys, which present an attractive balance between cost and resolution required for change detection on a regional scale.

Based on the results, we can conclude that the entire end-to-end pipeline for the generation of change layers can be implemented in a way that requires almost no user intervention. The main application of such methodology is for semi-automated updates of the landscape feature registry, where the user can set the desired granularity of detected changes before they are presented for manual validation. This allows the user to focus on major disruptions first in order to make an early assessment of trends.

There are several ways the proposed methodology could be improved in the future. In the segmentation stage, a better precision in detecting feature boundaries could be achieved by extending the loss function, or employing specialized neural network architectures. In the post-processing stage, the number of false positives could be reduced by explicit matching of overlapping shapes, i.e., detected and reference woody vegetation landscape features. The filtering thresholds could also be adjusted locally by using the canopy height model to estimate the possible amount of radial displacement due to orthorectification. These improvements would increase the precision of change detection without sacrificing recall, leading to an overall improved user experience.

From the end-user perspective, an important extension of the proposed workflow would include visual analytics in order to facilitate the supervision of how the national strategies for biodiversity and nature preservation are implemented. The proposed methodology can be used in a similar way to monitor other interesting types of habitats for ecological monitoring, such as meadows or marshes. Locating areas of fast vegetation expansion could also be used to indicate the potential progression of invasive species. Another possible application of the proposed approach is in planning and overseeing long-term periodic field operations, such as vegetation maintenance in transmission line corridors.

Author Contributions: Conceptualization, D.S. and Š.K.; methodology, D.S.; software, D.S. and Š.H.; validation, D.S., Š.K. and Š.H.; formal analysis, D.S. and Š.K.; writing—original draft preparation, D.S.; writing—review and editing, Š.K. and D.I.; visualization, Š.H.; supervision, D.M.; project administration, D.M.; funding acquisition, D.M. and D.I. All authors have read and agreed to the published version of the manuscript.

Funding: This research was funded by the Slovenian Research Agency through the research program P2-0041 and research project L7-2633. This research was also funded by the Slovenian Research Agency and the Slovenian Ministry of Agriculture, Forestry, and Food through research project V4-2018.

Data Availability Statement: The data used in this study are available publicly or upon free registration at the links provided in the paper. The sample layers generated by the current study are available from the corresponding author upon request.

Conflicts of Interest: The authors declare no conflict of interest.

References

1. Kokalj, Z.; Stančič, L.; Noumonvi, K.D.; Andrej, K. *Feasibility Study and Mapping of Vegetative Landscape Structures Important for Biodiversity and Climate Change Mitigation with Remote Sensing*; Final project report; Research Centre of the Slovenian Academy of Sciences and Arts & Slovenian Forestry Institute: Ljubljana, Slovenia, 2020.
2. Almalki, R.; Khaki, M.; Saco, P.M.; Rodriguez, J.F. Monitoring and Mapping Vegetation Cover Changes in Arid and Semi-Arid Areas Using Remote Sensing Technology: A Review. *Remote Sens.* **2022**, *14*, 5143. [CrossRef]
3. Decuyper, M.; Chávez, R.O.; Lohbeck, M.; Lastra, J.A.; Tsendbazar, N.; Hackländer, J.; Herold, M.; Vågen, T.G. Continuous monitoring of forest change dynamics with satellite time series. *Remote Sens. Environ.* **2022**, *269*, 112829. [CrossRef]
4. Moncrieff, G.R. Continuous Land Cover Change Detection in a Critically Endangered Shrubland Ecosystem Using Neural Networks. *Remote Sens.* **2022**, *14*, 2766. [CrossRef]
5. Pacheco-Pascagaza, A.M.; Gou, Y.; Louis, V.; Roberts, J.F.; Rodríguez-Veiga, P.; da Conceição Bispo, P.; Espírito-Santo, F.D.; Robb, C.; Upton, C.; Galindo, G.; et al. Near real-time change detection system using Sentinel-2 and machine learning: A test for Mexican and Colombian forests. *Remote Sens.* **2022**, *14*, 707. [CrossRef]
6. Wang, Z.; Wei, C.; Liu, X.; Zhu, L.; Yang, Q.; Wang, Q.; Zhang, Q.; Meng, Y. Object-based change detection for vegetation disturbance and recovery using Landsat time series. *GiScience Remote Sens.* **2022**, *59*, 1706–1721. [CrossRef]
7. Ben Abbes, A.; Bounouh, O.; Farah, I.R.; de Jong, R.; Martínez, B. Comparative study of three satellite image time-series decomposition methods for vegetation change detection. *Eur. J. Remote Sens.* **2018**, *51*, 607–615. [CrossRef]
8. Phil-Eze, P.; Wali, E.; Bosco-Abiahu, L.; Wizor, C.; Abdullahi, M.; Afolabi, O.; Ihunda, C. Vegetation Change Detection in the Niger Delta Region of Nigeria using Remote Sensing and GIS Techniques from 2000 to 2020. *Asian J. Environ. Ecol.* **2021**, *16*, 181–188. [CrossRef]
9. Aarfin, S.; Prakash, R.; Mishra, N. Land Vegetation Change Detection Using Remote Sensing and GIS. In *Advances in Civil Engineering and Infrastructural Development*; Springer: Singapore, 2021; pp. 265–274.
10. Morin, N.; Masse, A.; Sannier, C.; Siklar, M.; Kiesslich, N.; Sayadyan, H.; Faucqueur, L.; Seewald, M. Development and Application of Earth Observation Based Machine Learning Methods for Characterizing Forest and Land Cover Change in Dilijan National Park of Armenia between 1991 and 2019. *Remote Sens.* **2021**, *13*, 2942. [CrossRef]
11. Ahlswede, S.; Asam, S.; Röder, A. Hedgerow object detection in very high-resolution satellite images using convolutional neural networks. *J. Appl. Remote Sens.* **2021**, *15*, 018501. [CrossRef]
12. Magnússon, R.Í.; Limpens, J.; Kleijn, D.; van Huissteden, K.; Maximov, T.C.; Lobry, S.; Heijmans, M.M. Shrub decline and expansion of wetland vegetation revealed by very high resolution land cover change detection in the Siberian lowland tundra. *Sci. Total Environ.* **2021**, *782*, 146877. [CrossRef]
13. Papp, L.; Van Leeuwen, B.; Szilassi, P.; Tobak, Z.; Szatmári, J.; Árvai, M.; Mészáros, J.; Pásztor, L. Monitoring invasive plant species using hyperspectral remote sensing data. *Land* **2021**, *10*, 29. [CrossRef]
14. Ishida, T.; Kurihara, J.; Viray, F.A.; Namuco, S.B.; Paringit, E.C.; Perez, G.J.; Takahashi, Y.; Marciano, J.J., Jr. A novel approach for vegetation classification using UAV-based hyperspectral imaging. *Comput. Electron. Agric.* **2018**, *144*, 80–85. [CrossRef]
15. Durfee, N.; Ochoa, C.G.; Mata-Gonzalez, R. The use of low-altitude UAV imagery to assess western juniper density and canopy cover in treated and untreated stands. *Forests* **2019**, *10*, 296. [CrossRef]
16. Li, Y.; Fu, B.; Sun, X.; Fan, D.; Wang, Y.; He, H.; Gao, E.; He, W.; Yao, Y. Comparison of Different Transfer Learning Methods for Classification of Mangrove Communities Using MCCUNet and UAV Multispectral Images. *Remote Sens.* **2022**, *14*, 5533. [CrossRef]

17. He, J.; Lyu, D.; He, L.; Zhang, Y.; Xu, X.; Yi, H.; Tian, Q.; Liu, B.; Zhang, X. Combining Object-Oriented and Deep Learning Methods to Estimate Photosynthetic and Non-Photosynthetic Vegetation Cover in the Desert from Unmanned Aerial Vehicle Images with Consideration of Shadows. *Remote Sens.* **2022**, *15*, 105. [CrossRef]
18. Xie, L.; Meng, X.; Zhao, X.; Fu, L.; Sharma, R.P.; Sun, H. Estimating Fractional Vegetation Cover Changes in Desert Regions Using RGB Data. *Remote Sens.* **2022**, *14*, 3833. [CrossRef]
19. Olariu, H.G.; Malambo, L.; Popescu, S.C.; Virgil, C.; Wilcox, B.P. Woody Plant Encroachment: Evaluating Methodologies for Semiarid Woody Species Classification from Drone Images. *Remote Sens.* **2022**, *14*, 1665. [CrossRef]
20. Furukawa, F.; Laneng, L.A.; Ando, H.; Yoshimura, N.; Kaneko, M.; Morimoto, J. Comparison of RGB and multispectral unmanned aerial vehicle for monitoring vegetation coverage changes on a landslide area. *Drones* **2021**, *5*, 97. [CrossRef]
21. Sankey, T.T.; Leonard, J.M.; Moore, M.M. Unmanned aerial vehicle-based rangeland monitoring: Examining a century of vegetation changes. *Rangel. Ecol. Manag.* **2019**, *72*, 858–863. [CrossRef]
22. Bakó, G.; Molnár, Z.; Bakk, L.; Horváth, F.; Fehér, L.; Ábrám, Ö.; Morvai, E.; Biro, C.; Pápay, G.; Fűrész, A.; et al. Toward a high spatial resolution aerial monitoring network for nature conservation—how can remote sensing help protect natural areas? *Sustainability* **2021**, *13*, 8807. [CrossRef]
23. Meneguzzo, D.M.; Liknes, G.C.; Nelson, M.D. Mapping trees outside forests using high-resolution aerial imagery: A comparison of pixel-and object-based classification approaches. *Environ. Monit. Assess.* **2013**, *185*, 6261–6275. [CrossRef] [PubMed]
24. Daryaei, A.; Sohrabi, H.; Atzberger, C.; Immitzer, M. Fine-scale detection of vegetation in semi-arid mountainous areas with focus on riparian landscapes using Sentinel-2 and UAV data. *Comput. Electron. Agric.* **2020**, *177*, 105686. [CrossRef]
25. Aksoy, S.; Akçay, H.G.; Wassenaar, T. Automatic mapping of linear woody vegetation features in agricultural landscapes using very high resolution imagery. *IEEE Trans. Geosci. Remote Sens.* **2009**, *48*, 511–522. [CrossRef]
26. Nguyen, U.; Glenn, E.P.; Dang, T.D.; Pham, L.T. Mapping vegetation types in semi-arid riparian regions using random forest and object-based image approach: A case study of the Colorado River Ecosystem, Grand Canyon, Arizona. *Ecol. Inform.* **2019**, *50*, 43–50. [CrossRef]
27. Belcore, E.; Pittarello, M.; Lingua, A.M.; Lonati, M. Mapping riparian habitats of Natura 2000 network (91E0*, 3240) at individual tree level using UAV multi-temporal and multi-spectral data. *Remote Sens.* **2021**, *13*, 1756. [CrossRef]
28. O'Connell, J.; Bradter, U.; Benton, T.G. Wide-area mapping of small-scale features in agricultural landscapes using airborne remote sensing. *ISPRS J. Photogramm. Remote Sens.* **2015**, *109*, 165–177. [CrossRef]
29. Hellesen, T.; Matikainen, L. An object-based approach for mapping shrub and tree cover on grassland habitats by use of LiDAR and CIR orthoimages. *Remote Sens.* **2013**, *5*, 558–583. [CrossRef]
30. Jamil, A.; Bayram, B. Tree species extraction and land use/cover classification from high-resolution digital orthophoto maps. *IEEE J. Sel. Top. Appl. Earth Obs. Remote Sens.* **2017**, *11*, 89–94. [CrossRef]
31. Rusňák, T.; Halabuk, A.; Halada, L'.; Hilbert, H.; Gerhátová, K. Detection of Invasive Black Locust (Robinia pseudoacacia) in Small Woody Features Using Spatiotemporal Compositing of Sentinel-2 Data. *Remote Sens.* **2022**, *14*, 971. [CrossRef]
32. Blickensdörfer, L.; Schwieder, M.; Pflugmacher, D.; Nendel, C.; Erasmi, S.; Hostert, P. Mapping of crop types and crop sequences with combined time series of Sentinel-1, Sentinel-2 and Landsat 8 data for Germany. *Remote Sens. Environ.* **2022**, *269*, 112831. [CrossRef]
33. Merciol, F.; Faucqueur, L.; Damodaran, B.B.; Rémy, P.Y.; Desclée, B.; Dazin, F.; Lefèvre, S.; Masse, A.; Sannier, C. GEOBIA at the terapixel scale: Toward efficient mapping of small woody features from heterogeneous VHR scenes. *ISPRS Int. J. Geo-Inf.* **2019**, *8*, 46. [CrossRef]
34. Minaee, S.; Boykov, Y.; Porikli, F.; Plaza, A.; Kehtarnavaz, N.; Terzopoulos, D. Image segmentation using deep learning: A survey. *IEEE Trans. Pattern Anal. Mach. Intell.* **2021**, *44*, 3523–3542. [CrossRef] [PubMed]
35. Long, J.; Shelhamer, E.; Darrell, T. Fully convolutional networks for semantic segmentation. In Proceedings of the IEEE Conference on Computer Vision and Pattern Recognition, Boston, MA, USA, 7–12 June 2015; pp. 3431–3440.
36. Ronneberger, O.; Fischer, P.; Brox, T. U-net: Convolutional networks for biomedical image segmentation. In Proceedings of the Medical Image Computing and Computer-Assisted Intervention–MICCAI 2015: 18th International Conference, Munich, Germany, 5–9 October 2015; pp. 234–241.
37. Badrinarayanan, V.; Kendall, A.; Cipolla, R. Segnet: A deep convolutional encoder-decoder architecture for image segmentation. *IEEE Trans. Pattern Anal. Mach. Intell.* **2017**, *39*, 2481–2495. [CrossRef] [PubMed]
38. Chaurasia, A.; Culurciello, E. Linknet: Exploiting encoder representations for efficient semantic segmentation. In Proceedings of the 2017 IEEE Visual Communications and Image Processing (VCIP), St. Petersburg, FL, USA, 10–13 December 2017; pp. 1–4.
39. Deng, J.; Dong, W.; Socher, R.; Li, L.J.; Li, K.; Li, F.-F. Imagenet: A large-scale hierarchical image database. In Proceedings of the 2009 IEEE Conference on Computer Vision and Pattern Recognition, Miami, FL, USA, 20–25 June 2009; pp. 248–255.
40. Kattenborn, T.; Eichel, J.; Fassnacht, F.E. Convolutional Neural Networks enable efficient, accurate and fine-grained segmentation of plant species and communities from high-resolution UAV imagery. *Sci. Rep.* **2019**, *9*, 17656. [CrossRef] [PubMed]
41. Trenčanová, B.; Proença, V.; Bernardino, A. Development of semantic maps of vegetation cover from UAV images to support planning and management in fine-grained fire-prone landscapes. *Remote Sens.* **2022**, *14*, 1262. [CrossRef]
42. Flood, N.; Watson, F.; Collett, L. Using a U-net convolutional neural network to map woody vegetation extent from high resolution satellite imagery across Queensland, Australia. *Int. J. Appl. Earth Obs. Geoinf.* **2019**, *82*, 101897. [CrossRef]

43. Chen, L.C.; Zhu, Y.; Papandreou, G.; Schroff, F.; Adam, H. Encoder-decoder with atrous separable convolution for semantic image segmentation. In Proceedings of the European Conference on Computer Vision (ECCV), Munich, Germany, 8–14 September 2018; pp. 801–818.
44. Luo, H.; Chen, C.; Fang, L.; Zhu, X.; Lu, L. High-resolution aerial images semantic segmentation using deep fully convolutional network with channel attention mechanism. *IEEE J. Sel. Top. Appl. Earth Obs. Remote Sens.* **2019**, *12*, 3492–3507. [CrossRef]
45. Wurm, M.; Stark, T.; Zhu, X.X.; Weigand, M.; Taubenböck, H. Semantic segmentation of slums in satellite images using transfer learning on fully convolutional neural networks. *ISPRS J. Photogramm. Remote Sens.* **2019**, *150*, 59–69. [CrossRef]
46. Abdollahi, A.; Pradhan, B.; Alamri, A.M. An ensemble architecture of deep convolutional Segnet and Unet networks for building semantic segmentation from high-resolution aerial images. *Geocarto Int.* **2022**, *37*, 3355–3370. [CrossRef]
47. Yu, J.; Zeng, P.; Yu, Y.; Yu, H.; Huang, L.; Zhou, D. A Combined Convolutional Neural Network for Urban Land-Use Classification with GIS Data. *Remote Sens.* **2022**, *14*, 1128. [CrossRef]
48. Niu, R.; Sun, X.; Tian, Y.; Diao, W.; Chen, K.; Fu, K. Hybrid multiple attention network for semantic segmentation in aerial images. *IEEE Trans. Geosci. Remote. Sens.* **2021**, *60*, 1–18. [CrossRef]
49. Yue, K.; Yang, L.; Li, R.; Hu, W.; Zhang, F.; Li, W. TreeUNet: Adaptive tree convolutional neural networks for subdecimeter aerial image segmentation. *ISPRS J. Photogramm. Remote Sens.* **2019**, *156*, 1–13. [CrossRef]
50. Yu, D.; Ji, S.; Li, X.; Yuan, Z.; Shen, C. Earthquake crack detection from aerial images using a deformable convolutional neural network. *IEEE Trans. Geosci. Remote Sens.* **2022**, *60*, 1–12. [CrossRef]
51. Guo, H.; Bai, H.; Yuan, Y.; Qin, W. Fully deformable convolutional network for ship detection in remote sensing imagery. *Remote Sens.* **2022**, *14*, 1850. [CrossRef]
52. Diao, Q.; Dai, Y.; Zhang, C.; Wu, Y.; Feng, X.; Pan, F. Superpixel-based attention graph neural network for semantic segmentation in aerial images. *Remote Sens.* **2022**, *14*, 305. [CrossRef]
53. Tulczyjew, L.; Kawulok, M.; Longépé, N.; Le Saux, B.; Nalepa, J. Graph neural networks extract high-resolution cultivated land maps from Sentinel-2 image series. *IEEE Geosci. Remote Sens. Lett.* **2022**, *19*, 1–5. [CrossRef]
54. Yan, J.; Ji, S.; Wei, Y. A combination of convolutional and graph neural networks for regularized road surface extraction. *IEEE Trans. Geosci. Remote Sens.* **2022**, *60*, 1–13. [CrossRef]
55. Yan, L.; Huang, J.; Xie, H.; Wei, P.; Gao, Z. Efficient depth fusion transformer for aerial image semantic segmentation. *Remote Sens.* **2022**, *14*, 1294. [CrossRef]
56. Xu, Z.; Zhang, W.; Zhang, T.; Yang, Z.; Li, J. Efficient transformer for remote sensing image segmentation. *Remote Sens.* **2021**, *13*, 3585. [CrossRef]
57. Chen, X.; Qiu, C.; Guo, W.; Yu, A.; Tong, X.; Schmitt, M. Multiscale feature learning by transformer for building extraction from satellite images. *IEEE Geosci. Remote Sens. Lett.* **2022**, *19*, 1–5. [CrossRef]
58. Yi, S.; Liu, X.; Li, J.; Chen, L. UAVformer: A Composite Transformer Network for Urban Scene Segmentation of UAV Images. *Pattern Recognit.* **2023**, *133*, 109019. [CrossRef]
59. Khosravipour, A.; Skidmore, A.K.; Isenburg, M.; Wang, T.; Hussin, Y.A. Generating pit-free canopy height models from airborne lidar. *Photogramm. Eng. Remote Sens.* **2014**, *80*, 863–872. [CrossRef]
60. Roussel, J.R.; Auty, D.; Coops, N.C.; Tompalski, P.; Goodbody, T.R.; Meador, A.S.; Bourdon, J.F.; de Boissieu, F.; Achim, A. lidR: An R package for analysis of Airborne Laser Scanning (ALS) data. *Remote Sens. Environ.* **2020**, *251*, 112061. [CrossRef]
61. Tan, M.; Le, Q. EfficientNet: Rethinking model scaling for convolutional neural networks. In Proceedings of the International Conference on Machine Learning, PMLR, Long Beach, CA, USA, 9–15 June 2019; pp. 6105–6114.
62. Bric, V.; Oven, K.; Prešeren, P. Cyclical Aerial Photography of Slovenia—Digital Era. *Geod. Vestn.* **2022**, *66*, 77–84.
63. Cui, B.; Chen, X.; Lu, Y. Semantic segmentation of remote sensing images using transfer learning and deep convolutional neural network with dense connection. *IEEE Access* **2020**, *8*, 116744–116755. [CrossRef]
64. Liu, S.; Ye, H.; Jin, K.; Cheng, H. CT-UNet: Context-transfer-UNet for building segmentation in remote sensing images. *Neural Process. Lett.* **2021**, *53*, 4257–4277. [CrossRef]
65. Math, R.M.; Dharwadkar, N.V. Early detection and identification of grape diseases using convolutional neural networks. *J. Plant Dis. Prot.* **2022**, *129*, 521–532. [CrossRef]
66. Taori, R.; Dave, A.; Shankar, V.; Carlini, N.; Recht, B.; Schmidt, L. Measuring robustness to natural distribution shifts in image classification. *Adv. Neural Inf. Process. Syst.* **2020**, *33*, 18583–18599.
67 Liu, S.; Ding, W.; Liu, C.; Liu, Y.; Wang, Y.; Li, H. ERN: Edge Loss Reinforced Semantic Segmentation Network for Remote Sensing Images. *Remote Sens.* **2018**, *10*, 1339. [CrossRef]
68. Sladonja, B.; Damijanić, D. Remote Sensing in Invasive Species Detection and Monitoring. *Near East. Archaeol.* **2021**, *77*, 182–186.

remote sensing

MDPI

Article

Comparison of Phenological Parameters Extracted from SIF, NDVI and NIRv Data on the Mongolian Plateau

Cha Ersi [1,2], Tubuxin Bayaer [1,2,*], Yuhai Bao [1,2], Yulong Bao [1,2], Mei Yong [1,2], Quan Lai [1,2], Xiang Zhang [1,2] and Yusi Zhang [1,2]

1 College of Geographic Science, Inner Mongolia Normal University, Hohhot 010022, China
2 Inner Mongolia Key Laboratory of Remote Sensing and Geographic Information Systems, Inner Mongolia Normal University, Hohhot 010022, China
* Correspondence: tuvshin_b@imnu.edu.cn

Abstract: The phenological parameters estimated from different data may vary, especially in response to climatic factors. Therefore, we estimated the start of the growing season (SOS) and the end of the growing season (EOS) based on sunlight-induced chlorophyll fluorescence (SIF), the normalized difference vegetation index (NDVI) and the near-infrared reflectance of vegetation (NIRv). The SIF, NDVI and NIRv breakpoints were detected, and the trends and change-points of phenological parameters based on these data were analyzed. The correlations between the phenological parameters and snow-related factors, precipitation, temperature, soil moisture and population density were also analyzed. The results showed that SIF and NIRv could identify breakpoints early. SIF could estimate the latest SOS and the earliest EOS. NDVI could estimate the earliest SOS and the latest EOS. The change-points of SOS_{SIF} were mostly concentrated from 2001 to 2003, and those of SOS_{NDVI} and SOS_{NIRv} occurred later. The change-points of EOS_{SIF} and EOS_{NIRv} were mostly concentrated from 2001 to 2007, and those of EOS_{SIF} occurred later. Differently from the weak correlation with SOS_{SIF}, SOS_{NDVI} and SOS_{NIRv} were significantly correlated with snow-related factors. The correlation between the meteorological factors in the summer and autumn and EOS_{SIF} was the most significant. The population density showed the highest degree of interpretation for SOS_{NIRv} and EOS_{NDVI}. The results reveal the differences and potentials of different remote-sensing parameters in estimating phenological indicators, which is helpful for better understanding the dynamic changes in phenology and the response to changes in various influencing factors.

Keywords: vegetation phenology; Mongolian Plateau; SIF; NDVI; NIRv

1. Introduction

Phenology is a sensitive and key feature of the life history of plants. It can reflect how changes in climate affect vegetation growth, improve our understanding of vegetation ecological processes and allow us to quantify the impacts of meteorological changes on terrestrial ecosystems [1].

The Mongolian Plateau, located in arid and semiarid areas, is highly sensitive to climate change [2]. Its temperate continental climate forms large natural pastures [3]. Additionally, the grassland ecosystem of the Mongolian Plateau plays key roles in the carbon cycle in East Asia and globally [4]. The wide availability of low-cost remote sensing data facilitates the study of vegetation ecology on the Mongolian Plateau [5,6]. Previous studies have shown that the normalized difference vegetation index (NDVI) can be used to effectively monitor vegetation phenology [7]. Therefore, NDVI has frequently been used to estimate vegetation phenological parameters in studies of Mongolian Plateau phenology [1,2,8,9]. However, limitations may exist to using only NDVI when retrieving vegetation phenology information for evergreen forests [10,11]. Thus, evaluating the advantages and disadvantages of different types of remote sensing data in retrieving vegetation phenology and using a combination of approaches to obtain detailed observations are very important for high-precision phenology research [12].

In contrast to traditional remote sensing indicators based on changes in vegetation "greenness", sunlight-induced chlorophyll fluorescence (SIF) can directly reflect the photochemical reactions of plants [13,14]. Additionally, SIF is very sensitive to abiotic stress factors and can be effectively used to monitor vegetation stress [15,16]. SIF has unique technical advantages in the field of monitoring vegetation physiological conditions and changes in the surrounding environment [17–19]. Furthermore, it represents a new potential method for monitoring the physiological dynamics of vegetation from space [20].

Specifically, SIF is the energy flux released by plant chlorophyll molecules after absorbing sunlight in the wavelength range of 600–800 nm, which represents less than 5% of the light energy absorbed by plants [21]. Many studies have investigated the SIF products of the Greenhouse Gas Observation Satellite (GOSAT) and Global Ozone Monitoring Experiment-2 (GOME-2), but GOSAT and GOME-2 (GOME-2: 40 × 80 km; GOSAT: 10 km) cannot meet the need for high-accuracy research [22,23]. The SIF products estimated by the Orbital Carbon Observatory-2 (OCO-2) provide a better spatial resolution (1.3 × 2.2 km) than the SIF products of GOSAT and GOME-2 [24]. However, the OCO-2 SIF products with discontinuous OCO-2 still have limitations. Fortunately, Li and Xiao developed a new SIF product based on OCO-2 SIF and meteorological data, called "GOSIF", with a temporal resolution of 8 days and a spatial resolution of 0.05° [25]. This product has a better temporal and spatial resolution than existing products, and can greatly contribute to research using SIF to monitor vegetation physiology. Previous studies have shown the feasibility of GOSIF for retrieving vegetation phenology [26,27]. However, in arid and semiarid areas of the Mongolian Plateau, the characteristics of vegetation phenology estimated by SIF are not clear. At the same time, whether or not the high sensitivity of SIF to environmental stress has a higher advantage in terms of vegetation phenology also needs further exploration.

A recently developed vegetation index, the near-infrared reflectance of vegetation (NIRv), has the advantages of reducing the number of mixed pixels and minimizing the noise caused by soil pollution [28–30]. Previous studies have shown that NIRv can be used to monitor early changes in canopy development in grassland ecosystems. Thus, NIRv, as an index that can effectively reflect the growth status of vegetation, may also have great potential in retrieving vegetation phenology. Simultaneously, the advantages and disadvantages of sensitivity to climate factors are not clear.

Studies on the climatic factors affecting vegetation phenology have considered the effects of temperature, precipitation and soil moisture [31–34]. However, the effects of snow cover in winter and the snow melting period in spring at the start of the growing season (SOS) cannot be ignored. In addition, the phenological parameters of vegetation based on SIF (GOSIF), NDVI and NIRv and their responses to different climatic factors are also different. Understanding the response of phenological parameters estimated from different data to different climatic factors can enable us to more deeply understand the mechanism by which vegetation is affected by the environment. Concurrently, comparing the sensitivity of phenological parameters to influencing factors can help us to understand the advantages and disadvantages of the phenological retrieval abilities of different types of remote sensing data. The advantage of using different types of phenological parameters is of great value for retrieving phenological parameters with higher accuracy and improving our understanding of the relationship between vegetation phenology and climate factors in the future. The main objectives of this study were as follows:

1. To compare the characteristics of vegetation dynamics estimated by SIF, NDVI and NIRv;
2. To extract and compare phenological parameters and their dynamics based on SIF, NDVI and NIRv;
3. To determine the influence mechanism of snow-related factors on phenological parameters.

2. Materials and Methods

2.1. Study Area

The Mongolian Plateau is located in Central Asia (87°43′–126°04′E, 37°22′–53°23′N) [35], and it is surrounded by high-altitude mountains in the north and is adjacent to the Gobi Desert in the south [36,37]. Affected by the climate, the annual precipitation on the Mongolian Plateau is low, and drought is frequent. The average annual precipitation in most areas is 200 mm, and precipitation mainly occurs from June to August. Windy conditions are prevalent in winter and spring. The average temperature in January is −26 °C, and in July it is 17 °C [38]. The plateau has complex ecological zones, which mainly include meadow steppe, typical steppe, alpine steppe, desert steppe, coniferous forest, broadleaf forest and the Gobi desert (Figure 1). The vegetation-type data used to produce the figure are from the National Atlas of Mongolia. The vegetation map with a scale of 1:1,000,000 was rasterized to 0.083° [39]. Finally, we chose GOSIF, NDVI and NIRv to estimate the vegetation phenological parameters for the Mongolian Plateau.

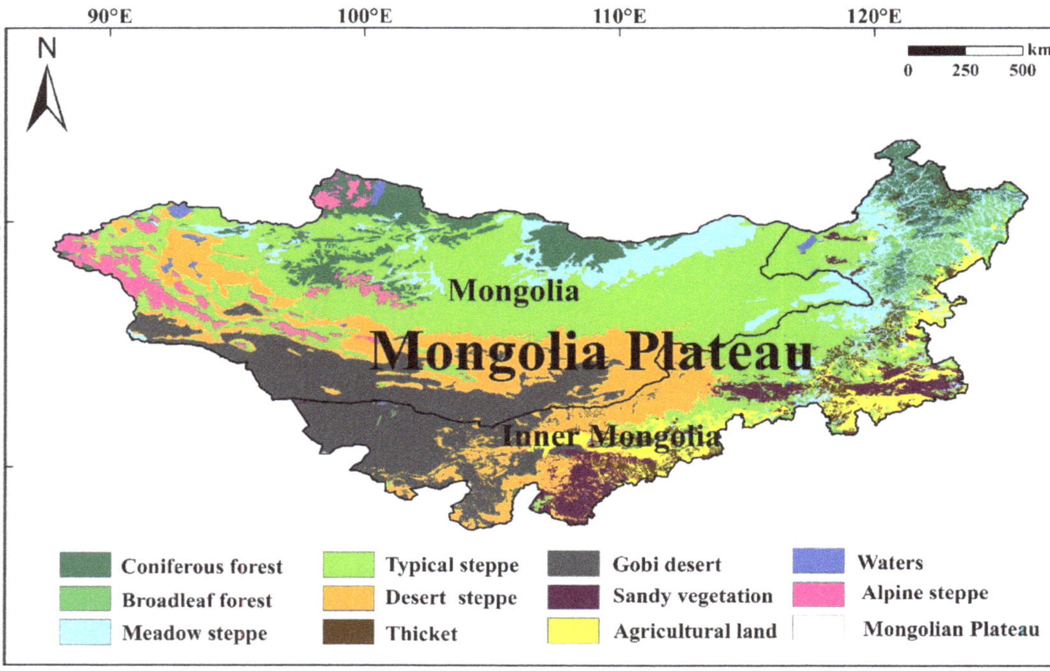

Figure 1. Land-type distribution map of the Mongolian Plateau.

2.2. Data Sources

2.2.1. Sunlight-Induced Chlorophyll Exposure

Li and Xiao developed a new global SIF data set with high spatiotemporal resolution (0.05° × 0.05°) based on OCO-2, using discrete OCO-2 SIF detection, remote sensing data from the moderate-resolution imaging spectroradiometer (MODIS) and meteorological reanalysis data. Compared with coarse-resolution SIF data aggregated directly from OCO-2 sounding, sunlight-induced chlorophyll exposure (GOSIF) has a higher spatial resolution and longer records, and it provides continuous global coverage [25].

The website used to access GOSIF data is as follows: http://globalecology.unh.edu/data/GOSIF.html (accessed on 3 April 2022).

In this study, we used GOSIF data sets from 2001 to 2020 with a spatial resolution of 0.05° × 0.05° and a temporal resolution of 8 d. To match the temporal resolution of NDVI and NIRv, we synthesized the GOSIF data of 8 days into a data set with a temporal resolution of 16 days, using the maximum-value composite (MVC) method.

2.2.2. MODIS Data Set

The vegetation indices we used included NDVI and near-infrared reflectance (NIR), both of which came from the MOD13A1 product of Terra MODIS version 6.1. The spatial resolution was 500 m, and the temporal resolution was 16 d. The website used to access these data is as follows: https://lpdaacsvc.cr.usgs.gov/appeears/ (accessed on 3 April 2022). NDVI and NIRv are calculated as follows:

$$\text{NDVI} = \frac{\rho_2 - \rho_1}{\rho_2 + \rho_1} \tag{1}$$

$$\text{NIRv} = (\text{NDVI} - C) \times \rho_2 \tag{2}$$

where ρ_1 is the red reflectance (620–670 nm), ρ_2 is the NIR (841–876 nm), and the parameter C was set to 0.08, according to Badgley [29]. In this study, we used NDVI and NIRv data sets from 2001 to 2020. We resampled the spatial resolution of the NDVI and NIRv data (the source data are 500 m) to $0.05° \times 0.05°$, to match the spatial resolution of GOSIF. (Downsampling of all high-resolution data in this study is realized through the aggregate tool of ArcGIS Pro3.0 software).

We used the daily snow-cover product of MOD10A1 from 2000 to 2020, with a spatial resolution of 500 m. The winter (December of the previous year and January and February of the current year) snow-cover data set (SFC) from 2001 to 2020 was calculated. In addition, we defined the snowmelt date as the period when the snow cover in spring had no value for five consecutive days, and we used Interactive Data Language (IDL, an application program and programming language for data analysis and visualization) to calculate the item-by-item snow-melting-period data set (SMD) from 2001 to 2020. To match the spatial resolution of the phenological data set, SFC and SMD were resampled to $0.05° \times 0.05°$. The data were acquired from https://www.environment.snu.ac.kr/bess-rad (accessed on 3 April 2022).

2.2.3. Extraction of Vegetation Phenological Parameters

Before extracting the phenological parameters, we discarded the areas with an annual average NDVI < 0.1, because such areas have been identified as unvegetated [40]. Additionally, we masked GOSIF and NIRv with NDVI of the discarded unvegetated area as the template to obtain the GOSIF and NIRv data sets of the discarded unvegetated area. Then, we used double logistic filtering to fit the smooth curve from the time series of the GOSIF, NDVI and NIRv data [41]. The threshold method is usually used to extract phenological vegetation parameters, and it mainly includes the fixed-threshold method and dynamic-threshold method [42,43]. However, the fixed-threshold method has limitations in areas with different land-cover types and soil backgrounds. Therefore, we used a more flexible dynamic-threshold method to extract the phenological parameters for the start of the growing season (SOS), the end of the growing season (EOS) and the length of the growing season (LOS, i.e., LOS = EOS − SOS). The dynamic threshold we selected was 20% [44–46]. For example, SOS was defined as the first day of the year that the GOSIF ratio exceeded 0.2 [47]. The GOSIF ratio is calculated as follows:

$$\text{GOSIF}_{ratio} = (\text{GOSIF}_t - \text{GOSIF}_{min})/(\text{GOSIF}_{max} - \text{GOSIF}_{min}) \tag{3}$$

where GOSIF_t is the value of GOSIF at time t and GOSIF_{min} and GOSIF_{max} are the minimum and maximum GOSIF values, respectively, in the annual GOSIF time series. The extraction of the above parameters was completed using TIMESAT 3.3 software. The software download website is http://web.nateko.lu.se/timesat/ (accessed on 4 April 2022).

In this study, we used SIF (GOSIF), NDVI and NIRv to extract phenological vegetation information, and obtained the following:

1. Phenological parameters based on SIF: SOS_{SIF}, LOS_{SIF} and EOS_{SIF};
2. Phenological parameters based on NDVI: SOS_{NDVI}, LOS_{NDVI} and EOS_{NDVI};

3. Phenological parameters based on NIRv: SOS_{NIRv}, LOS_{NIRv} and EOS_{NIRv}.

2.2.4. ERA5-Land Climate Data Set

ERA5-Land is a reanalysis data set that has a higher resolution than ERA5 and provides conventional longitude and latitude grid data with a spatial resolution of 0.1° × 0.1° (approximately 10 km). We obtained the data set of the average monthly values from 2001 to 2020, and the data were divided into the three types of climate data, as explained below.

We selected the total precipitation in m; the skin temperature, k; and the volume of the soil water layer 1 (0–7 cm) m^3/m^{-3}). The meteorological data sets of spring (March to May, PRE_{Spring}, TEM_{Spring}, and SW_{Spring}), summer (June to August, PRE_{Summer}, TEM_{Summer}, and SW_{summer}) and autumn (September to November, PRE_{Autumn}, TEM_{Autumn}, and SW_{Autumn}) were calculated. To match the spatial resolution of the phenological data, the data were resampled to 0.05° × 0.05°. The website providing detailed descriptions and access to the above ERA-Land climate data is as follows: https://cds.climate.copernicus.eu/cdsapp#!/dataset/reanalysis-era5-land-monthly-means?tab=overview (accessed on 3 April 2022).

2.2.5. Population Density Data

The population density data were obtained from open-space population data and previous research. We obtained the annual population-density data set from 2001 to 2020, with a spatial resolution of 1 km and a final resampling to 0.05° × 0.05°. The data were obtained from https://hub.worldpop.org/geodata/listing?id=77 (accessed on 3 April 2022).

2.2.6. Terrain Data Set

The terrain data were obtained from the Advanced Spaceborne Thermal Emission and Reflection Radiometer (ASTER) Global Digital Elevation Model (GDEM) V1 (https://www.gscloud.cn). Through an image mosaic, we obtained the DEM of the study area with a spatial resolution of 30 m (accessed on 3 April 2022).

2.3. Methods

2.3.1. The BFAST Package

The Breaks for Additive Season and Trend algorithm (BFAST) method was first proposed by Verbesselt in 2010 [48], and was initially used to identify disturbances in time-series models (with trend/season/regressor terms), using remote sensing data. The BFAST algorithm, which is widely applicable for meteorological, hydrological, economic and other fields, employs the parameters of fitted piecewise-linear trends and a seasonal trend model [49]. The formula of BFAST for long time-series is defined as follows:

$$Y_t = T_t + S_t + e_t, t = 1, 2, 3 \ldots \ldots n \tag{4}$$

$$T_t = \alpha_i + \beta_i t \tag{5}$$

$$M = (\alpha_{i-1} - \alpha_i) + (\beta_{i-1} - \beta_i)t \tag{6}$$

$$S_t = \sum_{k-1}^{j} \alpha_{j,k}\, sin\left(\frac{2\pi kt}{f}\right) + \delta_{j,k} \tag{7}$$

where t is the time step of a long time-series; Y_t is the 16-d SIF from 2001–2020 on the Mongolian Plateau, T_t is the trend component (trend), S_t is the seasonal component (seasonal), and e_t is the remainder (remainder). The long-term trend component, T_t, is segmented, has a specific slope (β_i) and intercepts (α_i) on $m + 1$ different segments. Thus, there are m breakpoints, $\tau_i - 1, \ldots, \tau_m$, where $i = 1, 2, \ldots \ldots m$, and we define $\tau_0 = 0$ and $\tau_m + 1 = n$.

The order of magnitude of the breakpoints, M, can be calculated using the slopes α_i and intercepts β_i of the trend component T_t between t_{i-1} and t_i. To fit the frequency, we define $t_0 = 0$ for $t_j < t \leq t_j + 1$, where j is the breakpoint position, $j = 1, \ldots m$, with m breakpoints in total; k is the number of harmonic terms; and α_j, k and $\delta_{j,k}$ are the segment-specific amplitude and phase, respectively. In this study, the SIF, NDVI and NIRv values during the growing season (from April to October) on the Mongolian Plateau from 2001 to 2020 were used for breakpoint analysis, and the differences in the vegetation-change trends retrieved from the three types of remote sensing data were evaluated. We also analyzed meadow-steppe, typical-steppe, broadleaf-forest and coniferous-forest data. The main purpose of this part of the study is to evaluate the sensitivity of three types of remote sensing data for retrieving the dynamic changes in vegetation and the ability to capture the dynamic changes in different vegetation types.

2.3.2. Theil–Sen Median Trend Analysis

Theil–Sen median-trend analysis and the Mann–Kendall analysis are effective for assessing the trends of long time-series data. The combination of these methods has the advantages of not requiring the data to obey a certain distribution, and having strong resistance to data errors [50]. Additionally, the Mann–Kendall method provides a significance test of linear trends. The Theil–Sen estimator calculates the median of the SOS data-combination slope as follows:

$$S_{SOS} = Median\left(\frac{SOS_i - SOS_j}{j - i}\right),\ 2001 \leq i < j \leq 2020 \tag{8}$$

$S_{SOS} > 0$ indicates an increasing SOS trend, while a negative value indicates a decreasing SOS trend.

The Mann–Kendall test can determine whether a trend is significant. Similar to the Theil–Sen median trend, this test does not require the sample to obey a certain distribution, and is robust to the presence of a few outliers [51,52].

$$Z = \begin{cases} \frac{S-1}{\sqrt{s(S)}}, & S > 0 \\ 0, & S > 0 \\ \frac{S+1}{\sqrt{s(S)}}, & S > 0 \end{cases},\ S = \sum_{j=1}^{n-1} \sum_{i=j+1}^{n} sgn(SOS_j - SOS_i) \tag{9}$$

$$sgn(SOS_j - SOS_i) = \begin{cases} 1, SOS_j - SOS_i > 0 \\ 0, SOS_j - SOS_i = 0 \\ -1, SOS_j - SOS_i < 0 \end{cases},\ s(S) = \frac{n(n-1)(2n+5)}{18} \tag{10}$$

where SOS_i and SOS_j represent the SOS in years *i* and *j*, respectively; *sgn* is the signum function; *n* represents the length of the time series, and the value range of the statistic *Z* is $(-\infty, +\infty)$. At a given significance level α, when $|Z| > u_{1-\alpha/2}$, the changes in the trend are considered significant. We evaluated the significance of a change in the SOS time-series trend at the 0.05 confidence level; thus, we applied $\alpha = 0.05$ [53]. In this study, we analyzed the trends of SOS_{SIF}, EOS_{SIF}, SOS_{NDVI}, EOS_{NDVI}, SOS_{NIRv}, and EOS_{NIRv} and the phenological parameters of different vegetation-types from 2001 to 2020. At the same time, we compared and evaluated the differences in the trends of the three phenological parameters.

2.3.3. Detection and Analysis of Change-Points of Phenological Parameters

We used the change-point detection tool in ArcGIS Pro3.0 to detect the change points of the three phenological parameters. The change-point detection tool can recognize the time step when some statistical attributes of a time-series change. The tool can detect changes in the average value, standard deviation or slope (linear trend) of continuous variables, and changes in the average value of the counting variables. In this study, we used the slope change of the phenological parameters over 20 years to detect the change points. To prevent all time steps from being detected as change points, one of two constraint types must be applied, using method parameters. Algorithms for the optimal identification of segment neighborhoods (SegNeigh) enable the user to determine the number of change points to be detected, using the number of change-points parameters This option uses the "segment neighborhood" [54] algorithm to find the lowest cost segment among all possible segments with the specified number of changepoints. The optimal detection of changepoints with a linear-computational-cost (PELT) option uses the pruned exact linear Ttme [55,56] algorithm to estimate the number and location of change points.

The idea behind PELT is that for a time step to be detected as a change point, the value of its segmentation cost reduction must be more than the added penalty value. If the cost reduction is less than the increased penalty value, the penalty cost will increase, and the time step will not be detected as a change point. The penalty value used in PELT is determined by the detection-sensitivity-parameter value. The sensitivity is provided in digital form between 0 and 1, where a higher sensitivity will detect more change points by using lower penalty values. For the change in slope (linear trend), a more conservative penalty formula is used:

$$\text{Penalty Value} = 6^{2-2\cdot sensitivity} \tag{11}$$

In this study, we used PELT to detect change points. Because change-point detection based on slope (step size 2) requires conservative control-sensitivity, we finally defined the sensitivity as 0.35. This sensitivity detects one change-point at most.

2.3.4. Pearson Correlation Analysis and Partial Correlation Analysis

The Pearson's correlation coefficient method can be used to analyze the degree of relatedness between two variables. By calculating the correlation coefficient, R, between the climate factors and SOS and EOS, we determined the relationships between them. The range of potential values of the correlation coefficient is $[-1, 1]$, and the calculation formula is as follows:

$$R_{SOS,SFC} = \frac{\sum_{i=1}^{n}(SOS_i - \overline{SOS})(SFC_i - \overline{SFC})}{\sqrt{\sum_{i=1}^{n}(SOS_i - \overline{SOS})^2}\sqrt{\sum_{i=1}^{n}(SFC_i - \overline{SFC})^2}} \tag{12}$$

where n is the number of years, n = 20 (2001–2020); i and j represent years i and j, respectively; $\overline{SOS}$ represents the average SOS value over the 20 years from 2001 to 2020, and $R_{SOS,SFC}$ are the correlation coefficients of SOS and SFC. In addition, we used a t-test to test whether the correlation coefficient was significant. A p value < 0.05 indicated a significant correlation.

Partial correlation analysis: In a multifactorial system, the relationship between two factors is inevitably disturbed by other factors. Therefore, excluding the influence of other factors is necessary to obtain a relatively pure relationship between the two variables. Suppose there are k ($k > 2$) variables $x_1, x_2 \ldots x_k$; then, after excluding the influence of other variables, the correlation coefficient of variables i and j is called the hth-order ($h \leq k - 2$) partial correlation coefficient, which is calculated as follows:

$$R_{ij,l_1\cdot l_2\ldots l_h} = \frac{R_{ij,l_1\cdot l_2\ldots l_{h-1}} - R_{il_h,l_1\cdot l_2\ldots l_{h-1}}\cdot R_{jl_h,l_1\cdot l_2\ldots l_{h-1}}}{\sqrt{(1 - R^2_{il_h,l_1\cdot l_2\ldots l_{h-1}})(1 - R^2_{jh,l_1\cdot l_2\ldots l_{h-1}})}} \tag{13}$$

where $R_{ij,l_1 l_2\ldots l_h}$ is the partial correlation coefficient of variables i and j, after excluding the influence of other factors. Suppose that there are five variables (k = 4): SOS(S), TEM_{Spring}(T), PRE_{Spring} (P) and SW_{Spring} (W). Consequently, the order of the partial correlation coefficient of the variable SOS and TEM is h = 2, and the formula is as follows:

$$R_{ST,PW} = \frac{R_{ST,P} - R_{SW,P}\cdot R_{TW,P}}{\sqrt{(1 - R^2_{SW,P})(1 - R^2_{TW,P})}} \tag{14}$$

In this study, we conducted a Pearson correlation analysis of three SOS values by SCF and SMD, and then we conducted a partial correlation analysis of spring precipitation, temperature and soil water and three SOS values, and a partial correlation analysis of summer and autumn precipitation, temperature and soil water, on three EOS values.

2.3.5. Path Analysis

Path analysis is an extension of multiple regression analysis. The causal relationship in path analysis is multileveled, and intermediate variables are added between causal variables, which makes the path-analysis model richer and more powerful than the general regression model in describing the actual causal relationship. Thus, path analysis is not a multiple regression equation, but rather a set of multiple regression equations. For example, the equations are as follows:

$$\text{SOS} = \beta_1 \cdot SFC + \beta_2 \cdot TEM + \beta_3 \cdot SW + \beta_4 \cdot SMD + \varepsilon_1 \tag{15}$$

$$\text{SMD} = \beta_5 \cdot SFC + \beta_6 \cdot TEM + \varepsilon_2 \tag{16}$$

$$\text{SW} = \beta_7 \cdot SFC + \beta_8 \cdot TEM + \beta_9 \cdot SMD + \varepsilon_3 \tag{17}$$

where $\beta_1 \ldots \beta_9$ are path coefficients, and $\varepsilon_1 \ldots \varepsilon_3$ are residuals. The main purpose of path analysis is to test the accuracy and reliability of a hypothetical causal model and to measure the strength of the causal relationship between variables. If a correlation exists between the two variables X and Y in the model, further study is required to determine if a causal relationship exists between them. This may be a direct or indirect relationship, or both. In this study, we assumed that a direct or indirect relationship exists between SMD and SOS. Path analysis can be used to more comprehensively understand the mechanism by which snow removal affects the SOS. For example, SCF and SMD affect SOS directly or through some variables, or some variables affect SOS through SMD. All path designs and path coefficients were calculated using IBM SPSS Amos 25 software.

2.3.6. Geographical-Detector Model

The geographical-detector model can analyze the relationship between factors based on spatial heterogeneity. In this study, the dependent variable Y was the phenological parameter of the Mongolian Plateau, and the independent variable X included natural factors and human factors. The geographical-detector model consists of four detectors: a factor detector, an interaction detector, a risk detector and an ecological detector [57]. In this study, we mainly used the factor detector to determine the explanatory degree of the independent variable to the dependent variable. The scope of interpretation is measured by the metric q [58], which is expressed as follows:

$$q = 1 - \frac{\sum_{h=1}^{L} N_h \sigma_h^2}{N\sigma^2} = 1 - \frac{SSW}{SST},\ SSW = \sum_{h=1}^{L} N_h \sigma_h^2,\ SST = N\sigma^2 \tag{18}$$

where q is the degree of interpretation of factor X to Y, and the value range of q is [0, 1]. $h = 1 \ldots, L$ represents the grade of the independent variable Y or the dependent variable X; σ_h^2 and σ^2 are the variance of Y in layer h and the variance in the entire region, respectively; N_h and N represent the number of layers, h, and the number of cells in the entire area, respectively.

3. Results

3.1. Comparison of Vegetation Trends Based on SIF, NDVI and NIRv

SIF, NDVI and NIRv all have the ability to restore vegetation changes, but their results may vary for vegetation dynamics. To explore the differences in the vegetation dynamics retrieved from the three types of remote-sensing data on the Mongolian Plateau, we used the BFAST package to analyze the SIF, NDVI and NIRv trends, and to explore the breakpoints. Additionally, we analyzed the trends and breakpoints in the meadow steppe, typical steppe, broadleaf forest and coniferous forest, over a large area of the Mongolian Plateau (Figure 2). The results show that the SIF, NDVI and NIRv trends and breakpoints on the Mongolian Plateau were obviously different.

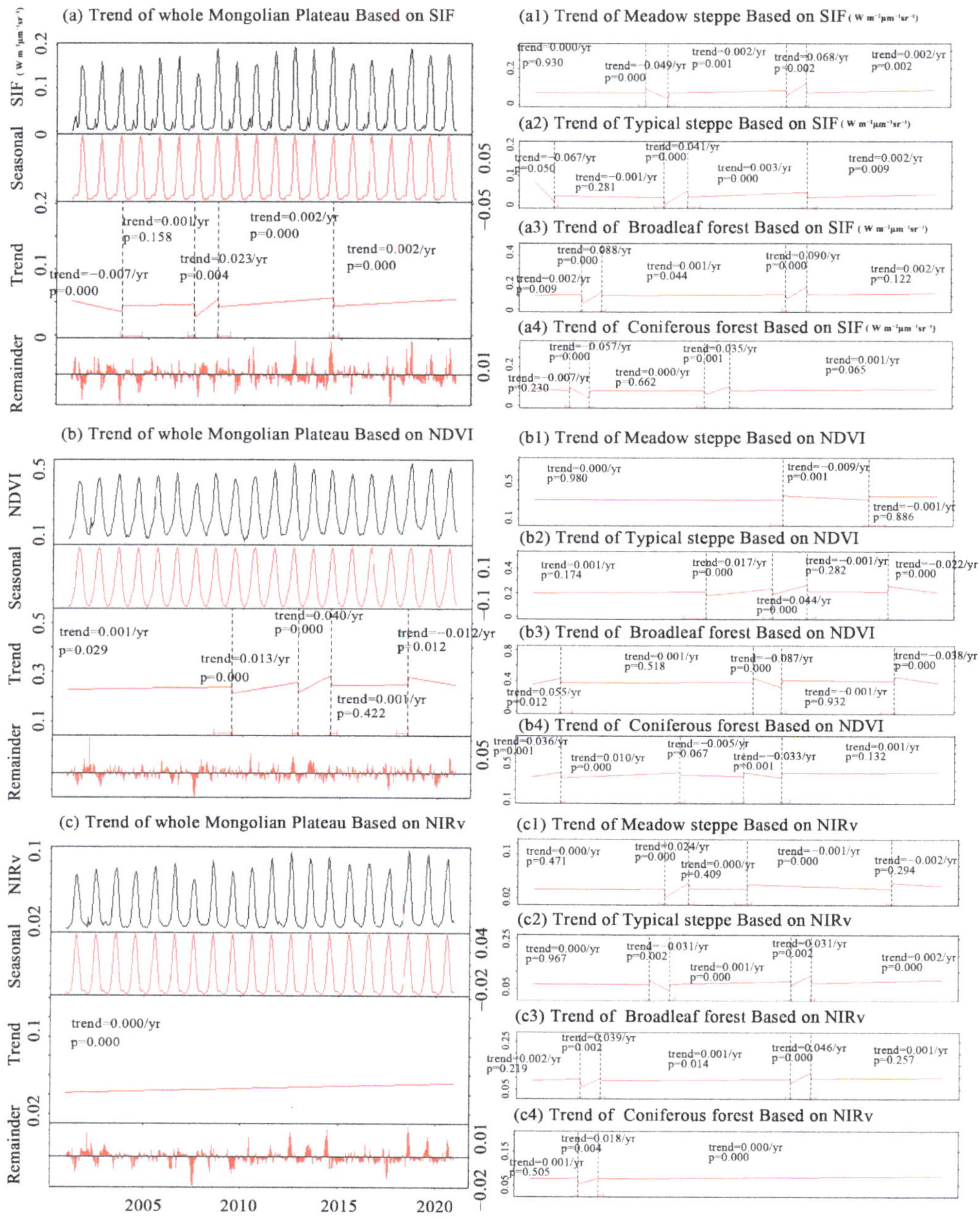

Figure 2. Vegetation trends and breakpoints retrieved from three types of remote-sensing data: (**a–c**) Mongolian Plateau results based on SIF, NDVI and NIRv; (**a1**,**b1**,**c1**) meadow-steppe results based on SIF, NDVI and NIRv; (**a2**,**b2**,**c2**) typical-steppe results based on SIF, NDVI and NIRv; (**a3**,**b3**,**c3**) broadleaf-forest results based on SIF and NDVI; (**a4**,**b4**,**c4**) coniferous-forest results based on SIF, NDVI and NIRv.

SIF breakpoints were found on day 225 in 2003, day 145 in 2007, day 225 in 2008 and day 209 in 2014. Moreover, SIF decreased significantly from 2003 to 2007, increased sharply from 2007 to 2008, and increased significantly from 2008 to 2014 and from 2014 to 2020 (Figure 2a). NDVI breakpoints were found on day 161 in 2009, day 305 in 2012, day 193 in 2014 and day 177 in 2018 (Figure 2b). NDVI experienced a continuous surge from 2009, 2012 and 2012 to 2014 and a decline from 2018 to 2020. However, no NIRv breakpoint was detected on the Mongolian Plateau. The comparison showed that the SIF and NDVI breakpoint years overlapped greatly (2007 to 2014), showing significant growth, and the SIF breakpoint years appeared earlier.

In addition, our detection results for different vegetation types were very different. SIF of meadow steppe experienced a sharp increase from 2006 to 2007 and a sharp decrease from 2013 to 2014. Both meadow steppe and forest experienced a sharp increase or a decrease from 2002 to 2007, but NDVI did not detect these changes. The drastic changes in NDVI mostly occurred from 2008 to 2011. During this period, SIF also underwent drastic changes. Therefore, SIF may capture more mutation details than NDVI, especially for mutations within a short period of time that may be caused by human activities. NIRv and SIF were similar, especially in forest areas. Both could detect breakpoints in earlier years, but the NIRv breakpoints occurred later than those for SIF. This result may indicate that SIF and NIRv have more advantages in detecting short-term upheaval.

3.2. Temporal and Spatial Variability of Vegetation Phenological Parameters

In the past 20 years, the SOS_{NDVI} and SOS_{NIRv} values of the Mongolian Plateau have been relatively close, and have not increased significantly, with rates of 0.143 days/year and 0.192 days/year, respectively (Figure 3a). The average values over the past 20 years were 103.6 day and 97.4 day, respectively. The earliest SOS occurred in 2016 and 2013, and the latest SOS occurred in 2006 and 2008. In the past 20 years, SOS_{SIF} was late, with an insignificant increase of 0.038 days/year, and the annual average value was the 132.3nd day. The earliest and latest SOS occurred in 2001 and 2005, respectively, indicating that this difference may be due to the low chlorophyll concentration and weak SIF of the returning vegetation at higher latitudes on the Mongolian Plateau. Concurrently, the earliest and latest years of SOS are breakpoint years.

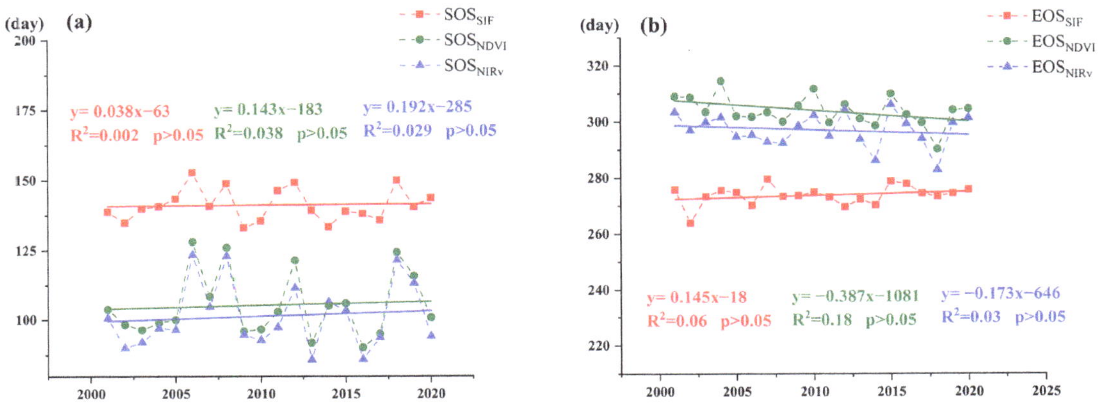

Figure 3. Temporal variation in phenological parameters of the three vegetation types on the Mongolian Plateau, over 20 years; (**a**) is the linear trend of the three SOS; (**b**) is the linear trend of the three EOS.

The EOS_{NDVI} and EOS_{NIRv} values on the Mongolian Plateau are also close, and EOS_{NDVI} is slightly later than EOS_{NIRv}. EOS_{NDVI} and EOS_{NIRv} decreased at the rates of 0.387 days/year and 0.173 days/year, respectively, but not significantly, with mean values of the 303.9nd day and the 297nd day over 20 years. The earliest and latest SOS were in

2001 and 2019, respectively. However, EOS_{SIF} increased at a rate of 0.145 days/year, but not significantly. The average value from the past 20 years was 274nd day, and the earliest and latest EOSs were in 2001 and 2006, respectively. LOS_{SIF} was the shortest, and LOS_{NDVI} was the longest. The three LOSs were 112.3 days, 198.6 days and 195.6 days, respectively.

As shown in Figure 4, the spatial distributions of the three SOS trends in central and eastern Mongolia were similar. Central and eastern Mongolia showed an upward trend, while western Mongolia and central and eastern Inner Mongolia showed a downward trend. However, SOS_{SIF} had a higher significance; 10.29% of pixels passed the significance test ($p < 0.05$), and the agricultural land in south-central Inner Mongolia and the broadleaf forest in the northeast showed a significant downward trend (Figure 4a).

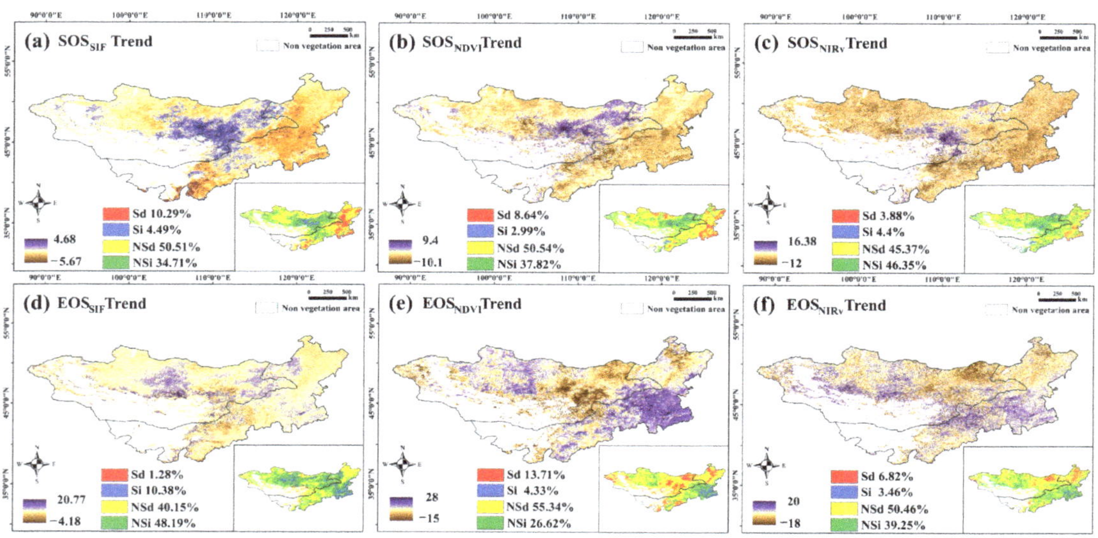

Figure 4. Comparison of the spatial-distribution trends of the phenological parameters extracted based on SIF, NDVI and NIRv. Sd: significant decrease; Si: significant increase; NSd: nonsignificant decrease; NSi: nonsignificant increase.

Obvious differences were observed among the three EOSs. Approximately 58.57% of the EOS_{SIF} pixels showed an increasing trend that was mainly distributed in the middle of Mongolia and the middle of Inner Mongolia. Approximately 10.38% of the pixels passed the significance test ($p < 0.05$), and these pixels were distributed in the coniferous forest belt in the middle of Mongolia and on agricultural land and desert steppe in the middle and south of Inner Mongolia (Figure 4d). The growth of EOS_{NDVI} and EOS_{NIRv} in this region was also detected, but it was not significant. Approximately 69.05% of the EOS_{NDVI} pixels showed a decreasing trend, and these pixels, of which 13.71% passed the significance test ($p < 0.05$), were mainly distributed in the typical steppe belt in the southeast and the meadow steppe belt in the northeast of Mongolia, as well as some coniferous and broadleaf forest-areas in Inner Mongolia. The spatial distribution trend of EOS_{NIRv} was similar to that of EOS_{NDVI}, but the decreasing trend of the coniferous and broadleaf forest belts detected by EOS_{NIRv} was not significant (Figure 4f). However, EOS_{SIF} did not detect significant changes in the forest region. In general, the spatial distribution of the three EOS trends on the Mongolian Plateau was generally consistent, and the difference was mainly reflected in the sensitivity of the trend to different vegetation types.

Figure 5 shows that SOS_{SIF} tends to be significantly delayed by approximately 2 days/100 m of elevation. The difference is that SOS_{NDVI} and SOS_{NIRv} significantly advance at rates of 0.6 days/100 m and 1.3 days/100 m, respectively. EOS_{SIF} and EOS_{NDVI}

advanced by approximately 0.9 days/100 m and 0.04 days/100 m, respectively, although the trend for EOS_{NDVI} was not significant. EOS_{NIRv} was significantly delayed by approximately 0.5 days/100 m. As the altitude increased, the change trends of the three phenological parameters varied. Therefore, as the altitude increased, LOS_{SIF} was significantly shorter, LOS_{NDVI} slowly increased, and LOS_{NIRv} was significantly longer.

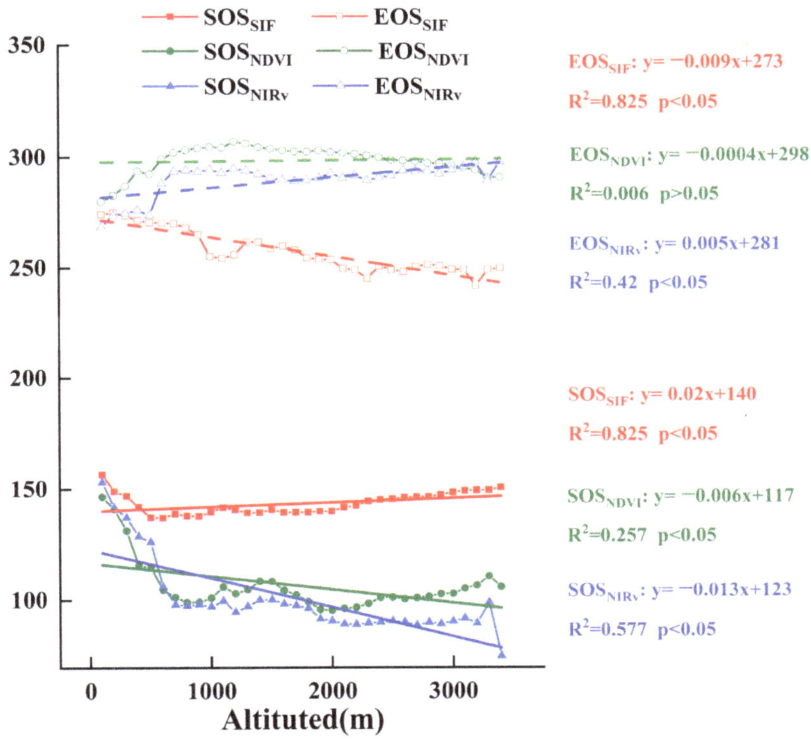

Figure 5. Characteristics of vegetation phenology changes with altitude.

This may be due to the difference in chlorophyll concentration of different vegetation types, which affects the observation of phenology by NDVI and NIRv, especially in high altitude areas, where the greenness reflectance of forest vegetation is high. SIF can directly express the photosynthesis inflection-point, and may better show the relationship between vegetation phenology and altitude, which also highlights the advantages of SIF in retrieving phenological parameters.

We analyzed the occurrence of the change trend of the vegetation phenological parameters on the Mongolian Plateau over the past 20 years (Figures 6 and 7). The results show that the years that SOS_{SIF} experienced the most changes were 2001 to 2003 and 2016 to 2019. A total of 43.55% and 24.41% of the changes occurred in these two periods, respectively. The years with the least change were from 2011 to 2013, and only 2.24% of the change occurred during this period. The change points were mainly distributed in the broadleaf forest area in northeastern Inner Mongolia, the desert steppe belt in southern Mongolia, and the coniferous forest areas in northern and middle Mongolia (Figure 6a). The decreasing trends of SOS_{SIF} before and after the change accounted for 70.95% and 48.28%, respectively. The increasing trends before and after the change accounted for 21.97% and 51.65%, respectively (Figure 6b,c). This result shows that the change was mainly reflected in the change from the original decreasing trend to the increasing trend, especially in the central and western regions of Mongolia and some regions in southern Inner Mongolia. In northeastern Inner Mongolia, the original increasing trend changed to a decreasing trend.

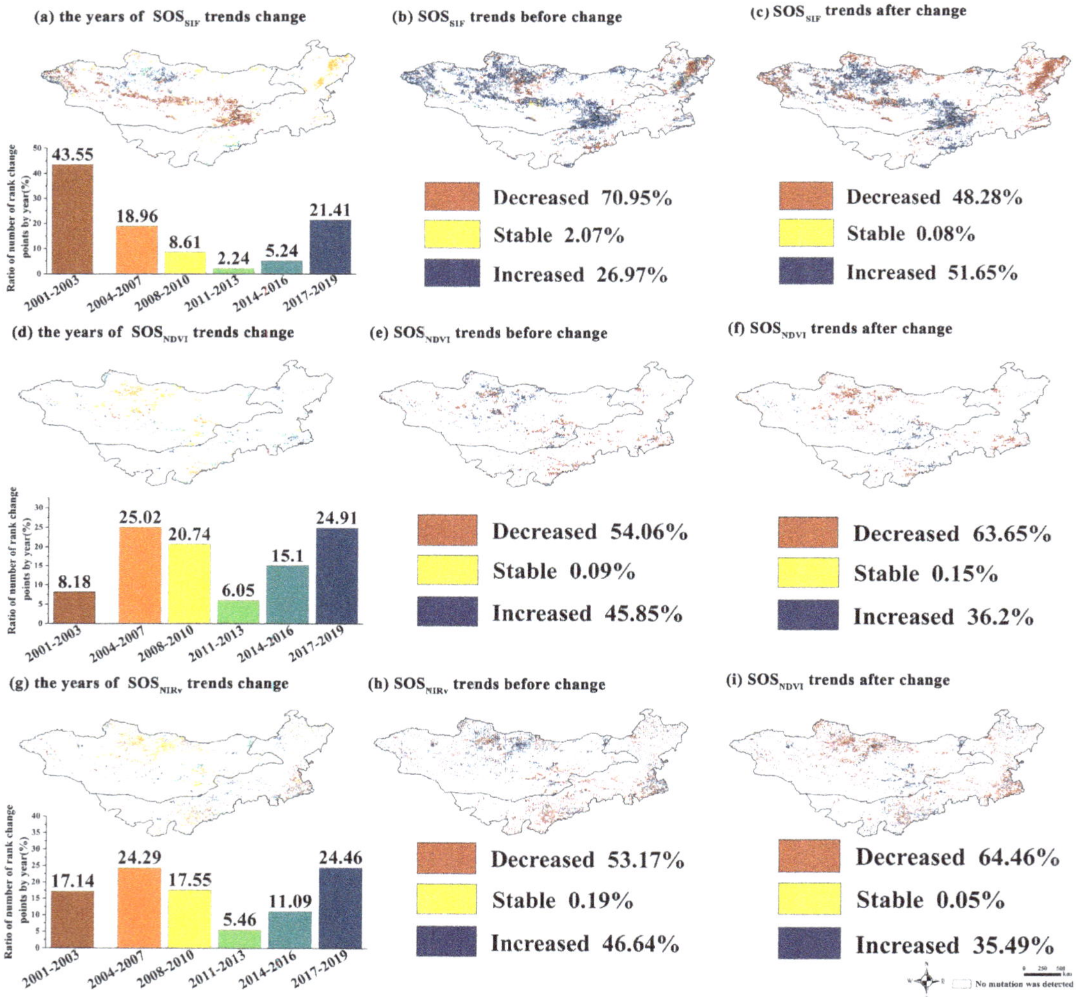

Figure 6. The years when the SOS trends changed during the past 20 years, and a trend comparison before and after the change: (**a**,**d**,**g**) are the change-points years and the change proportions of SOS_{SIF}, SOS_{NDVI} and SOS_{NIRv}, respectively; (**b**,**c**), (**e**,**f**) and (**h**,**i**) are the trends before and after the changes in SOS_{SIF}, SOS_{NDVI} and SOS_{NIRv}, respectively.

The years that SOS_{NDVI} experienced the most changes were 2004 to 2007 and 2016 to 2019. A total of 25.02% and 24.91% of the changes occurred in these two periods. The years with the least change were from 2011 to 2013, and only 4.54% of the changes occurred then (Figure 6d). The change points were mainly distributed in the coniferous forest area in northern Mongolia and in central and southern Inner Mongolia. The decreasing trend of SOS_{NDVI} before and after the change accounted for 48.82% and 72.07%, respectively. The proportion of the increasing trend before and after the change was 53.55% and 27.58%, respectively (Figure 6e,f). This result shows that the original increasing trend changed into a decreasing trend, which was mainly distributed in the coniferous forest in northern Mongolia and in the sandy steppe and agricultural land in central Inner Mongolia.

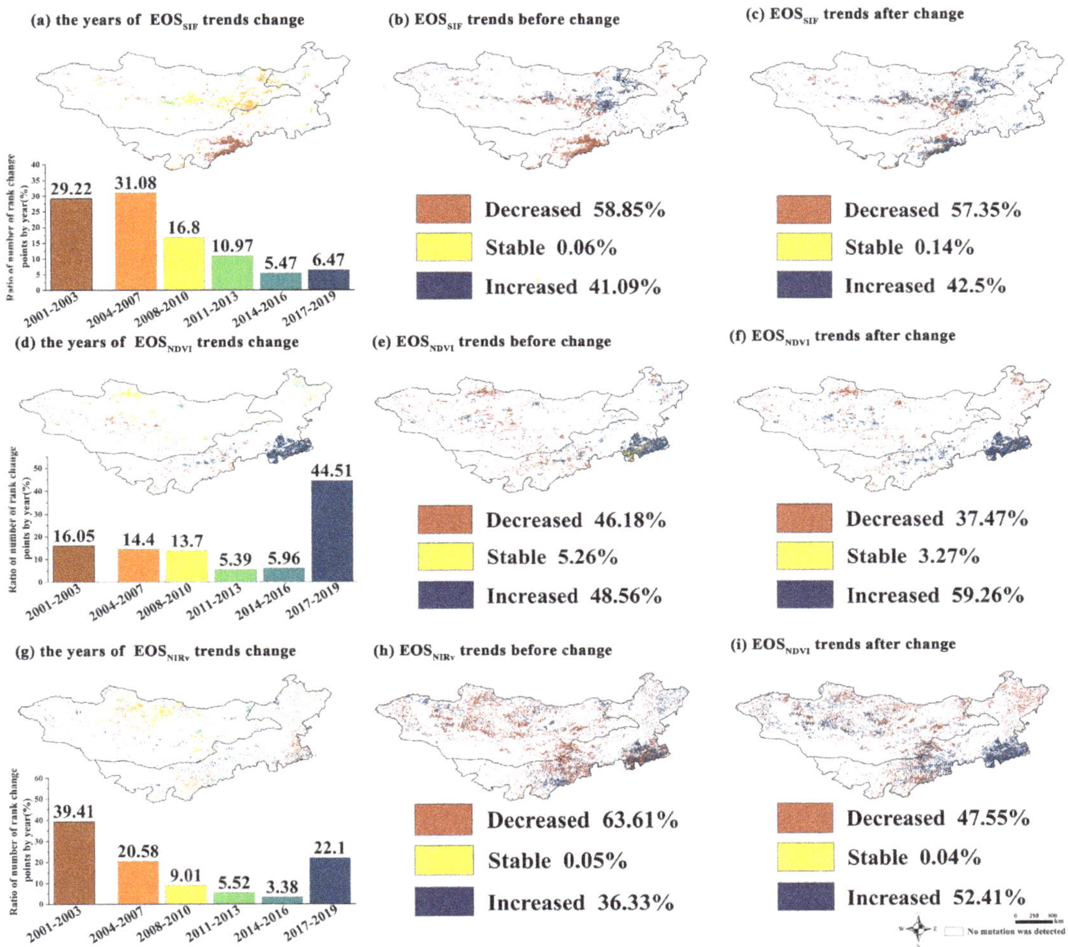

Figure 7. The years of the EOS change-trends in the past 20 years and a trend comparison before and after the change: (**a**,**d**,**g**) are the years of the change-points and the change proportions of EOS_{SIF}, EOS_{NDVI} and EOS_{NIRv}, respectively; (**b**,**c**), (**e**,**f**) and (**h**,**i**) are the trends before and after the changes of EOS_{SIF}, EOS_{NDVI} and EOS_{NIRv}, respectively.

The years that SOS_{NIRv} experienced the most changes were 2017 to 2019 and 2004 to 2007. A total of 24.46% and 24.25% of the changes occurred in these two periods. The years with the least change were from 2011 to 2013, and only 5.46% of the change occurred then (Figure 6g). The change years of SOS_{SIF} and SOS_{NIRv} were relatively close. The distribution was closer to SOS_{NDVI}, and more pixels were detected. The decreasing trend of SOS_{NIRv} before and after the trend-change accounted for 53.17% and 64.46%, respectively. The increasing trend before and after the change accounted for 46.64% and 35.49%, respectively (Figure 6h,i). This result shows that the change was mainly manifested as the original increasing trend changing to a decreasing trend. The change points were mainly distributed in the coniferous forest in northern Mongolia and most areas of Inner

Mongolia. The number of change-point pixels detected in the forest area of northeast Inner Mongolia was greater than that of SOS_{NDVI}. In general, the change years of the three SOSs were similar to the BFSAT breakpoint-monitoring years for the three data sources. In terms of the years and the distribution, SOS_{NDVI} and SOS_{NIRv} were closer, and the changes in NDVI occurred later than the changes in SIF and NIRv. SOS_{SIF} was more sensitive to the change in the forest area, and SOS_{NDVI} was more sensitive to the change in the steppe area. SOS_{NIRv} detected the most change points.

In the past 20 years, the years that EOS_{SIF} experienced the most changes were 2004 to 2007 and 2001 to 2003. A total of 31.08% and 29.22% of the changes occurred in these two periods. The years with the least change were from 2014 to 2016, with only 5.47% of changes. The trend change-points were mainly distributed in the grassland belt in eastern Mongolia and the desert steppe in southern Inner Mongolia (Figure 7a). The decreasing trend of EOS_{SIF} before and after the change point accounted for 58.85% and 57.35%, respectively. The proportion of the increasing trend before and after the trend change was 41.52% and 51.66%, respectively (Figure 7b,c). This result shows that the change was mainly reflected in the change from the original decreasing trend to the increasing trend, especially in the central and western regions of Mongolia and in some regions in southern Inner Mongolia. In northeastern Inner Mongolia, the original increasing trend changed to a decreasing trend.

The years in which EOS_{NDVI} experienced the most change were from 2017 to 2019, and 44.51% of the changes occurred during this period. The years with the least change were from 2011 to 2013, with only 5.39% of the change (Figure 7d). This change mainly occurred in the coniferous forest belt in the north of Mongolia and the artificial vegetation area in the middle of Inner Mongolia. The decreasing trend of EOS_{NDVI} before and after the change accounted for 46.18% and 37.47%, respectively. The proportion of the increasing trend before and after the change was 48.56% and 59.26%, respectively (Figure 7e,f). This result shows that the overall trend is from decreasing to increasing. The trends mainly changed in some areas in the middle of Mongolia, and the trends in the sandy steppe and agricultural land in the middle of Inner Mongolia changed from decreasing to increasing. The slope of the decreasing trend in northeastern Inner Mongolia also changed.

The years that EOS_{NIRv} experienced the most changes were 2001 to 2003 and 2016 to 2019, and 39.41% and 22.1% of the changes occurred in these two periods, respectively. The years with the least change in trends were from 2014 to 2016, and only 3.38% of the change occurred during this period (Figure 7g). The year and spatial distribution of the EOS_{NIRv} change points were close to those of EOS_{NDVI}, and EOS_{NIRv} detected more pixels. The decreasing trend of EOS_{NIRv} before and after the change accounted for 63.61% and 47.55%, respectively. The increasing trend before and after the change accounted for 36.33% and 52.41%, respectively (Figure 7h,i). This result shows that the change in the trend was mainly manifested in the change from the original decreasing trend to the increasing trend. The change points were mainly distributed in the coniferous forest in northern Mongolia and in central and southern Inner Mongolia. In addition, the slope of the decreasing trend in the middle of Inner Mongolia changed. In terms of the year and location, EOS_{SIF} and EOS_{NIRv} were closer. The change in EOS_{NDVI} occurred later than that of EOS_{NDVI} and EOS_{NIRv}. EOS_{NIRv} detected the most change points.

3.3. Correlation Analysis between Vegetation Phenological Parameters and Meteorological Factors

The area of snow-cover in winter showed a significant negative correlation with SOS_{NDVI} and SOS_{NIRv} (Figure 8a–c). Approximately 41.56% and 34.87% of the pixels passed the significance test ($p < 0.05$), and were mainly distributed in most regions of Mongolia and the central region of Inner Mongolia. This indicated that the larger snow-cover area in winter will lead to the advance of the SOS. This is related to the supplementation of snow to soil moisture in spring. However, SOS_{SIF} was not sensitive to the change in the snow-cover area (Figure 8a). This result shows that snow-cover in winter had a significant effect on the advance of the SOS. The spring snow-melting period showed a more significant negative correlation with SOS_{NDVI} and SOS_{NIRv} (Figure 8d–f). Approximately 54.48% and 47.65% of the pixels, respectively, passed the significance test ($p < 0.05$), and were mainly distributed in most regions of Mongolia and parts of the middle and north of Inner Mongolia. Specifically, the late snowmelt-period may lead to the advance of the SOS; in contrast, if the snowmelt period is relatively early, it will lead to the delay of the SOS. However, SOS_{SIF} was still not sensitive to changes in the snowmelt period (Figure 8a).Therefore, the larger snow-cover in winter and late snow-melting in spring positively contributed to the advance of the SOS. This result shows that snow-melting may play an important role in supplementing soil water and in the supplementation time for the growth of vegetation in spring. However, the mechanism of the snow-cover and the snowmelt period on the SOS is complex and indirect, especially when the soil water is supplemented by melted-snow cover, which plays an important role here; thus, further analysis is needed.

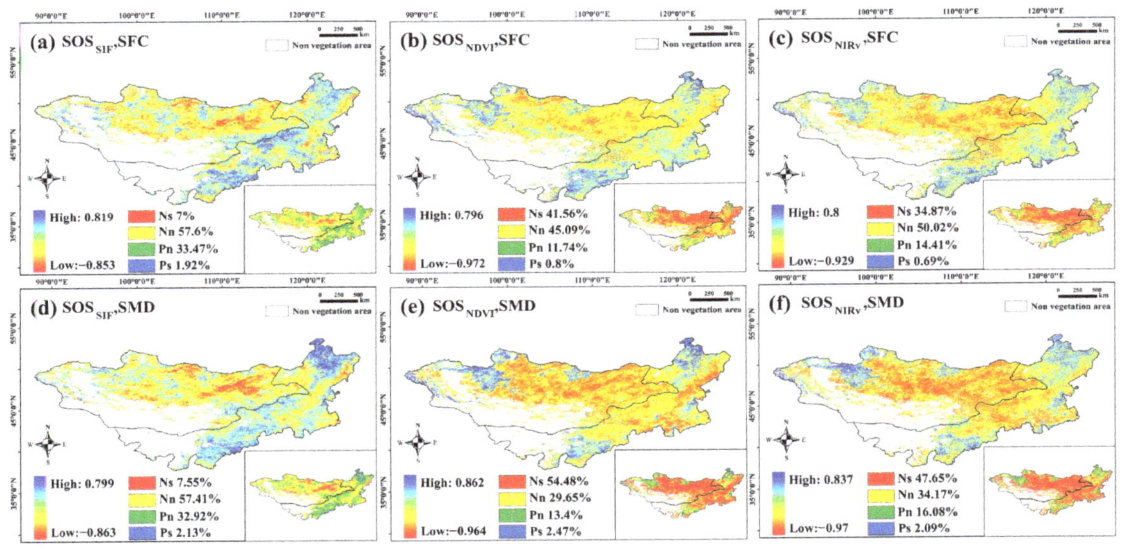

Figure 8. Spatial distribution of the correlations between winter snow-related climatic factors (SFC, SMD) and the three types of phenological parameters. Note: Ns: significant ($p < 0.05$) negative correlation; Nn: nonsignificant ($p > 0.05$) negative correlation; Pn: nonsignificant ($p > 0.05$) positive correlation; Ps: significant ($p < 0.05$) positive correlation. The same applies below.

To eliminate the interference between the meteorological factors in spring, we conducted a partial correlation analysis on the precipitation, temperature and soil moisture of the three SOSs (Figure 9). SOS_{SIF}, SOS_{NDVI} and SOS_{NIRv} all showed a positive correlation with PRE_{Spring}, and 5.26%, 6.09% and 6.69% of the pixels, respectively, passed the significance test ($p < 0.05$). Consequently, compared with the other two SOSs, SOS_{NIRv} and PRE were more significant. SOS_{SIF}, SOS_{NDVI} and SOS_{NIRv} showed a negative correlation with TEM_{Spring}. Approximately 11.83%, 12.8% and 8.07%, respectively, of the pixels passed the significance test ($p < 0.05$); thus, SOS_{NDVI} is highly significant, and mainly distributed in the forest areas in the northeast of Inner Mongolia and the west of Mongolia. SOS_{SIF}, SOS_{NDVI} and SOS_{NIRv} all showed a significant negative correlation with SW_{Spring}. Approximately 20.41%, 24.71% and 17.52%, respectively, of the pixels passed the significance test ($p < 0.05$) and were mainly distributed in the northern coniferous forest of Mongolia, the eastern meadow steppe and the northeast broadleaf forest of Inner Mongolia. After comparison, the significance of SOS_{NDVI} was the highest. The results show that the effect of spring precipitation on the SOS was relatively small, and due to the decreased precipitation in spring. In addition, the early precipitation in spring may lead to low temperatures, thus delaying the SOS; thus, precipitation and the SOS were positively correlated. In the meantime, the water source for vegetation in spring is mostly from the soil water formed after the melting of snow or frozen soil. Therefore, the temperature and soil moisture in spring showed a high negative correlation with the SOS.

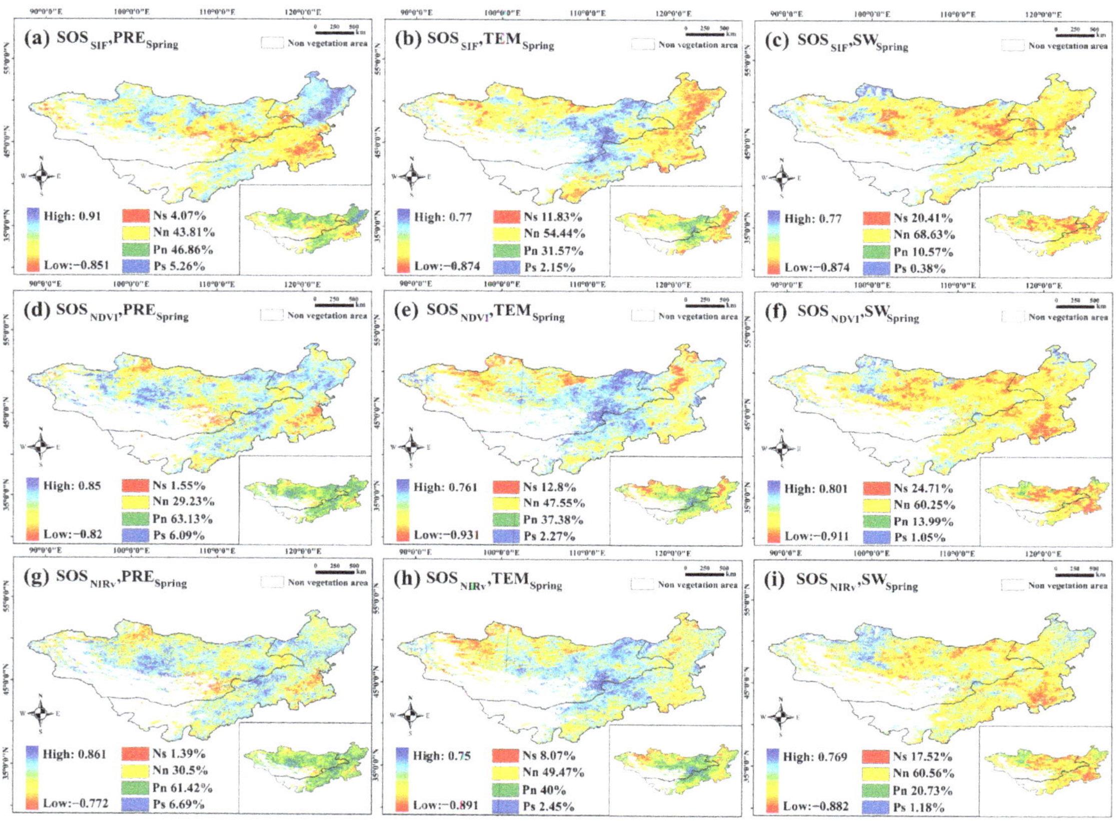

Figure 9. Spatial distributions of the correlations between temperature (TEM_{Spring}), precipitation (PRE_{Spring}), soil moisture (SW_{Spring}) and three types of SOS in (spring).

Due to the complex impact of water and heat synergy on the vegetation in spring, we used the path-analysis method to analyze the direct or indirect effect of the snow-cover area in winter and snow-melting date on SOS under the synergistic effect of spring temperature and soil moisture (the precipitation in spring is small, and the influence is low; thus, no analysis was made). SOS_{SIF} is not sensitive to snow-cover and snowmelt periods. Therefore, in this study, we only analyzed SOS_{NDVI} and SOS_{NIRv} (Figure 10). The results show that a large snow area will delay the snow-melting date. The temperature in spring has a high negative correlation with snow-melting date and soil moisture. Consequently, an increase in temperature may lead to earlier snowmelt. Simultaneously, the advance of the snow-melting date means that the time of soil water supplementation is also advanced. The temperature rise will further lead to the evaporation of soil water and thus be consumed. Hence, the later the snow-melting date is, the higher the amount of soil moisture preserved, which plays an important role in advancing the vegetation rejuvenation period. However, the earlier melting of snow will lead to an increase in evaporation time, and the water eventually stored in the soil will be insufficient to supply vegetation. Moreover, according to the path coefficient, SW_{Spring} has little effect on the SOS, which indicates that the soil moisture content is low during the spring drought. In this period, soil moisture is more important to the supply time of vegetation, which depends on the content of snow and the influence of temperature on snow-melting date. In other words, even if snow-melting supplies much soil water, because the snow-melting date is too early, the soil water is not sufficient for vegetation growth but rather is consumed by evaporation, which means that higher spring soil water may not have the effect of advancing the SOS. This also shows that the path coefficients of SMD and SOS are higher than those of SW_{Spring}. The response of SOS_{NDVI} and SOS_{NIRv} to snow-melting conforms to this mechanism.

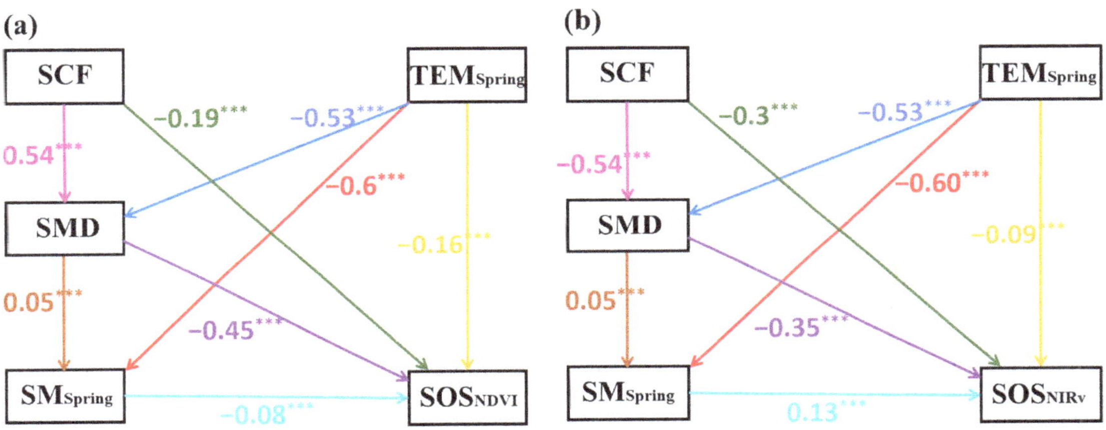

Figure 10. Path analysis of SOS_{NDVI} (**a**), and SOS_{NIRv} (**b**), for each influencing factor. *** indicates that the path coefficient is extremely significant, $p < 0.01$.

On the Mongolian Plateau, the summer precipitation is positively correlated with the three EOSs, and the spatial distribution is similar; however, EOS_{SIF} is more significant (Figure 11a). A total of 21.19% of the pixels passed the significance test ($p < 0.05$), and they were mainly distributed in the coniferous forest area in the central part of Mongolia, the eastern border area and the central area of Inner Mongolia. The correlation between the EOS and summer temperature was not significant. The positive correlation areas are mainly distributed in Mongolia and northeastern Inner Mongolia; this may be related to the low temperature at high latitudes (Figure 11b,e,h). The three EOSs showed a negative correlation with soil water in summer, with a similar spatial distribution, and EOS_{SIF} was more significant. A total of 16.44% of the pixels passed the significance test, and they were mainly distributed in southern Inner Mongolia, the meadow steppe belt at the border between central Inner Mongolia and Mongolia, and some forest areas in northeastern Inner Mongolia. The results show that the EOS is delayed by precipitation in summer, and the effect of temperature on the EOS is weak. In addition, too much soil water will lead to the advance of the EOS, and the significant-negative-correlation spatial distribution of EOS_{SIF} and SW_{Summer} overlaps with the positive-correlation spatial distribution of EOS_{SIF} and PRE_{Summer}, mainly distributed in the meadow steppe at the border between Inner Mongolia and Mongolia and the forest area in northeastern Inner Mongolia. However, in the coniferous forest region in western Mongolia, precipitation is significantly positively correlated with EOS_{SIF}; meanwhile, in this region, soil water and EOS_{SIF} were not significantly negatively correlated. This shows that the precipitation in western Mongolia is lower than that in central and northeastern Inner Mongolia, and will not lead to excessive soil moisture. The precipitation on the arid and semiarid Mongolian Plateau is mostly concentrated in summer; thus, excessive precipitation will lead to high soil-moisture in areas with high precipitation, which will affect the growth of vegetation. In general, EOS_{SIF} is more sensitive to meteorological factors than EOS_{NDVI} and EOS_{NIRv}.

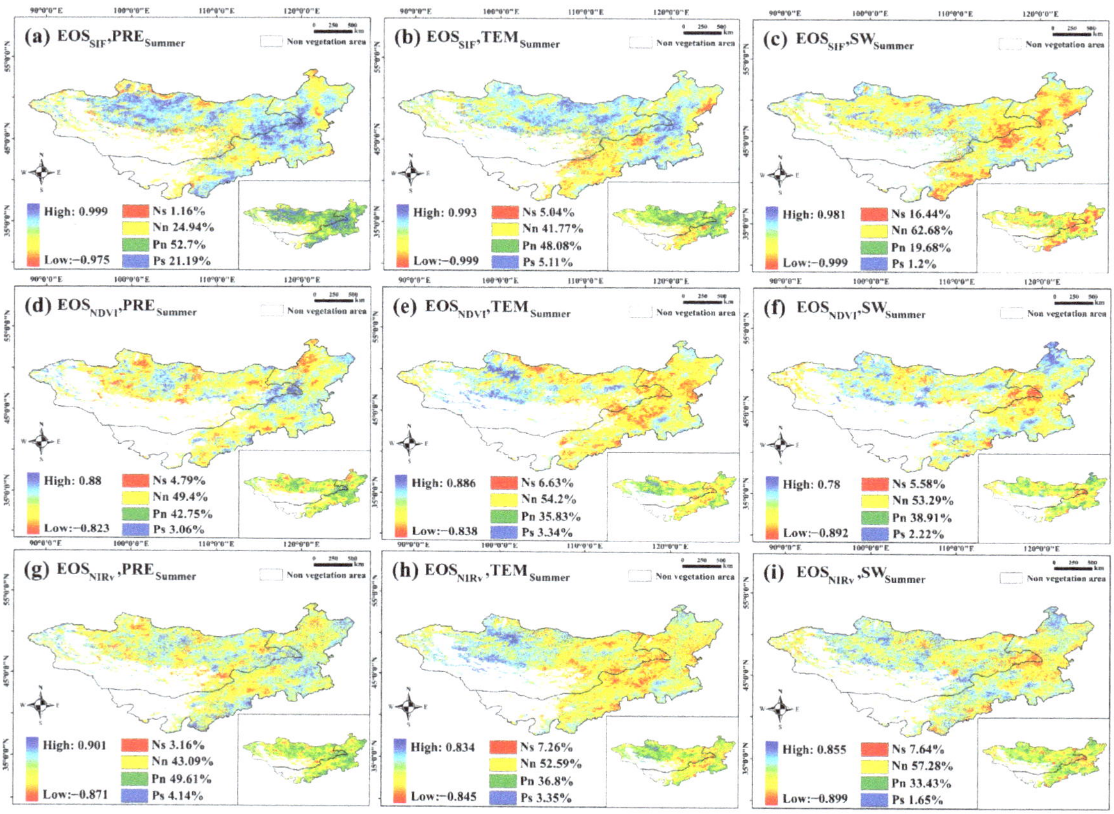

Figure 11. Spatial distributions of the correlations between temperature (TEM_{Summer}), precipitation (PRE_{Summer}), soil moisture (SW_{Summer}) and three types of EOS in summer.

The impact of autumn precipitation on the three EOSs was not significant, which may be due to the low precipitation in autumn (Figure 12a,d,g). TEM_{Autumn} and EOS_{SIF} have a significant positive correlation (Figure 12b). A total of 11.49% of the pixels passed the significance test, mainly distributed in the middle and south of Inner Mongolia, which is mostly sandy vegetation or desert grassland and agricultural land. The first two vegetation types may have strong resistance to drought, while agricultural land may be affected by human irrigation. A 65.17% positive correlation was found between autumn soil-moisture and EOS_{SIF}, and 14.07% of the pixels passed the significance test ($p < 0.05$), which indicates that soil moisture is most important for vegetation growth in autumn, with less precipitation (Figure 12c). This further explains the negative correlation between soil moisture and EOS in the rainy season. Both EOS_{NDVI} and EOS_{NIRv} are less sensitive to meteorological factors than EOS_{SIF}. This shows that SIF has a stronger feedback ability toward environmental changes.

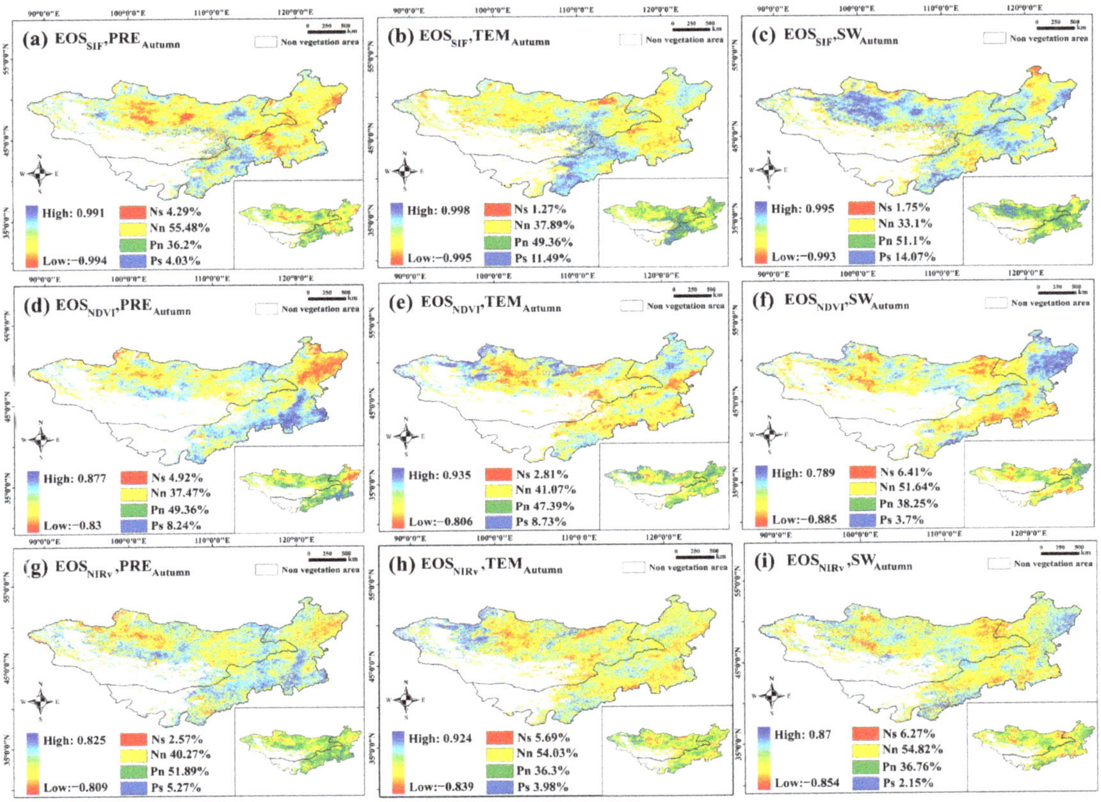

Figure 12. Spatial distributions of the correlations between temperature (TEM$_{Autumn}$), precipitation (PRE$_{Autumn}$), soil moisture (SW$_{Autumn}$) and three types of EOS in (autumn).

3.4. Geographical-Detection Model for SOS and EOS Drivers

From the perspective of spatial differentiation, we analyzed the explanatory power of each factor for the three SOS and EOS. Furthermore, we used the interpretation of the population density on the SOS and EOS to quantify the impact of human activities on the vegetation phenology (Figure 13).

According to the results of the factor detector, the factors with the highest interpretation degree (q value) for SOS$_{SIF}$ were PRE$_{Spring}$ and SW$_{Spring}$, with q values of 0.0964 and 0.0699, respectively (Figure 13a). This result shows that the water factor had a high degree of interpretation of SOS$_{SIF}$ in spatial differentiation, but the SFC had the lowest degree of interpretation of SOS$_{SIF}$ among all factors. The factors with the highest interpretation of SOS$_{NDVI}$ were SMD and SFC, and the q values were 0.3362 and 0.2462, respectively (Figure 13b). The degree of interpretation of the population density to SOS$_{NDVI}$ was higher than that of spring precipitation. SFC and SMD were the factors with the highest interpretation of SOS$_{NIRv}$, with q values of 0.3425 and 0.2442, respectively (Figure 13c). Unlike SOS$_{NDVI}$, SFC had a higher interpretation of SOS$_{NIRv}$ than SMD. The degree of interpretation of the population density to SOS$_{NDVI}$ was higher than that of spring precipitation, temperature and soil moisture. The results show that all influencing factors had the best explanatory power for SOS$_{NDVI}$ (the highest q value), human activities had the highest explanatory power for SOS$_{NIRv}$, and SOS$_{SIF}$ had the lowest explanatory power.

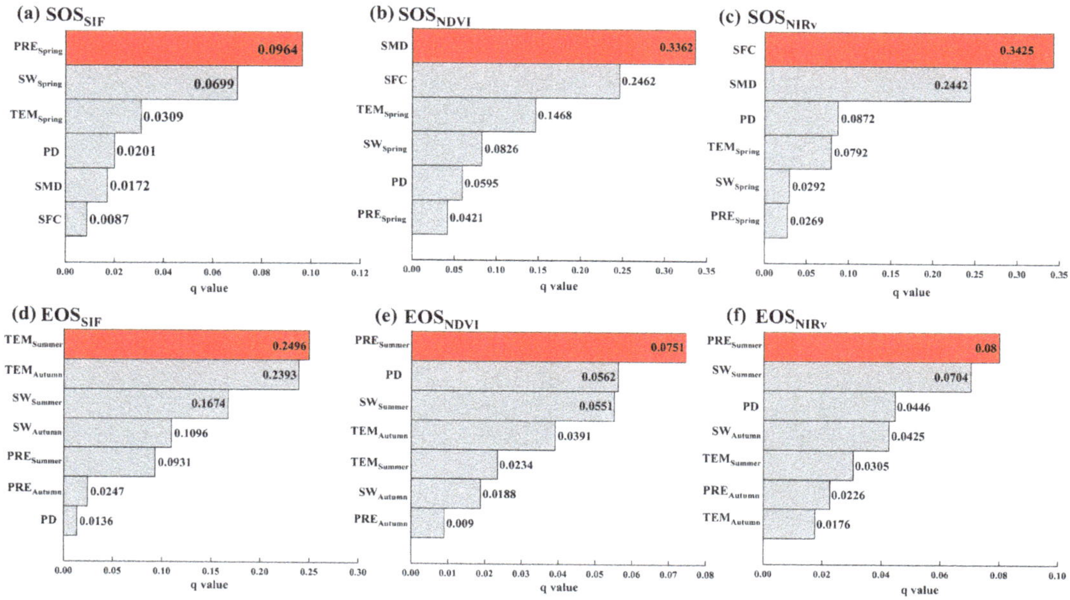

Figure 13. Interpretations of various influencing factors on SOS and EOS: (**a**–**c**) are SOS_{SIF}, SOS_{NDVI} and SOS_{NIRv}, respectively; (**d**–**f**) are EOS_{SIF}, EOS_{NDVI} and EOS_{NIRv}, respectively.

The factors with the highest degree of interpretation (q value) for EOS_{SIF} were TEM_{Summer} and TEM_{Autumn}. The q values were 0.2496 and 0.2393, respectively. This result shows that the temperature factor had a high degree of interpretation of EOS_{SIF} in terms of spatial differentiation, but the population density had the lowest degree of interpretation of EOS_{SIF} among all factors. The factors with the highest interpretation of SOS_{NDVI} were PRE_{Summer} and PD, and the q values were 0.0751 and 0.0562, respectively. The factors with the highest interpretation of EOS_{NIRv} were PRE_{Summer} and SW_{Summer}, and the q values were 0.08 and 0.0704, respectively. The interpretation of the population density to EOS_{NDVI} was higher than that of SW_{Autumn}, TEM_{Summer}, PRE_{Autumn} and TEM_{Autumn}. The results show that each influencing factor had the best explanatory power for EOS_{SIF} (the highest q value), human activities had the highest explanatory power for EOS_{NIRv}, and EOS_{NDVI} had the lowest explanatory power.

4. Discussion

At the beginning of the study, we analyzed the SIF, NDVI and NIRv trends and breakpoints. The difference in the three types of data is mainly shown in the sensitivity difference of the breakpoint occurrence time to different vegetation types. The breakpoints detected by SIF and NIRv were later than those detected by NDVI. SIF and NIRv can detect mutations in a short time and add more details of changes in trend. This finding shows that reflecting the advantages and disadvantages of vegetation growth based on the data may lead to some vegetation physiological information being neglected. Therefore, the retrieval of vegetation phenological parameters from singular data may have limitations.

From the perspective of time differentiation, the phenological parameters estimated by NDVI and NIRv were closer, while those estimated by SIF were quite different. In addition, SIF is related to the chlorophyll concentration and directly characterizes photosynthesis. At higher latitudes on the Mongolian Plateau, SIF signals are weak in spring and autumn; thus, the estimated SOS is later, and the EOS is earlier. The difference between NDVI and NIRv is small, and the time trend of EOS_{SIF} is also opposite to that of EOS_{NDVI} and

EOS_{NIRv}. This result may be because the phenological changes of different vegetation types are different, and SIF is strongly affected by some vegetation types; thus, the time-trend is different from that of NDVI and NIRv. This finding is related to the difference in the spatial distribution of the phenological trend changes (Figure 4d–f) and the difference in the spatial distribution of the change-points (Figure 7). The relationships between the three phenological parameters and altitude were also different. SOS_{NDVI} and SOS_{NIRv} decrease significantly as the altitude increases. The opposite is true for SOS_{SIF}. At the same time, EOS_{NIRv} increases significantly with elevation, EOS_{SIF} decreases significantly, and NDVI decreases slightly. These results shows that photosynthesis is negatively correlated with altitude, and the duration of vegetation photosynthesis is also shortened with increasing altitude. However, this situation is not necessarily reflected in vegetation greenness. The NDVI and NIRv which were obtained based on spectral reflection do not have this obvious feature. In contrast, the LOS estimated by NIRv increases with elevation, which may be caused by higher chlorophyll concentrations and deeper vegetation colors in high-altitude areas. However, the change in color alone may not directly represent the change in vegetation photosynthesis. For example, the lag in NDVI changes caused by precipitation or temperature has been widely reported [59–61]. Vegetation with high greenness may still retain a certain chlorophyll concentration during the EOS. At this moment in time, the activity of photosynthesis-related enzymes is inhibited, due to the decrease in temperature. This may make some difference. SIF directly represents the change in photosynthesis. When photosynthesis is affected by environmental stress, it will immediately show a change in chlorophyll fluorescence. The study of this part still needs to compare and analyze the relationship between phenology and the elevation gradient of different types of vegetation.

In terms of time, the change-point of SOS_{SIF} is earlier than that of SOS_{NDVI} and SOS_{NIRv} (the most change-point is from 2001 to 2003). However, the time of SIF and NIRv is closer (mainly from 2001 to 2003). In addition, in the analysis of EOS change-points, NDVI and NIRv are more similar in time and space. The reason may be that SIF is usually sensitive to changes in forest areas, desert vegetation and agricultural land. NDVI is more sensitive to changes in grassland ecosystems. In addition, the results based on the change-point detection showed that SOS_{SIF} mainly showed a reversal of the decreasing trend to an increasing trend. The opposite was true for SOS_{NDVI} and SOS_{NIRv}. The greenness of the vegetation returned earlier, but photosynthesis was not enhanced and the inflection-point based on the SIF curve was delayed. The early greening of vegetation may not be beneficial to vegetation growth, but lead to a decrease in photosynthetic capacity. Meanwhile, the trend-change of SOS_{SIF} was earlier than that of SOS_{NDVI} and SOS_{NIRv}, which may indicate that SIF can signal the trend-change of vegetation phenology in advance, in terms of the change of phenology.

Additionally, phenological parameters extracted from different observation angles may represent different meanings, and their responses to climate factors may also be different. Because SOS_{SIF} is too late, the correlation between SOS_{SIF} and snow-cover is weak. However, the correlation between the SOS and EOS which were estimated based on SIF and meteorological factors in the spring, summer and autumn, was the most significant. This result shows that the phenological parameters estimated by SIF have the highest sensitivity to environmental changes within a comparison of the three phenological parameters. In general, the impact of environmental factors on vegetation physiology is significant. This may be due to the rapid response of SIF to environmental stress and the advantage of direct characterization of photosynthesis. Reflectivity-based NDVI and NIRv have a certain time-lag in response to environmental changes, which may also be reflected in the extraction of vegetation phenology.

The results also show that the snowmelt period and soil moisture were the most important factors in spring, but their effects on the SOS were very complex. Based on the path-analysis results, the change in the SOS is directly or indirectly caused by multiple factors, or even superposition. The increase in SCF will delay the SMD but also increase the soil water content. The effect of precipitation in spring is less. Therefore, the soil moisture in relatively dry springs may mainly come from the melting supply of snow. The soil moisture content in spring may still not be high overall. Therefore, for the advance of the SOS, the soil moisture must be fully supplied to the spring vegetation instead of being consumed by evaporation. Because the temperature in spring has a strong negative correlation with the snow-melting period and soil moisture, the effect of temperature on the SOS is indirect. This result mainly shows that the temperature causes the SMD to advance and the melted snow to supply soil water prematurely. Simultaneously, the temperature will further lead to the consumption of soil water. During the time interval between the SOS and the snow-melting period, a large amount of soil water will be consumed, and the soil water will not be fully used by growing vegetation. Thus, in a dry spring, the right time for soil water to supply vegetation may be even more important than the total amount of soil water. The delay of the SMD can have critical effects. Of course, soil water is not only supplied by snow; frozen soil is also very important [62,63]. In addition, the positive effect of snow-cover on vegetation is not limited to water supplementation; nitrogen from melted-snow supplementation is also important [64–67]. The SMD showed a significant correlation with SOS_{NDVI} and SOS_{NIRv}. Additionally, from the perspective of regional differentiation, the interpretation of winter snow cover to SOS_{NIRv} is the highest. The interpretation of EOS_{NDVI} by natural factors in summer and autumn is low, and, except for summer precipitation, the interpretation of SOS_{NDVI} by population-density is high. Moreover, the q value of natural factors and EOS_{SIF} is the highest. Accordingly, NDVI in the Mongolian Plateau may be more suitable for extracting the SOS, and SIF is more suitable for extracting the EOS.

5. Conclusions

We used SIF, NDVI and NIRv to retrieve vegetation changes and detect discontinuities. Based on three types of remote-sensing data, we estimated the phenological parameters of the Mongolian Plateau for 20 years, and conducted a trend analysis and change-point detection. Then, we analyzed the potential impact of meteorological factors in each season and the human factors, on phenological parameters.

The main conclusions of this study are as follows:

1. SIF and NIRv can retrieve more details of vegetation dynamics and detect breakpoints earlier than NDVI;
2. On the Mongolian Plateau, SOS_{SIF} is the latest, and EOS_{SIF} is the earliest. The time-trends of the phenological parameters estimated based on NIRv and NDVI were similar, and SIF can better express the relationship between altitude and vegetation phenology. The year of main change-points of SOS_{NDVI} and SOS_{NIRv} is later than that of SOS_{SIF}; the change-point of EOS_{SIF} and EOS_{NIRv} is earlier than that of EOS_{NDVI}. The trend change-points of EOS_{SIF} and EOS_{NIRv} were mostly concentrated from 2001 to 2007, and the change-points of EOS_{NDVI} occurred later. The difference in different phenological parameters is mostly due to the difference in the sensitivity of different data to different types of vegetation changes, as well as the different data characteristics;
3. The contribution of snow-cover to SOS is mainly shown in the late snow-melting date, which can reduce the evaporation loss of melted snow-cover, and fully supplement the vegetation in the green-returning period as soil water;
4. The correlation between spring climate factors and SOS_{NDVI} is most significant. The correlation between summer and autumn meteorological factors and EOS_{SIF} is most significant. From the perspective of spatial differentiation, each influencing factor has the highest degree of interpretation for SOS_{NDVI} and EOS_{SIF}. The population density has the highest interpretation of SOS_{NDVI} and EOS_{NIRv}. On the Mongolian Plateau, NDVI may be more suitable for extracting the SOS, while SIF is more suitable for extracting the EOS. According to the analysis results from different angles, NIRv did not show the most prominent advantages, and SIF is more predictable in the trend analysis of phenology.

Author Contributions: C.E. analyzed the data and wrote the paper. C.E. conceived and designed the experiments. T.B., Y.B. (Yuhai Bao), Y.B. (Yulong Bao), M.Y., Q.L., X.Z. and Y.Z. revised the manuscript. All authors have read and agreed to the published version of the manuscript.

Funding: This study was supported by Early Warning and Information Sharing of Forest and Grassland Fire Risk on the Mongolian Plateau under the Background of Climate Change (Funder: Yuhai Bao; No. 4191101037), Research on Remote Sensing Monitoring and Propagation Path Prediction of Typical Cross-border Disasters in China and Mongolia (Funder: Yuhai Bao; No. 61631011), Research on Monitoring and Early Warning Methods of Black Dzud in Pastoral Areas Based on Daily Snow Products (Funder: Yulong Bao; No. 2021MS04016) and the National Natural Science Foundation of China (Funder: Mei Yong; No. 41867070).

Data Availability Statement: Not applicable.

Acknowledgments: We thank the reviewers for their suggestions, and we are also very grateful to Li and Xiao for sharing their GOSIF data, which were very helpful for our research.

Conflicts of Interest: The authors declare no conflict of interest.

References

1. Zhang, G.; Zhang, Y. Green-up dates in the Tibetan Plateau have continuously advanced from 1982 to 2011. *Proc. Natl. Acad. Sci. USA* **2013**, *110*, 4309–4314. [CrossRef] [PubMed]
2. Bao, G.; Chen, J.; Chopping, M.; Bao, Y.; Qin, Z. Dynamics of net primary productivity on the Mongolian Plateau: Joint regulations of phenology and drought. *Int. J. Appl. Earth Obs. Geoinf.* **2019**, *81*, 85–97. [CrossRef]
3. Nandintsetseg, B.; Shinoda, M. Seasonal change of soil moisture in Mongolia: Its climatology and modelling. *Int. J. Climatol.* **2011**, *31*, 1143–1152. [CrossRef]
4. Gang, C.; Zhou, W.; Chen, Y.; Wang, Z.; Sun, Z.; Li, J.; Qi, J.; Odeh, I. Quantitative assessment of the contributions of climate change and human activities on global grassland degradation. *Environ. Earth Sci.* **2014**, *72*, 4273–4282. [CrossRef]
5. Guo, E.; Wang, Y.; Wang, C.; Sun, Z.; Bao, Y.; Mandula, N.; Jirigala, B.; Bao, Y.; Li, H. NDVI Indicates Long-Term Dynamics of Vegetation and Its Driving Forces from Climatic and Anthropogenic Factors in Mongolian Plateau. *Remote Sens.* **2021**, *13*, 688. [CrossRef]
6. Cza, B.; Yan, Y.A.; Wm, C.; Xu, S.A.; Jc, B.; Yr, A.; Tian, X.A.; Yuan, Q.A. RESTREND-based assessment of factors affecting vegetation dynamics on the Mongolian Plateau. *Ecol. Model.* **2021**, *440*, 109415.
7. Zhang, X.; Friedl, M.A.; Schaaf, C.B.; Strahler, A.H.; Hodges, J.C.; Gao, F.; Reed, B.C.; Huete, A. Monitoring vegetation phenology using MODIS. *Remote Sens. Environ.* **2003**, *84*, 471–475. [CrossRef]
8. Miao, L.; Daniel, M.; Cui, X.; Ma, M.; Sylvain, D. Changes in vegetation phenology on the Mongolian Plateau and their climatic determinants. *PLoS ONE* **2017**, *12*, e0190313. [CrossRef]
9. Bao, G.; Jin, H.; Tong, S.; Chen, J.; Du, L. Autumn Phenology and Its Covariation with Climate, Spring Phenology and Annual Peak Growth on the Mongolian Plateau. *Agric. For. Meteorol.* **2021**, 108312. [CrossRef]
10. Hmimina, G.; Dufrene, E.; Pontailler, J.Y.; Delpierre, N.; Aubinet, M.; Caquet, B.; Grandcourt, A.D.; Flechard, C.; Gross, P.; Heinesch, B. Evaluation of the potential of MODIS satellite data to predict vegetation phenology in different biomes: An investigation using ground-based NDVI measurements. *Remote Sens. Environ.* **2013**, *132*, 145–158. [CrossRef]
11. Wu, C.; Gonsamo, A.; Gough, C.M.; Chen, J.M.; Xu, S. Modeling growing season phenology in North American forests using seasonal mean vegetation indices from MODIS. *Remote Sens. Environ.* **2014**, *147*, 79–88. [CrossRef]
12. Meng, F.; Huang, L.; Chen, A.; Zhang, Y.; Piao, S. Spring and autumn phenology across the Tibetan Plateau inferred from normalized difference vegetation index and solar-induced chlorophyll fluorescence. *Big Earth Data* **2021**, *5*, 182–200. [CrossRef]
13. Zarco-Tejada, P.J.; Morales, A.; Testi, L.; Villalobos, F.J. Spatio-temporal patterns of chlorophyll fluorescence and physiological and structural indices acquired from hyperspectral imagery as compared with carbon fluxes measured with eddy covariance. *Remote Sens. Environ.* **2013**, *133*, 102–115. [CrossRef]
14. Meroni, M.; Rossini, M.; Guanter, L.; Alonso, L.; Rascher, U.; Colombo, R.; Moreno, J. Remote sensing of solar-induced chlorophyll fluorescence: Review of methods and applications. *Remote Sens. Environ.* **2009**, *113*, 2037–2051. [CrossRef]
15. Tubuxin, B.; Rahimzadeh-Bajgiran, P.; Ginnan, Y.; Hosoi, F.; Omasa, K. Estimating chlorophyll content and photochemical yield of photosystem II (Φ PSII) using solar-induced chlorophyll fluorescence measurements at different growing stages of attached leaves. *J. Exp. Bot.* **2015**, *66*, 5595–5603. [CrossRef]
16. Rahimzadeh-Bajgiran, P.; Tubuxin, B.; Omasa, K. Estimating chlorophyll fluorescence parameters using the joint Fraunhofer line depth and laser-induced saturation pulse (FLD-LISP) method in different plant species. *Remote Sens.* **2017**, *9*, 599. [CrossRef]
17. Joiner, J.; Guanter, L.; Lindstrot, R.; Voigt, M.; Vasilkov, A.; Middleton, E.; Huemmrich, K.; Yoshida, Y.; Frankenberg, C. Global monitoring of terrestrial chlorophyll fluorescence from moderate-spectral-resolution near-infrared satellite measurements: Methodology, simulations, and application to GOME-2. *Atmos. Meas. Tech.* **2013**, *6*, 2803–2823. [CrossRef]

18. Guanter, L.; Aben, I.; Tol, P.; Krijger, J.; Hollstein, A.; Köhler, P.; Damm, A.; Joiner, J.; Frankenberg, C.; Landgraf, J. Potential of the TROPOspheric Monitoring Instrument (TROPOMI) onboard the Sentinel-5 Precursor for the monitoring of terrestrial chlorophyll fluorescence. *Atmos. Meas. Tech.* **2015**, *8*, 1337–1352. [CrossRef]
19. Frankenberg, C.; O'Dell, C.; Berry, J.; Guanter, L.; Joiner, J.; Köhler, P.; Pollock, R.; Taylor, T.E. Prospects for chlorophyll fluorescence remote sensing from the Orbiting Carbon Observatory-2. *Remote Sens. Environ.* **2014**, *147*, 1–12. [CrossRef]
20. Guanter, L.; Zhang, Y.; Jung, M.; Joiner, J.; Voigt, M.; Berry, J.A.; Frankenberg, C.; Huete, A.R.; Zarco-Tejada, P.; Lee, J.-E. Global and time-resolved monitoring of crop photosynthesis with chlorophyll fluorescence. *Proc. Natl. Acad. Sci. USA* **2014**, *111*, E1327–E1333. [CrossRef]
21. Baker, N.R. Chlorophyll fluorescence: A probe of photosynthesis in vivo. *Annu. Rev. Plant Biol.* **2008**, *59*, 89–113. [CrossRef] [PubMed]
22. Sun, Y.; Frankenberg, C.; Jung, M.; Joiner, J.; Guanter, L.; Köhler, P.; Magney, T. Overview of Solar-Induced chlorophyll Fluorescence (SIF) from the Orbiting Carbon Observatory-2: Retrieval, cross-mission comparison, and global monitoring for GPP. *Remote Sens. Environ.* **2018**, *209*, 808–823. [CrossRef]
23. De Smedt, I.; Van Roozendael, M.; Stavrakou, T.; Müller, J.F.; Lerot, C.; Theys, N.; Valks, P.; Hao, N.; Van Der, A.R. Improved retrieval of global tropospheric formaldehyde columns from GOME-2/MetOp-A addressing noise reduction and instrumental degradation issues. *Atmos. Meas. Tech.* **2012**, *5*, 2933–2949. [CrossRef]
24. Sun, Y.; Frankenberg, C.; Wood, J.D.; Schimel, D.; Jung, M.; Guanter, L.; Drewry, D.; Verma, M.; Porcar-Castell, A.; Griffis, T.J. OCO-2 advances photosynthesis observation from space via solar-induced chlorophyll fluorescence. *Science* **2017**, *358*, eaam5747. [CrossRef] [PubMed]
25. Li, X.; Xiao, J. A Global, 0.05-Degree Product of Solar-Induced Chlorophyll Fluorescence Derived from OCO-2, MODIS, and Reanalysis Data. *Remote Sens.* **2019**, *11*, 517. [CrossRef]
26. Ren, P.; Liu, Z.; Zhou, X.; Peng, C.; Xiao, J.; Wang, S.; Li, X.; Li, P. Strong controls of daily minimum temperature on the autumn photosynthetic phenology of subtropical vegetation in China. *For. Ecosyst.* **2021**, *8*, 31. [CrossRef]
27. Li, X.; Fu, Y.H.; Chen, S.; Xiao, J.; Yin, G.; Li, X.; Zhang, X.; Geng, X.; Wu, Z.; Zhou, X. Increasing importance of precipitation in spring phenology with decreasing latitudes in subtropical forest area in China. *Agric. For. Meteorol.* **2021**, *304*, 108427. [CrossRef]
28. Wong, C.; D'Odorico, P.; Arain, M.A.; Ensminger, I. Tracking the phenology of photosynthesis using carotenoid-sensitive and near-infrared reflectance vegetation indices in a temperate evergreen and mixed deciduous forest. *New Phytol.* **2020**, *226*, 1682–1695. [CrossRef]
29. Badgley, G.; Field, C.B.; Berry, J.A. Canopy near-infrared reflectance and terrestrial photosynthesis. *Sci. Adv.* **2017**, *3*, e1602244. [CrossRef]
30. Fernández-Martínez, M.; Yu, R.; Gamon, J.; Hmimina, G.; Peuelas, J. Monitoring Spatial and Temporal Variabilities of Gross Primary Production Using MAIAC MODIS Data. *Remote Sens.* **2019**, *11*, 874. [CrossRef]
31. Sha, Z.; Zhong, J.; Bai, Y.; Tan, X.; Li, J. Spatio-temporal patterns of satellite-derived grassland vegetation phenology from 1998 to 2012 in Inner Mongolia, China. *J. Arid Land* **2016**, *8*, 462–477. [CrossRef]
32. Ren, S.; Ao, Q.; Han, M.; Xin, F.; Tian, J.L. A semi-analytical model for sound propagation in sintered fiber metals. *Compos. Part B Eng.* **2017**, *126*, 17–26. [CrossRef]
33. Luo, M.; Meng, F.; Sa, C.; Duan, Y.; Bao, Y.; Liu, T.; De Maeyer, P. Response of vegetation phenology to soil moisture dynamics in the Mongolian Plateau. *Catena* **2021**, *206*, 105505. [CrossRef]
34. Jeong, S.J.; Chang-Hoi, H.O.; Gim, H.J.; Brown, M.E. Phenology shifts at start vs. end of growing season in temperate vegetation over the Northern Hemisphere for the period 1982–2008. *Glob. Chang. Biol.* **2011**, *17*, 2385–2399. [CrossRef]
35. Peel, M.C.; Finlayson, B.L.; Mcmahon, T.A. Updated world map of the Köppen-Geiger climate classification. *Hydrol. Earth Syst. Sci.* **2007**, *11*, 259–263. [CrossRef]
36. Shi, H.; Gao, Q.; Qi, Y.; Liu, J.; Hu, Y. Wind erosion hazard assessment of the Mongolian Plateau using FCM and GIS techniques. *Environ. Earth Sci.* **2010**, *61*, 689–697. [CrossRef]
37. Liu, Y.; Zhuang, Q.; Chen, M.; Pan, Z.; Tchebakova, N.; Sokolov, A.; Kicklighter, D.; Melillo, J.; Sirin, A.; Zhou, G. Response of evapotranspiration and water availability to changing climate and land cover on the Mongolian Plateau during the 21st century. *Glob. Planet. Chang.* **2013**, *108*, 85–99. [CrossRef]
38. Zhang, X.; Hu, Y.; Zhuang, D.; Qi, Y.; Ma, X. NDVI spatial pattern and its differentiation on the Mongolian Plateau. *J. Geogr. Sci.* **2009**, *19*, 403–415. [CrossRef]
39. Bao, G.; Bao, Y.; Qin, Z.; Xin, X.; Bao, Y.; Bayarsaikan, S.; Zhou, Y.; Chuntai, B. Modeling net primary productivity of terrestrial ecosystems in the semi-arid climate of the Mongolian Plateau using LSWI-based CASA ecosystem model. *Int. J. Appl. Earth Obs. Geoinf.* **2016**, *46*, 84–93. [CrossRef]
40. Jiang, L.; Yao, Z.; Huang, H.Q. Climate variability and change on the Mongolian Plateau: Historical variation and future predictions. *Clim. Res.* **2016**, *67*, 1–14. [CrossRef]
41. Hird, J.N.; McDermid, G.J. Noise reduction of NDVI time series: An empirical comparison of selected techniques. *Remote Sens. Environ.* **2009**, *113*, 248–258. [CrossRef]
42. Richardson, A.D.; Keenan, T.F.; Migliavacca, M.; Ryu, Y.; Sonnentag, O.; Toomey, M. Climate change, phenology, and phenological control of vegetation feedbacks to the climate system. *Agric. For. Meteorol.* **2013**, *169*, 156–173. [CrossRef]

43. Lloyd, D. A phenological classification of terrestrial vegetation cover using shortwave vegetation index imagery. *Remote Sens.* **1990**, *11*, 2269–2279. [CrossRef]
44. Liu, X.; Chen, Y.; Li, Z.; Li, Y.; Zhang, Q.; Zan, M. Driving Forces of the Changes in Vegetation Phenology in the Qinghai–Tibet Plateau. *Remote Sens.* **2021**, *13*, 4952. [CrossRef]
45. Jonsson, P.; Eklundh, L. Seasonality extraction by function fitting to time-series of satellite sensor data. *IEEE Trans. Geosci. Remote Sens.* **2002**, *40*, 1824–1832. [CrossRef]
46. Hufkens, K.; Friedl, M.; Sonnentag, O.; Braswell, B.H.; Milliman, T.; Richardson, A.D. Linking near-surface and satellite remote sensing measurements of deciduous broadleaf forest phenology. *Remote Sens. Environ.* **2012**, *117*, 307–321. [CrossRef]
47. Shao, Y.; Wang, J.; Yan, X. The phenological characteristics of Mongolian vegetation and its response to geographical elements. *Geogr. Res.* **2021**, *40*, 3029–3043. [CrossRef]
48. Verbesselt, J.; Hyndman, R.; Newnham, G.; Culvenor, D. Detecting trend and seasonal changes in satellite image time series. *Remote Sens. Environ.* **2010**, *114*, 106–115. [CrossRef]
49. Verbesselt, J.; Hyndman, R.; Zeileis, A.; Culvenor, D. Phenological change detection while accounting for abrupt and gradual trends in satellite image time series. *Remote Sens. Environ.* **2010**, *114*, 2970–2980. [CrossRef]
50. Fensholt, R.; Langanke, T.; Rasmussen, K.; Reenberg, A.; Prince, S.D.; Tucker, C.; Scholes, R.J.; Le, Q.B.; Bondeau, A.; Eastman, R. Greenness in semi-arid areas across the globe 1981–2007—An Earth Observing Satellite based analysis of trends and drivers. *Remote Sens. Environ.* **2012**, *121*, 144–158. [CrossRef]
51. Lunetta, R.S.; Knight, J.F.; Ediriwickrema, J.; Lyon, J.G.; Worthy, L.D. Land-cover change detection using multi-temporal MODIS NDVI data. *Remote Sens. Environ.* **2006**, *105*, 142–154. [CrossRef]
52. Kendall, M.G. Rank Correlation Methods. *Br. J. Psychol.* **1990**, *25*, 86–91. [CrossRef]
53. Verbesselt, J.; Zeileis, A.; Herold, M. Near real-time disturbance detection using satellite image time series. *Remote Sens. Environ.* **2012**, *123*, 98–108. [CrossRef]
54. Lawrence, A. Algorithms for the optimal identification of segment neighborhoods. *Bull. Math. Biol.* **1989**, *51*, 39–54.
55 Killick, R.; Fearnhead, P.; Eckley, I.A. Optimal detection of changepoints with a linear computational cost. *J. Am. Stat. Assoc.* **2012**, *107*, 1590–1598. [CrossRef]
56. Killick, R.; Eckley, I.A.; Pebesma, E.; Zeileis, A. Changepoint: An R Package for Changepoint Analysis. *J. Stat. Softw.* **2014**, *3*, 1–19.
57. Su, Y.; Li, T.X.; Cheng, S.K.; Wang, X. Spatial distribution exploration and driving factor identification for soil salinisation based on geodetector models in coastal area. *Ecol. Eng.* **2020**, *156*, 18. [CrossRef]
58. Wang, J.F.; Li, X.H.; Christakos, G.; Liao, Y.L.; Zhang, T.; Gu, X.; Zheng, X.Y. Geographical Detectors-Based Health Risk Assessment and its Application in the Neural Tube Defects Study of the Heshun Region, China. *Int. J. Geogr. Inf. Sci.* **2010**, *24*, 107–127. [CrossRef]
59. Zhe, M.; Zhang, X. Time-lag effects of NDVI responses to climate change in the Yamzhog Yumco Basin, South Tibet. *Ecol. Indic.* **2021**, *124*, 107431. [CrossRef]
60. Li, P.; Wang, J.; Liu, M.; Xue, Z.; Bagherzadeh, A.; Liu, M. Spatio-temporal variation characteristics of NDVI and its response to climate on the Loess Plateau from 1985 to 2015. *Catena* **2021**, *203*, 105331. [CrossRef]
61. Ji, L.; Peters, A.J. Lag and seasonality considerations in evaluating AVHRR NDVI response to precipitation. *Photogramm. Eng. Remote Sens.* **2005**, *71*, 1053–1061. [CrossRef]
62. Sturm, M.; Schimel, J.; Michaelson, G.; Welker, J.M.; Oberbauer, S.F.; Liston, G.E.; Fahnestock, J.; Romanovsky, V.E. Winter biological processes could help convert arctic tundra to shrubland. *Bioscience* **2005**, *55*, 17–26. [CrossRef]
63. Li, Y.; Fu, Q.; Li, T.; Liu, D.; Hou, R.; Li, Q.; Yi, J.; Li, M.; Meng, F. Snow melting water infiltration mechanism of farmland freezing-thawing soil and determination of meltwater infiltration parameter in seasonal frozen soil areas. *Agric. Water Manag.* **2021**, *258*, 107165. [CrossRef]
64. Warner, K.A.; Saros, J.E.; Simon, K.S. Nitrogen Subsidies in Glacial Meltwater: Implications for High Elevation Aquatic Chains. *Water Resour. Res.* **2017**, *53*, 9791–9806. [CrossRef]
65. Morse, J.L.; Durán, J.; Groffman, P.M. Soil Denitrification Fluxes in a Northern Hardwood Forest: The Importance of Snowmelt and Implications for Ecosystem N Budgets. *Ecosystems* **2015**, *18*, 520–532. [CrossRef]
66. Harpold, A.A.; Molotch, N.P.; Musselman, K.N.; Bales, R.C.; Kirchner, P.B.; Litvak, M.; Brooks, P.D. Soil moisture response to snowmelt timing in mixed-conifer subalpine forests. *Hydrol. Process.* **2015**, *29*, 2782–2798. [CrossRef]
67. Ambadan, J.T.; Berg, A.A.; Merryfield, W.J.; Lee, W.S. Influence of snowmelt on soil moisture and on near surface air temperature during winter-spring transition season. *Clim. Dyn.* **2018**, *51*, 1295–1309. [CrossRef]

remote sensing

MDPI

Article

Monitoring of Plant Ecological Units Cover Dynamics in a Semiarid Landscape from Past to Future Using Multi-Layer Perceptron and Markov Chain Model

Masoumeh Aghababaei [1], Ataollah Ebrahimi [1,*], Ali Asghar Naghipour [1], Esmaeil Asadi [1] and Jochem Verrelst [2]

[1] Department of Nature Engineering, Faculty of Natural Resources and Earth Sciences, Shahrekord University, Shahrekord 8818634141, Iran; ma.aghababaeei@stu.sku.ac.ir (M.A.); aa.naghipour@sku.ac.ir (A.A.N.); asadi-es@sku.ac.ir (E.A.)

[2] Image Processing Laboratory (IPL), Parc Científic, Universitat de València, 46980 Paterna, Spain; jochem.verrelst@uv.es

* Correspondence: ataollah.ebrahimi@sku.ac.ir; Tel.: +98-9132808343

Abstract: Anthropogenic activities and natural disturbances cause changes in natural ecosystems, leading to altered Plant Ecological Units (PEUs). Despite a long history of land use and land cover change detection, the creation of change detection maps of PEUs remains problematic, especially in arid and semiarid landscape. This study aimed to determine and describe the changes in PEUs patterns in the past and present, and also predict and monitor future PEUs dynamics using the multi-layer perceptron-Markov chain (MLP-MC) model in a semiarid landscape in Central Zagros, Iran. Analysis of PEUs classification maps formed the basis for the identification of the main drivers in PEUs changes. First, an optimal time-series dataset of Landsat images were selected to derive PEUs classification maps in three periods, each separated by 16 years. Then, PEUs multi-temporal maps classified for period 1 (years 1986–1988) period 2 (years 2002–2004), and period 3 (years 2018–2020) were employed to analyze and predict PEUs dynamics. The dominant transitions were identified, and the transition potential was determined by developing twelve sub-models in the final change prediction process. Transitions were modeled using a Multi-Layer Perceptron (MLP) algorithm. To predict the PEU map for period 3, two PEUs classification maps of period 1 and period 2 were used using the MLP-MC method. The classified map and the predicted map of period 3 were used to evaluate and validate the predicted results. Finally, based on the results, transitions of future PEUs were predicted for the year 2036. The MLP-MC model proved to be a powerful model that can predict future PEUs dynamics that are the result of current human and managerial activities. The findings of this study demonstrate that the impact of anthropogenic processes and management activities will become visible in the natural environment and ecosystem in less than a decade.

Keywords: plant ecological unit's changes; land change modeler; time-series dataset; Markov chain model

1. Introduction

Land cover is the combination of biotic and abiotic physical substances on the earth's surface. Typically, it includes natural vegetation (forests, grassland), water, soil, and man-made surfaces [1,2]. Mapping and monitoring changes in land cover types are pivotal for land cover sustainability planning and environmental management needs [3–5]. However, the benefits and importance of land cover maps based on plant ecological units (PEUs) are less understood [6]. PEUs represent the potential plant communities that can occur on a site. Because of differences in elevation, soil, historic background, and land abandonment processes, a variety of plant communities with distinctive amounts and types of vegetation can be found in an area. Thereby, the identification of PEUs provides a reference for the interpretation of land cover data and research, monitoring, and land management [7]. PEUs mapping through classification techniques is extremely important, and their accuracy will directly affect the extraction of the final prediction results and change detection maps [8]. Therefore, documenting and evaluating the temporal dynamics of PEUs and making predictions of plausible future PEUs dynamics is vital for plant community and land cover monitoring.

Optical remote sensing data offer a valuable source of information for studying change detection. Several studies in change detection have been related to land use/cover changes [9,10]. The key areas in these land use studies included urban areas [11,12], drylands [13], protected areas [14], and wetlands [8,15]. Modeling PEUs changes, on the other hand, not only helps evaluate the current condition of land cover but also helps managers and natural resource planners prevent or reduce negative consequences of undesirable future PEUs changes. Meanwhile, PEUs are hardly distinguishable due to their similar spectral behavior and low inter-class separability, especially in arid and semiarid areas. Therefore, these PEUs impose challenges to classification and predictions of future changes. In a related study about improving land cover map accuracy, Aghababaei et al. [7] found that the development of an accurate land cover map is feasible in a semiarid rangeland when a dataset of time-series images is entered into the classification. So far, publicly available satellite data have been used in change detection studies, such as MODIS [16], SPOT [3], and Landsat [17,18]. Nevertheless, the most notable applications of Landsat data are land cover change analyses. Landsat is widely used in these studies due to its long-range time-series data and free access [19]. By exploiting the temporal dimension of Landsat data, the effects of poor-quality observations can be minimized in the classification and change detection [20]. Regarding the processing of multi-temporal images, the Google Earth Engine (GEE) platform allows researchers to classify and process large volumes of satellite images, including Landsat [21–23]. The combination of an unprecedented data catalog with powerful processing possibilities caused a paradigm shift in Optical Earth observation (EO) data analysis that moved away from traditional image analysis using desktop software to cloud-based processing.

Various models have been used in predicting LULC changes, such as the Markov Chain model (MC), Cellular Automata (CA), and Multi-Layer Perceptron–Markov Chain (MLP-MC). However, integrated models such as MLP-MC are implemented in the integration of the Multi-Layer Perceptron Neural Network (NN-MLP) model with the MC model as an effective approach in prediction studies [8]. This model is used to predict geospatial changes with the use of previous changes [24]. Studies show that the MC model gives better results when predicting long-term temporal LULC variations, future scenarios, and landscape changes [15,25–27].

Altogether, the objectives of the study presented here are the following. Firstly, to develop a classification strategy using a multi-temporal dataset that achieves accurate and efficient PEUs mapping. Secondly, using the MLP-MC to monitor changes in PEUs dynamics from 1986 to 2020 with 16-year time intervals and forecasting future changes for 16 years later, i.e., 2036, in a semiarid rangeland landscape in Southwest Iran, where the low vegetation covers make it a challenging task. This work can help natural resource managers and planners make a sustainable result for natural resource protection, climate regulation, and erosion control by understanding current and future PEUs changes.

2. Materials and Methods

2.1. Study Area

The study presented here is based on the Marjan watershed case study area located in the Chaharmahal-Va-Bakhtiari province in Southwest Iran. As shown in Figure 1, PEUs boundaries can be straightforwardly observed in this area due to the relatively narrow ecotones and sharp borders between them. The study area is characterized by a heterogeneous, semiarid landscape with a predominance of different PEUs. The covered area is 7736.58 ha. There are some effects of anthropogenic processes and management activities present, leading to spatial changes in PEUs maps in different periods. We are not aware of work focusing on modeling these PEUs dynamics.

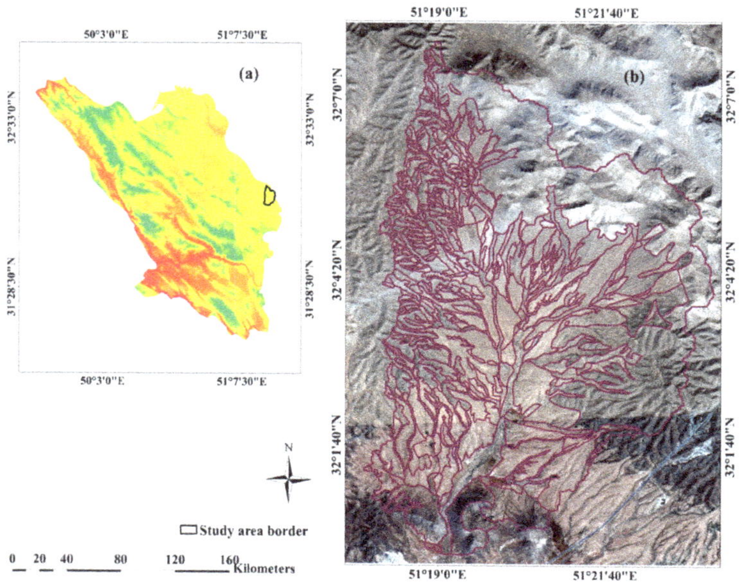

Figure 1. Location of Marjan in the Chaharmahal-Va-Bakhtiari province: (**a**) Chaharmahal-Va-Bakhtiari border; (**b**) study area border (Marjan).

2.2. Field Measurements

Four PEUs groups were distinguished in the area, including (1) PEU1 (*Astragalus verus Olivier* (As ve)), (2) PEU2 (*Bromus tomentellus Boiss* (Br to)), (3) PEU3 (*Scariola orientalis Sojak* (Sc or)), and (4) PEU4 (*Astragalus verus Olivier—Bromus tomentellus Boiss* (As ve—Br to)). Canopy cover data could potentially be used to identify structural and compositional PEUs or a combination of both, the so-called physiognomic–floristic classification, to gain a sound and accurate perspective on PEUs. We sampled the four identified PEUs using three replicates, in each of which canopy cover was sampled along three transects of 100 m that were evenly distributed throughout the study area. The sampling was randomly systematic (the first node was selected randomly but the rest were systematically distributed along the transects). We collected species-based canopy cover within each quadrat. In each PEUs, the canopy cover percentage was calculated and the PEUs were named according to their dominant floristic composition (Figure 2). For this purpose, first, the dominant plant species of each PEU was identified and then its accompanying species was determined, with the dominant species having 50% or more canopy cover than the previously dominant species. Thus, each PEU was named based on a physiognomic–floristic method.

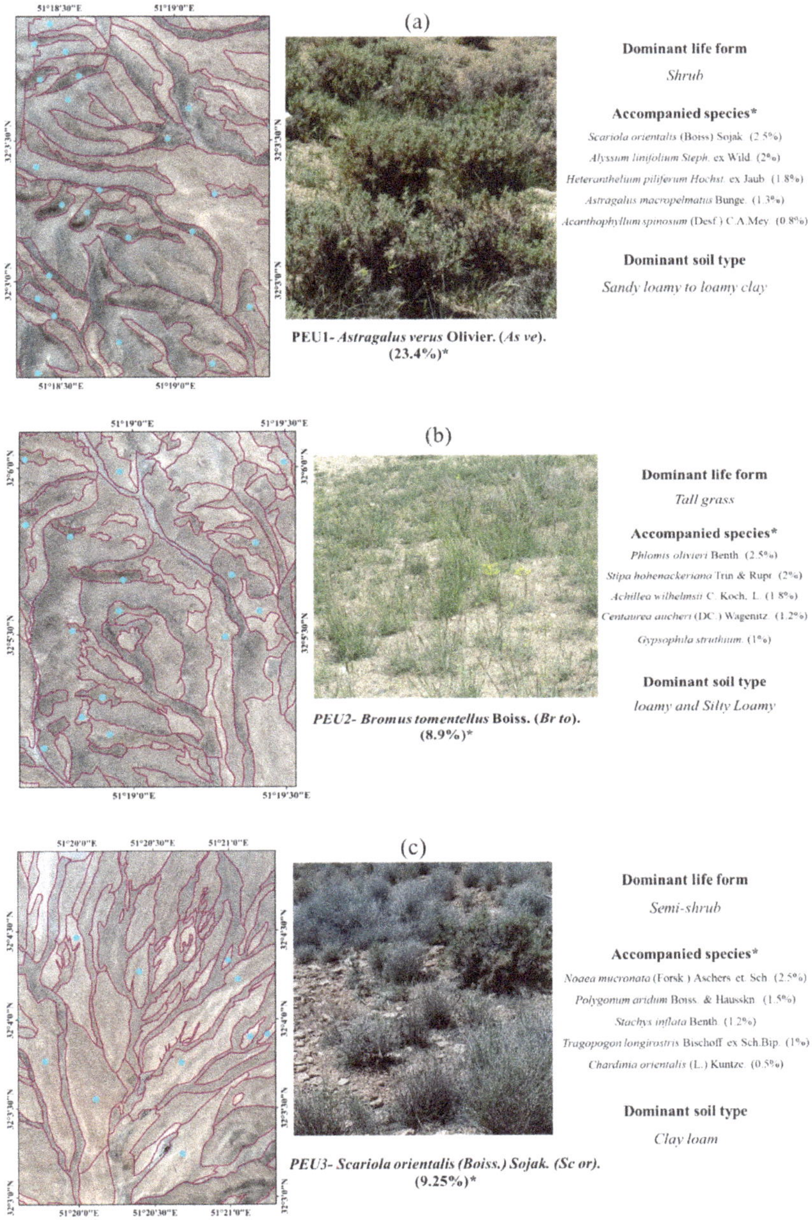

Figure 2. *Cont.*

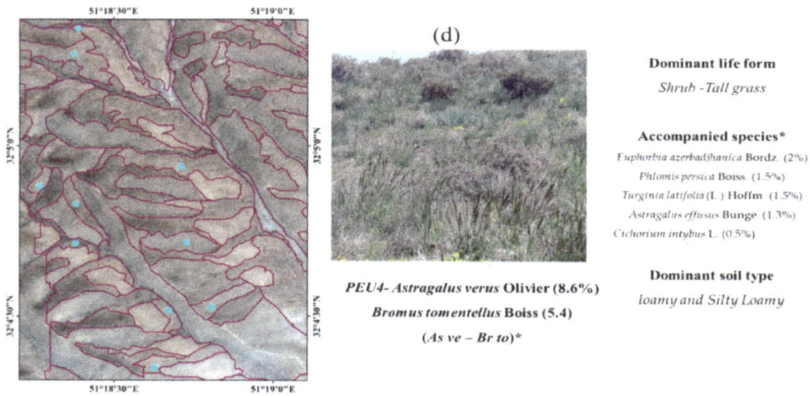

Figure 2. The location of PEUs in Google Earth images and the corresponding field photos. (**a**) PEU1 (As ve); (**b**) PEU2 (Br to); (**c**) PEU3 (Sc or); and (**d**) PEU4 (As ve—Br to). * Canopy cover percentage of dominant and accompanied species that was calculated on transects.

2.3. NDVI Spectral Curve and Selection of Optimal Time-Series Datasets

This study used multi-temporal satellite data. The satellite data consisted of 3 periods with 16-year time intervals acquired by Landsat 5 TM for period 1 (1986, 1987, 1988), Landsat 7 ETM+ for period 2 (2002, 2003, 2004), and Landsat 8 OLI for period 3 (2018, 2019, 2020). With the support of Google Earth Engine (GEE), we used the Top of Atmosphere (TOA) reflectance (ee. Image Collection ('LANDSAT/LT05/C01/T1_TOA')), (ee. Image Collection ('LANDSAT/LE07/C01/T1_TOA')), and (ee. Image Collection ('LANDSAT/LC08/C01/T1_TOA')) for Landsat 5, Landsat 7, and Landsat 8 images, respectively. Likewise, using the cloud filter (ee. Filter. Less than ('CLOUD_COVER', 5)), only satellite images with less than 5% cloud cover were selected. So, some images were filtered due to persistent cloudiness. Finally, the NDVI time-series profile was calculated from January to December of each year (Equation (1)).

By analyzing the NDVI temporal profile, we identified a multi-temporal dataset as an input for PEUs classification for each period [6].

$$\text{NDVI} = \frac{(NIR - Red)}{(NIR + Red)} \tag{1}$$

where *RED* is the reflectance in the red band, and *NIR* is the reflectance in the near-infrared band [24].

2.4. Methodology

Figure 3 shows the conducted workflow of PEUs change detection and prediction using the MLP-MC model. First, using the physiognomic–floristic method, we selected 4 dominant PEUs in the region, and using GPS, training and testing samples were collected by field survey in the study area. We used the RF classification algorithm and an optimal combination of multi-temporal images to produce PEUs maps for period 1 (years 1986–1988), period 2 (years 2002–2004), and period 3 (years 2018–2020), respectively. Detection of PEUs changes was based on the Markov Chain model (MLP-MC model) that was obtained from change analysis in PEUs classes from two dates.

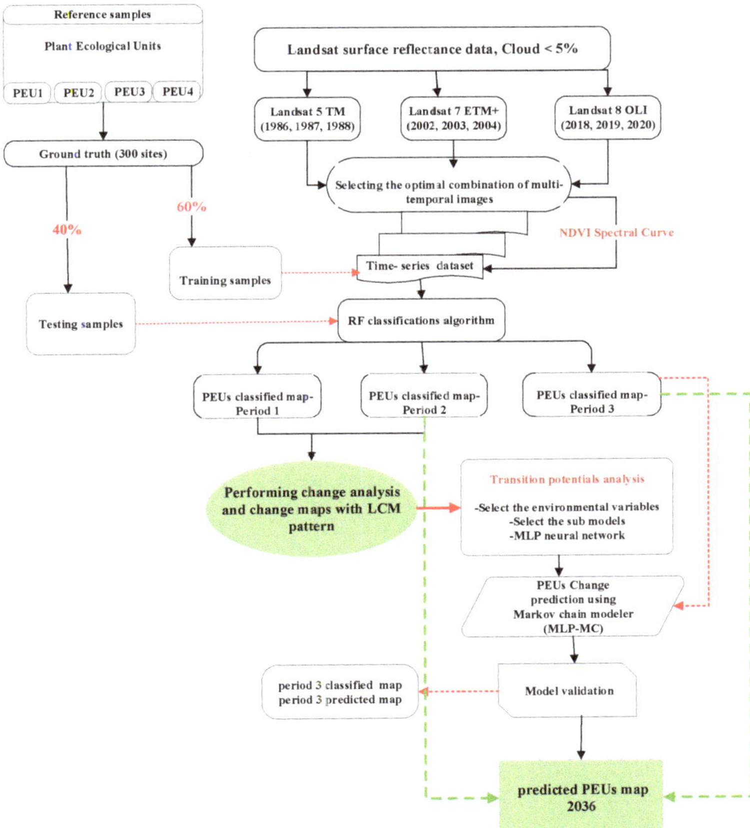

Figure 3. The workflow for PEUs change detection (1986–2020) and future prediction of PEUs (2020–2036) in this study.

For period 3, a prediction map was created using the forecast model and the PEUs classification maps for period 1 and period 2. To validate the accuracy of the predicted map, we used the PEU classification map of the same period (period 3 classified map and period 3 predicted map). Finally, using the MLP-MC model, we obtained a PEU prediction map for the year 2036.

2.4.1. Training and Verification Points

After identifying the dominant PEUs in the study area, in total, 300 points were recorded for the PEUs using a Garmin eTrex 32× GPS. Then, by examining aerial photographs from each period, it was ensured that each sample point was selected in the PEUs center. Finally, the points were divided into two groups, the "training samples" (60%) and "verification samples" (40%), used for the classification and validation of results.

2.4.2. PEUs Multi-Temporal Classification and Maps Validation

We used random forests (RF) to produce PEU maps for period 1 (years 1986–1988), period 2 (years 2002–2004), and period 3 (years 2018–2020). After selecting the multi-temporal dataset for each period, RF algorithm, as one of the most accurate machine learning algorithms [28–31], was used to perform PEUs classification. The mapping accuracy was evaluated by means of a confusion matrix (including Overall Accuracy (OA), Overall Kappa

(OK), User's Accuracy (UA), Producer's Accuracy (PA), and Kappa Index of Agreement (KIA)) for all years [32].

2.4.3. PEUs Change Detection and Analysis Using the LCM Model

Change detection and prediction mapping were achieved by the Land Change Modeler (LCM). The LCM consists of a suite of tools for evaluating losses and gains and assessment of land cover changes and transformations between the land cover classes in the map [33]. The PEUs maps based on the RF classification algorithm and an optimal combination of multi-temporal images of Landsat 5 TM, Landsat 7 ETM+, and Landsat 8 OLI images were used in order to quantify the PEUs. The recognition of changes can facilitate the determination of future PEUs changes and scenarios. The detection of PEUs changes are based on the Markov Chain matrix that is obtained from change analysis in PEUs classes from two dates (period 1 to period 2). In addition, this matrix can predict future PEUs changes [26]. Based on the detection and analysis of PEUs changes, changes from one class type to another were characterized. Cross-tabulation analysis was performed to quantify PEUs changes from period 1 to period 2, from period 2 to period 3, and from period 1 to period 3. The PEUs changes in areas from one period to another period can be determined quantitatively and spatially by this analysis. The gains and losses experienced by various PEUs classes and analysis of the spatial trend of these changes for the study area were investigated for period 1, period 2, and period 3.

2.4.4. Selection of PEUs Transitions

Multiple mechanisms of transitions may occur in PEUs between two periods. All the PEUs changes and transitions were collected into a set of sub-models using the NN-MLP algorithm. The land cover change prediction was based on dependent and independent variables. In this work, six dependent variables (environmental factors), i.e., slope, aspect, DEM, distance from roads, temperature, and precipitation, were considered. These variables were selected according to the results of other studies in this field [8] (Dey, Nataraj Narayan, 2021) and according to the Kramer correlation coefficient. PEUs maps were considered independent variables. PEUs transition potential maps for each sub-model were produced based on PEUs transitions, as well as environmental factors with the help of an NN-MLP integrated in the LCM model.

2.4.5. Prediction of Future PEUs Changes

The MLP-MC model was used to predict and simulate the PEUs dynamics for a specified future date. Using the NN-MLP model, it is possible to determine the transitions that will be integrated for future PEUs change prediction. This model includes an input layer, many hidden layers, and an output layer, thereby known as a reputed change prediction model for its higher accuracy [34]. The MLP-MC model for the period 3 predicted map was created using the forecast model and the PEUs classification maps of period 1 and period 2. To validate the accuracy of the predicted map, we used the PEU classification map of the same period (period 3 classified map and period 3 predicted map). Finally, using the MLP-MC model, we obtained a PEU prediction map for the year 2036.

3. Results

3.1. Selecting the Optimal Time-Series Datasets

Figure 4a–c show the extracted NDVI temporal profiles for three periods. In these profiles, patterns and trends in NDVI changes can be observed for selected periods. The highest NDVI value belongs to the spring season, as this season is the peak of PEUs vegetative growth. In contrast, the lowest NDVI value can be observed in autumn and winter. As shown in Figure 4a–c, generally, the highest NDVI value occurs for selected periods between April and June. The multi-temporal images with the maximum NDVI values were selected for PEUs classification for each period (Table 1).

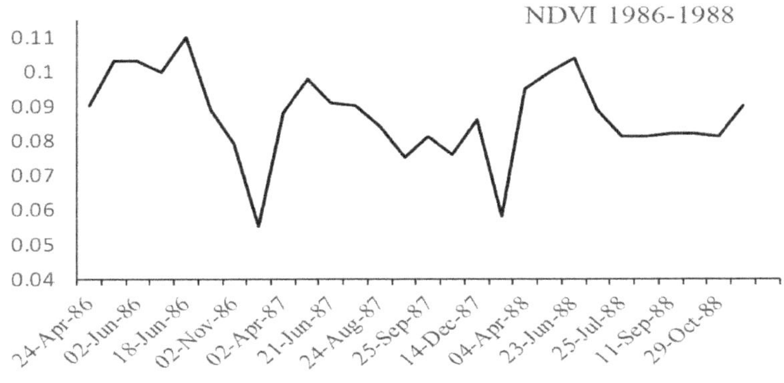

(a)

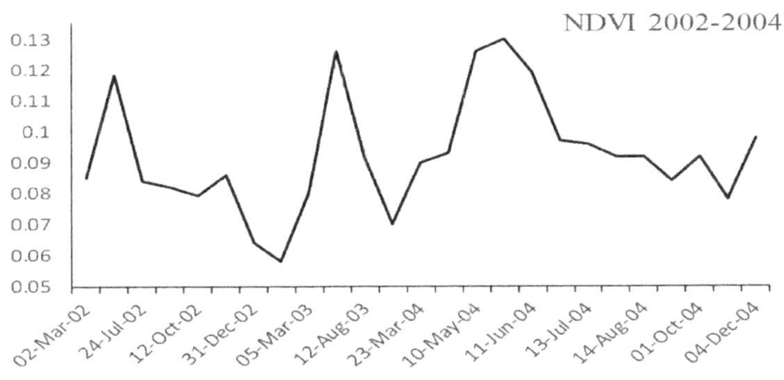

(b)

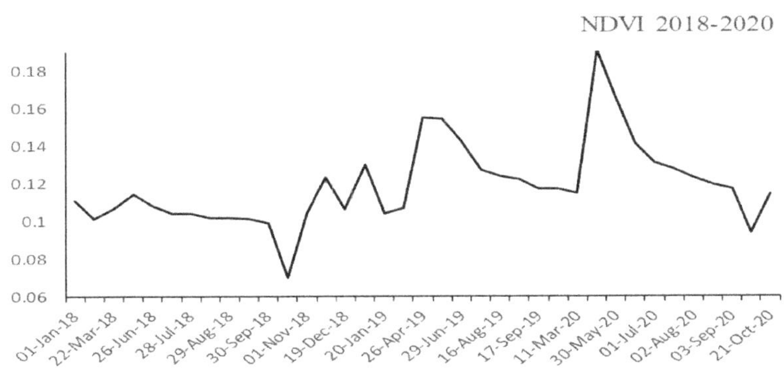

(c)

Figure 4. (**a**) The NDVI time-series profile of Landsat 5 TM (1986–1988), images for the periods of January to December in each year. (**b**) The NDVI time-series profile of Landsat 7 ETM+ (2002–2004), images for the periods of January to December in each year. (**c**) The NDVI time-series profile of Landsat 8 OLI (2018–2020), images for the periods of January to December in each year.

Table 1. Details of multi-temporal images used in this study for PEUs classification.

Sensor	Year	Month/Day	Year	Month/Day	Year	Month/Day
Landsat 5 TM	1986	24 April 17 May 2 June 18 June	1987	2 April 4 May 20 May 5 June 21 June	1988	4 April 7 June 23 June
Landsat 7 ETM+	2002	5 May 14 May 17 June	2003	24 May	2004	8 April 10 May 26 May 11 June 27 June
Landsat 8 OLI	2018	25 May 10 June 26 June	2019	26 April 28 May 13 June 29 June	2020	12 April 28 April 14 May 30 May 15 June

3.2. PEUs Classification and Validation

Using the RF classification algorithm, PEUs classification was performed for selected periods. Figure 5 shows three PEUs classification maps. The accuracy of the PEUs maps were was estimated using confusion matrices. In Table 2, the OA and OK of each period are presented. Also, the PA, UA, and KIA of PEUs are reported. The OA of the maps was 0.73, 0.77, and 0.81 for period 1, period 2, and period 3, respectively.

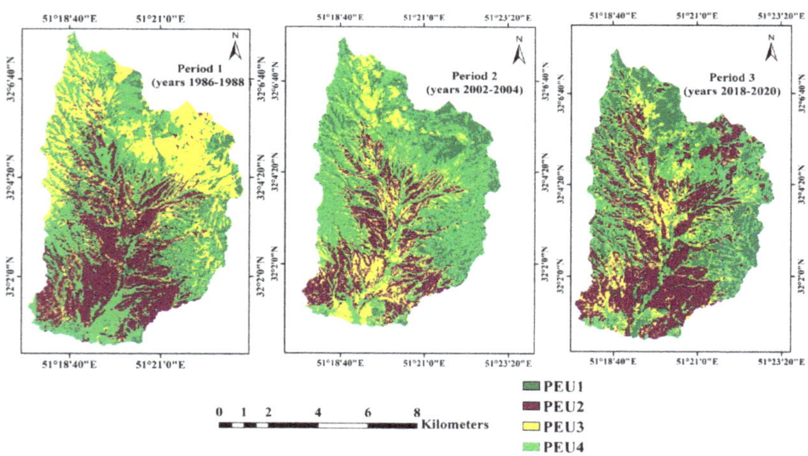

Figure 5. PEUs maps of period 1, period 2, and period 3.

Table 2. Confusion matrix results of PEUs classification of periods 1, 2, and 3.

PEUs Class	Period 1			Period 2			Period 3		
	PA1	UA1	KIA1	PA2	UA2	KIA2	PA3	UA3	KIA3
PEU1	0.84	0.74	0.84	0.91	0.77	0.87	0.91	0.91	0.88
PEU2	0.79	0.64	0.50	0.84	0.72	0.76	0.75	0.82	0.67

Table 2. *Cont.*

PEUs Class	Period								
	Period 1			Period 2			Period 3		
	PA1	**UA1**	**KIA1**	**PA2**	**UA2**	**KIA2**	**PA3**	**UA3**	**KIA3**
PEU3	0.58	0.75	0.63	0.67	0.89	0.58	0.84	0.67	0.75
PEU4	0.70	0.78	0.56	0.67	0.73	0.56	0.75	0.90	0.68
	OK = 0.63, OA = 0.73			OK = 0.68, OA = 0.77			OK = 0.74, OA = 0.81		

PA: Producer's Accuracy %; UA: User's Accuracy %; and KIA: Kappa Index of Agreement %.

3.3. Change Analysis in PEUs Classes

As shown in Table 3(a), significant changes occurred in all PEUs over the three periods. These changes were analyzed from period 1 to period 2, period 2 to period 3, and period 1 to period 3 (Table 3(b)). PEU1 in period 1 covered an area of 1224.27 ha (15.82%) and it increased to 1582.12 ha (20.48%) and 1779.84 ha (23%) in period 2 and period 3, respectively. From period 1 to period 2, period 2 to period 3, and period 1 to period 3, PEU1 grew by 4.66%, 2.51%, and 7.18%, respectively.

Table 3. (a) Area of PEUs of period 1, period 2, and period 3. (b) Amount of changes in PEUs during period 1, period 2, and period 3 in the study area.

(a) Area						
	Period 1		**Period 2**		**Period 3**	
PEU Class	**Area (ha)**	**Area (%)**	**Area (ha)**	**Area (%)**	**Area (ha)**	**Area (%)**
PEU1	1224.27	15.82	1582.12	20.48	1779.84	23
PEU2	2089.62	27.08	1389.57	17.96	2567.79	33.19
PEU3	2314.17	29.91	1444.58	18.67	1189.26	15.37
PEU4	2108.7	27.25	3317.31	42.87	2199.69	28.43
Total class	7736.58	100	7736.58	100	7736.58	100
(b) Amount of Changes						
	Period 1–Period 2		**Period 2–Period 3**		**Period 2–Period 3**	
PEU Class	**Area (ha)**	**Area (%)**	**Area (ha)**	**Area (%)**	**Area (ha)**	**Area (%)**
PEU1	360.95	4.66	194.72	2.51	555.67	7.18
PEU2	−699.97	−9.04	1178.22	15.22	478.25	6.18
PEU3	−869.59	−11.23	−255.32	−3.3	−1124.91	−14.54
PEU4	1208.61	15.62	−1117.62	−14.4	90.99	1.17

PEU2 covered an area of 2089.62 ha (27.08%) in period 1, and it decreased to 1389.57 ha (17.96%) in period 2, while it increased to 2567.79 ha (33.19%) in period 3. From period 1 to period 2, PEU2 decreased by 9.04%, while from period 2 to period 3 and period 1 to period 3, PEU2 increased by 15.22% and 6.18%, respectively. Also, it was determined that in period 1, the PEU3 area covered an area of 2314.17 ha (29.91%) and it reduced to 1444.58 ha (18.67%) and 1189.26 ha (15.37%) in period 2 and period 3, respectively. A continuous decrease in PEU3 was observed by 11.23%, 3.30%, and 14.54% from period 1 to period 2, period 2 to period 3, and period 1 to period 3, respectively. The area covered by PEU4 in period 1 was 2108.7 ha (27.25%), and it increased to 3317.31 ha (42.87%) in period 2, while it decreased to 2199.69 ha (28.43%) in period 3. From period 1 to period 2, PEU4 increased by 15.62%, while during period 2 to period 3, PEU4 decreased by 14.4% and again increased by 1.17% from period 1 to period 3.

3.4. Transition Potential Modeling

3.4.1. Testing and Selecting Environmental Variables

This step identifies the environmental variables that have the ability to describe changes in the area. As shown in Figure 6, a total of six environmental variables, including the Digital Elevation Model (DEM), aspect, slope, distance from roads, precipitation, and temperature, were used for the transition potential modeling. These variables were selected according to other studies in this field [25,35], and also according to Cramer's V statistics. Cramer's V statistics were used to evaluate the potential power of these variables (as an independent variable) on the sub-model in which change was observed (as a dependent variable). The variables with a Cramer's V value of about 0.15 or higher are regarded as effective [36]. As presented in Table 4, the best Overall Cramer's coefficient of 0.44 was obtained using the DEM variable.

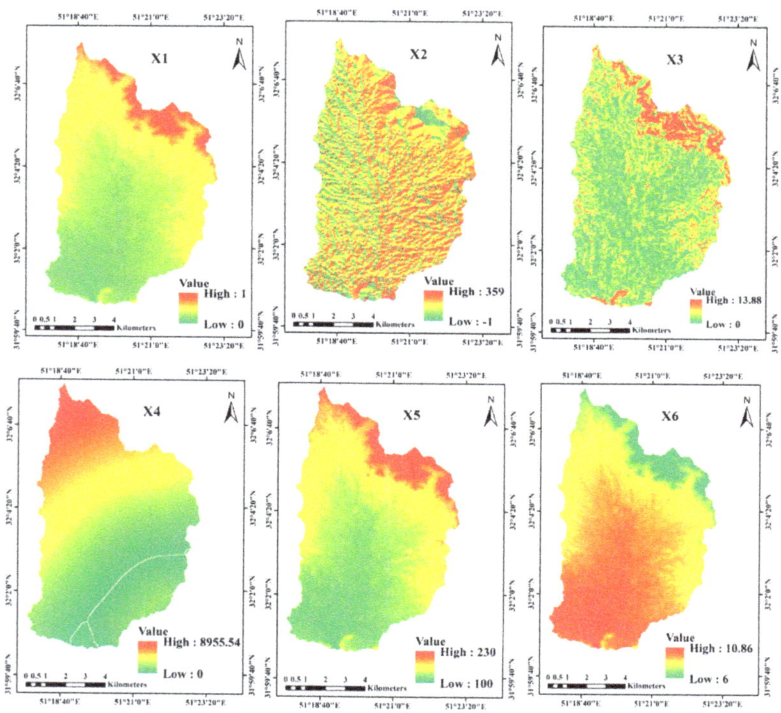

Figure 6. Environmental variable maps. (X1) DEM; (X2) aspect; (X3) slope; (X4) distance from roads; (X5) precipitation; (X6) temperature.

Table 4. Cramer's V statistics results for environmental variables.

Environmental Variables	Overall Cramer's
Digital Elevation Model (DEM)	0.4458
Aspect	0.182
Slope	0.194
Distance from roads	0.21
Precipitation	0.396
temperature	0.264

3.4.2. Transition Sub-Models

After reaching acceptable Cramer's V values for all environmental factors, subsequently, the transfer potential model was run. This model expresses the tendency of each image cell to receive a change from one PEU type to another type with respect to environmental variables. All the observed PEUs transitions were collected into a set of sub-models. Likewise, the transition potential was determined by developing twelve sub-models in the NN-MLP algorithm. Table 5 gives the results of transfer potential modeling using the NN-MLP algorithm. To evaluate the transfer potential modeling, three factors of Accuracy Rate, Testing RMS, and Training RMS were determined. The results in all sub-models show an accuracy of 61 to 89%.

Table 5. Results of accuracy assessing of transfer potential modeling using NN-MLP algorithm.

Sub-Models	Testing RMS	Training RMS	Accuracy Rate
PEU1 to PEU2	0.27	0.27	89.14
PEU1 to PEU3	0.39	0.39	78.16
PEU1 to PEU4	0.43	0.43	71.51
PEU2 to PEU1	0.41	0.43	73.26
PEU2 to PEU3	0.42	0.42	75.27
PEU2 to PEU4	0.4	0.39	77
PEU3 to PEU1	0.41	0.4	76
PEU3 to PEU2	0.39	0.4	79
PEU3 to PEU4	0.44	0.44	70.53
PEU4 to PEU1	0.37	0.37	82
PEU4 to PEU2	0.41	0.4	75.29
PEU4 to PEU3	0.47	0.48	61.18

3.5. *Prediction and Validation of PEUs Changes with Markov Chain Modeling*

To predict a PEU map for period 3 (years 2018–2020), two different PEUs maps of period 1 and period 2 were used to create the prediction map (Figure 7) and transition probability matrix (Table 6) using the MC prediction process. As shown in Table 6, the probability of a change of PEU4 into PEU1 and PEU2 in period 3 from period 1 to period 2 is 33.91% and 17.22%, respectively. Also, the probability that PEU3 converts into PEU2 in period 3 is 26.99%.

Table 6. Markov Chain transition probability matrix of changes among PEUs (period 1–period 2) for period 3.

PEU Class	PEU1	PEU2	PEU3	PEU4
PEU1	0.7563	0	0.1149	0.1228
PEU2	0	0.7958	0.1282	0.0761
PEU3	0.028	0.2699	0.5162	0.1853
PEU4	0.3391	0.1722	0.1411	0.3476

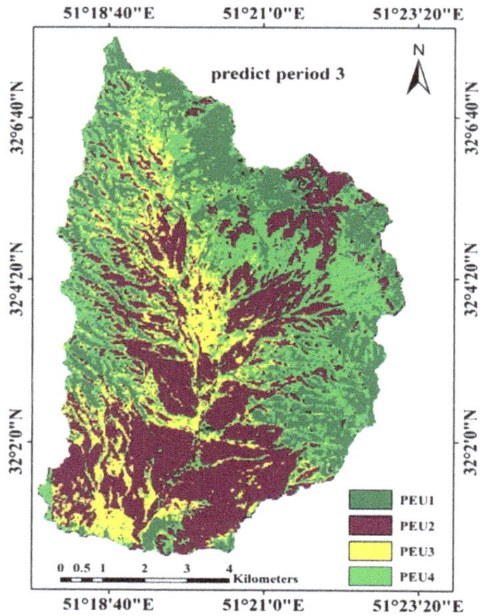

Figure 7. Predicted PEUs map for period 3 (years 2018–2020).

To evaluate and validate the MC prediction process, the period 3 classified map and the period 3 predicted map were used. Cross-tabulation analysis shows that the Overall Cramer's value is 68.53% and the Overall Kappa is 77.6%.

In addition, Figure 8 shows the side-by-side comparison of PEUs in two classified and predicted maps of period 3. In this map, the PEUs that are classified in both maps in the correct pixels are marked with 1-1, 2-2, 3-3, and 4-4. This means that there are pixels that are classified in the same class in both maps. The parts that are white on the map show the wrong pixels. This means that this model has incorrectly predicted the white parts.

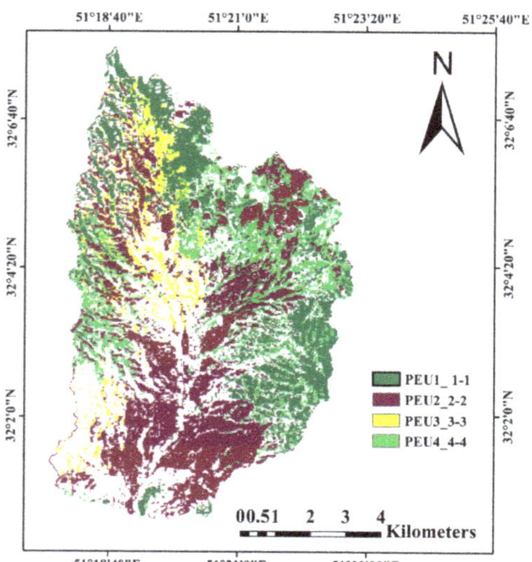

Figure 8. The map resulting from the comparison of the classified and predicted maps of period 3 (years 2018–2020).

3.6. Prediction of Future PEUs Changes

The PEUs maps of period 2 and period 3 were used to predict the future PEU scenarios for the year 2036 using the MLP-MC model. The predicted PEUs map (Figure 9) and transition probability matrices (Table 7) were produced using the PEUs maps of period 2 and period 3. By utilizing the PEU of period 3 as the base map, the transition potential maps, and the transition probability matrices of period 2 to period 3, the future PEU scenarios were predicted for the year 2036. The probability of PEU4 turning into PEU1 and PEU2 in the year 2036 is 32.10% and 15.88%, respectively. Also, the probability of PEU3 turning into PEU2 and PEU4 in 2036 is 19.17% and 20.93%, respectively.

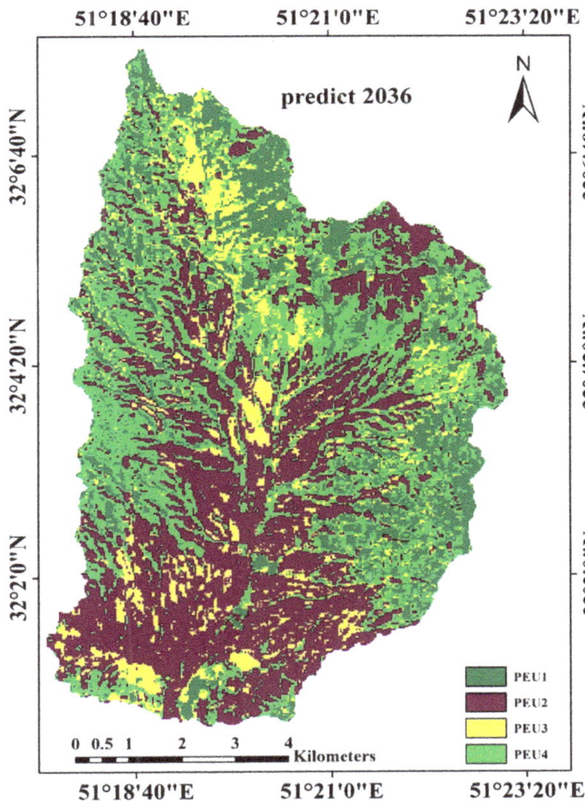

Figure 9. Predicted PEUs map for year 2036.

Table 7. Markov Chain transition probabilities matrix of changing among PEUs (period 2–period 3) for 2036.

PEU Class	PEU1	PEU2	PEU3	PEU4
PEU1	0.7437	0	0.0935	0.1628
PEU2	0	0.8369	0.1482	0.0150
PEU3	0	0.1917	0.5990	0.2093
PEU4	0.3210	0.1588	0.1119	0.4084

The areas of the PEUs classes of the predicted map of 2036 are presented in Table 8. A comparison of the prediction map of the year 2036 and the classification map of period 3 (2018–2020) reveals that PEU2 and PEU4 will increase by 333.63 ha (4.30%) and 60.46 ha

(0.07%), respectively. Meanwhile, PEU1 and PEU3 will decrease by 153.07 ha (1.98%) and 241.02 ha (3.12%), respectively.

Table 8. Area distribution of PEUs classes for predicted year 2036.

	Area (ha)	Area %
PEU1	1626.77	21.02
PEU2	2901.42	37.50
PEU3	948.24	12.25
PEU4	2260.15	29.21
Total area	7736.58	100

4. Discussion

Knowledge and understanding of historical PEUs patterns, changes, and future trends are important in proactive environmental management. To enable the study and prediction of long-term PEUs dynamics, it is critical to analyze and understand the various changes shaping the land cover, which, among other factors, can help to minimize undesired changes in plant communities [6]. PEUs change is the result of complex interactions between human activities and environmental factors from the past to the present [33]. We started with the four dominant PEU classes in the study area: PEU1 (*As ve*), PEU2 (*Br to*), PEU3 (*Sc or*), and PEU4 (*As ve—Br to*).

With the purpose of understanding the spatial–temporal dynamics of PEUs, we divided this work into four sections: (1) Preparation of PEU multi-temporal classification maps and accuracy assessment for three different periods. (2) Change analysis of periods 1–2, periods 2–3, and periods 1–3. (3) Prediction of period 3 by the MLP-MC model, and comparison of the prediction results with the classified PEU map of period 3. (4) Prediction of future scenarios of PEUs for the year 2036 using the MLP-MC model that produced the best results in the prediction of period 3. These sections are further discussed below.

4.1. PEUs Multi-Temporal Classification Maps for Three Periods

This study used time series of Landsat data in three periods of three years that include period 1 (Landsat 5 TM images for 1986–1988), period 2 (Landsat 7 ETM+ images for 2002–2004), and period 3 (Landsat 8 OLI images for 2018–2020). According to the NDVI temporal profile, we selected the optimal multi-temporal images for each period. As shown in Figure 5, the most informative temporal windows were observed in spring for the period of April through June. A total of 12 cloudless images were extracted for period 1, 9 cloudless images were extracted for period 2, and 12 cloudless images were extracted for period 3 (Table 1).

The GEE platform allows the synchronization of all the Landsat data and then the creation of a high-quality multi-temporal dataset using codes already provided [37]. The RF algorithm was chosen for PEUs classification. The RF algorithm is a tree-based machine learning method that leverages the power of multiple decision trees for making decisions and is suitable for situations when we have a large dataset [1]. At this stage, three multi-temporal PEU maps were obtained with an OA of 73%, 77%, and 81% for period 1, period 2, and period 3, respectively. The increasing OA from period 1 to period 3 can be related to the development of Landsat generations' capabilities specifically in terms of radiometric properties and the consequently decrease in the noise-to-signal ratio. PEUs as a sub-class of rangeland cover are involved. Sub-classes of vegetation cover are more similar in terms of their spectral reflectance than that of a higher hierarchical land cover classification [7]. Thereby, PEUs classification process using an optimal time-series dataset is required to accurately identify and discriminate the past and current trends as well as to predict future trends of PEUs.

4.2. Analysis of PEUs Changes

The study area during period 1 (1986–1988), period 2 (2002–2004), and period 3 (2018–2020) experienced intense changes in PEUs. There are significant changes that occurred in all PEUs classes. Each of the PEU represents the different succession stages of each period in the study area. Natural disturbances and human activities such as fire, drought, land tillage, and irregular grazing can change the plant community's successional stages. Thereby, it will lead to a decrease in vegetation and a descending trend in plant community succession. The PEUs classification maps of period 1, period 2, and period 3 demonstrate PEUs changes well (Figure 5). In period 1, PEU3 covered most of the study area and was the dominant PEU with 2314.17 ha (29.91%). PEU3 is dominated by semi-shrub species (*Sc or*). This plant species is ecologically an invasive species, and in areas where the plant community has been destroyed by tilling, it becomes dominant in the area [38]. Examination of 1986-1988 documents shows that this area was heavily tilled and cultivated but was abandoned due to low yields. Thereby, PEU3 became the dominant PEU in the area, which shows plant succession's declining trend and the severe destruction of the plant community. Meanwhile, due to the implementation of appropriate management practices in the region, such as tillage inhibition, PEU3 reduced to 18.67% and 15.37% in period 2 and period 3, respectively. Thereby, after 16 years (period 2) and 32 years (period 3), much of the area has recovered its vegetation with a suitable canopy cover, whereby shrubs (PEU1) and perennial grasses (PEU2) became dominant. Likewise, the plant succession trend is upward and positive. In period 2, PEU2 had the lowest coverage in the study area, with 1389.57 ha (17.96%). Meanwhile, shrubs and non-palatable species (e.g., PEU1) have become more widespread. PEU2 is dominated by grass species (*Br to*). These plant species are palatable to livestock and are destroyed by grazing before their growth is completed [38]. Examination of 2002–2004 documents shows that this area was affected by severe grazing. Therefore, PEU2 decreased in period 2. After 16 years, PEU2 increased to 33% in period 3, mainly due to appropriate management practices and preventing severe grazing. Much of the area is dominated by perennial grasses (i.e., PEU2). This suggests that, with the implementation of proper management policies and the prevention of destructive processes such as tilling and heavy grazing, the plant community will approach a climax stage.

4.3. Transition Probabilities of Change among PEUs

For modeling future PEUs dynamics, the transition probability assessment among different PEUs is an important aspect. The NN-MLP algorithm was used to determine the weights of transitions for two periods that were included in the transition probability matrix using the MC model for future prediction. The MC model is very effective in determining the behaviors and extent of land cover changes by analyzing land cover maps of two periods [8]. Based on a projection of the transition potentials, the method determines exactly how many PEUs would transition from the specified date to the predicted date in the future [33]. In this study, in the first step, two different PEUs maps of period 1 and period 2 were used to create the transition potential maps for period 3 (Table 6). For validation of the MC prediction process, the PEUs prediction map of period 3 was compared with the PEUs classification map of period 3 using cross-tabulation analysis. It was found that the Kappa Index of Agreement value is 77.6%, showing good agreement between the observed and predicted PEUs maps. Then, in the next step, the PEUs map of period 2 and period 3 were used to predict the future PEUs changes for 2036 (Table 7). In both stages, the probability of the change of PEU4 into PEU1 and PEU2 in period 3 and 2036. Also, the probability of change of PEU3 into PEU2 is significant in both predicted dates. This suggests that the MC is a robust model and can predict future PEUs changes.

4.4. Prediction of Future PEUs for Year 2036

The last step is the prediction of future scenarios of PEUs for the year 2036 using the MLP-MC model that produced the best results in the prediction of period 3. By utilizing the PEU of period 3 as the base map, the transition potential maps, and the transition probability matrices of period 2 to period 3, the future PEU scenarios were predicted for 2036. The prediction results for 2036 are shown in Table 8; the dominant vegetation of the area for 2036 will be PEU2 and PEU4 with 37.50% and 29.21%, respectively. The predicted results further show that by 2036, PEU3 will have the lowest coverage with 12.25%. Comparing the 2036 predicted map with the period 3 classified map shows that the PEU1 area decreases (153.07 ha). PEU1 is dominated by shrub species (*As ve*). Field investigations in recent years have shown that tragacanth gum extraction from *As ve* has dramatically increased. However, little information is available about the best method of gum harvesting, which is economically efficient and maintains the health of the plant. This study's results demonstrate that the MLP-MC method has the ability to predict and describe future PEU changes that are the result of current activities. Also, by aiming to compare three prediction models, Cellular Automata–Markov Chain (CA-MC), Stochastic Markov Chain (ST-MC), and MLP-MC, to predict land cover changes in the Varanasi district, ref. [25] reported that the MLP-MC prediction model had the best results for an accurate understanding of changes and predicting future landscape scenarios. The implementation of an arsenal of management activities and natural resource conservation policies will reduce pressure on plant communities such as PEUs, thereby contributing to natural environmental sustainability even in the future.

5. Conclusions

This study aimed to reveal and analyze the dynamics in PEUs patterns. Nowadays, land use change patterns and urban area development and construction are considered the main variables in the analysis and monitoring of the terrestrial environment and natural system changes. In many areas, however, land cover and plant communities have declined for multiple reasons. The main cause is human activities, such as intense grazing, fire, tilling, and drought. But what is often unsaid is that anthropogenic processes and management activities are the most important drivers in plant community changes. It is therefore essential to identify these activities for the sake of future planning and natural resource management. This study evaluated the MLP-MC model to identify the main processes in PEUs changes from the past to the future and predicted future PEUs dynamics. We first selected three periods with 16-year time intervals. PEUs information extracted for period 1 and period 2 was employed to predict PEUs changes for period 3 using the MLP-MC prediction model. The predicted results were compared with the classified PEUs information for period 3 to appraise the validity of models through cross-tabulation analysis and kappa index statistics (Overall Kappa: 77.6%). Finally, we predicted future PEUs changes 16 years ahead of period 3, for the year 2036. The results demonstrated that the MLP-MC model is efficient in predicting PEUs patterns in the future. The basis for this model is the anthropogenic and natural processes of the past. Our study suggests that the impact of anthropogenic processes and management activities will become visible in the natural environment and ecosystem in less than a decade.

Author Contributions: Methodology, A.E. and M.A.; conceptualization, A.E. and M.A.; software, M.A.; validation, A.E.; investigation, M.A.; resources, M.A.; formal analysis, M.A.; data curation, A.E. and M.A.; writing—original draft preparation, M.A. and J.V.; writing—review and editing, A.E., J.V., A.A.N. and E.A.; supervision, A.E.; visualization, M.A. and A.E.; project administration, A.E.; funding acquisition: J.V. All authors have read and agreed to the published version of the manuscript.

Funding: This work was supported by Shahrekord University. J.V. was funded by the European Union (ERC, FLEXINEL, 101086622). Views and opinions expressed are however those of the author(s) only and do not necessarily reflect those of the European Union or the European Research Council. Neither the European Union nor the granting authority can be held responsible for them.

Data Availability Statement: Data supporting this study may be provided by the first author upon a reasonable request. The data are not publicly available due to the privacy of parts of the data.

Conflicts of Interest: The authors declare no conflict of interest.

References

1. Feng, D.; Yu, L.; Zhao, Y.; Cheng, Y.; Xu, Y.; Li, C.; Gong, P. A multiple dataset approach for 30-m resolution land cover mapping: A case study of continental Africa. *Int. J. Remote Sens.* **2018**, *39*, 3926–3938. [CrossRef]
2. Pflugmacher, D.; Rabe, A.; Peters, M.; Hostert, P. Mapping pan-European land cover using Landsat spectral-temporal metrics and the European LUCAS survey. *Remote Sens. Environ.* **2019**, *221*, 583–595. [CrossRef]
3. McCarthy, M.J.; Radabaugh, K.R.; Moyer, R.P.; Muller-Karger, F.E. Enabling efficient, large-scale high-spatial resolution wetland mapping using satellites. *Remote Sens. Environ.* **2018**, *208*, 189–201. [CrossRef]
4. Xie, Z.; Chen, Y.; Lu, D.; Li, G.; Chen, E. Erxue, Classification of Land Cover, Forest, and Tree Species Classes with ZiYuan-3 Multispectral and Stereo Data. *Remote Sens.* **2019**, *11*, 164. [CrossRef]
5. Macintyre, P.D.; Van Niekerk, A.; Dobrowolski, M.P.; Tsakalos, J.L.; Mucina, L. Impact of ecological redundancy on the performance of machine learning classifiers in vegetation mapping. *Ecol. Evol.* **2018**, *8*, 6728–6737. [CrossRef] [PubMed]
6. Aghababaei, M.; Ebrahimi, A.; Naghipour, A.A.; Asadi, E.; Verrelst, J. Classification of Plant Ecological Units in Heterogeneous Semi-Steppe Rangelands: Performance Assessment of Four Classification Algorithms. *Remote Sens.* **2021**, *13*, 3433. [CrossRef] [PubMed]
7. Aghababaei, M.; Ebrahimi, A.; Naghipour, A.A.; Asadi, E.; Verrelst, J. Vegetation Types Mapping Using Multi-Temporal Landsat Images in the Google Earth Engine Platform. *Remote Sens.* **2021**, *13*, 4683. [CrossRef]
8. Dey, N.N.; Al Rakib, A.; Kafy, A.A.; Raikwar, V. Geospatial modelling of changes in land use/land cover dynamics using Multi-layer perception Markov chain model in Rajshahi City, Bangladesh. *Environ. Chall.* **2021**, *4*, 100148. [CrossRef]
9. Tolessa, T.; Senbeta, F.; Kidane, M. The impact of land use/land cover change on ecosystem services in the central highlands of Ethiopia. *Ecosyst. Serv.* **2017**, *23*, 47–54. [CrossRef]
10. Naboureh, A.; Bian, J.; Lei, G.; Li, A. A review of land use/land cover change mapping in the China-Central Asia-West Asia economic corridor countries. *Big Earth Data* **2020**, *5*, 237–257. [CrossRef]
11. Dou, P.; Chen, Y. Dynamic monitoring of land-use/land-cover change and urban expansion in Shenzhen using Landsat imagery from 1988 to 2015. *Int. J. Remote Sens.* **2017**, *38*, 5388–5407. [CrossRef]
12. Somvanshi, S.S.; Bhalla, O.; Kunwar, P.; Singh, M.; Singh, P. Monitoring spatial LULC changes and its growth prediction based on statistical models and earth observation datasets of Gautam Budh Nagar, Uttar Pradesh, India. *Environ. Dev. Sustain.* **2018**, *22*, 1073–1091. [CrossRef]
13. Osaliya, R.; Wasonga, O.V.; Majaliwa Mwanjalolo, J.G.; MacOpiyo, L.; Kironchi, G.; Adipala, E. Predicted land use and land cover outlook for semi-arid Lokere and Lokok catchments in Karamoja region, Uganda. *Afr. Crop Sci. J.* **2020**, *28*, 595–616. [CrossRef]
14. Gambo, J.; Mohd Shafri, H.Z.; Shaharum, N.S.N.; Abidin, F.A.Z.; Rahman, M.T.A. Monitoring and Predicting Land Use-Land Cover (Lulc) Changes within and around Krau Wildlife Reserve (Kwr) Protected Area in Malaysia Using Multi-Temporal Landsat Data. *Geoplanning J. Geomat. Plan.* **2018**, *5*, 17. [CrossRef]
15. Bunyangha, J.; Majaliwa, M.J.G.; Muthumbi, A.W.; Gichuki, N.N.; Egeru, A. Past and future land use/land cover changes from multi-temporal Landsat imagery in Mpologoma catchment, eastern Uganda. *Egypt. J. Remote Sens. Space Sci.* **2021**, *24*, 675–685. [CrossRef]
16. Al-Hamdan, M.Z.; Oduor, P.; Flores, A.I.; Kotikot, S.M.; Mugo, R.; Ababu, J.; Farah, H. Evaluating land cover changes in Eastern and Southern Africa from 2000 to 2010 using validated Landsat and MODIS data. *Int. J. Appl. Earth Obs. Geoinf.* **2017**, *62*, 8–26. [CrossRef]
17. Phiri, D.; Morgenroth, J. Developments in Landsat Land Cover Classification Methods: A Review. *Remote Sens.* **2017**, *9*, 967. [CrossRef]
18. Ettehadi Osgouei, P.; Kaya, S. Analysis of land cover/use changes using Landsat 5 TM data and indices. *Environ. Monit. Assess* **2017**, *189*, 136. [CrossRef] [PubMed]
19. Phiri, D.; Morgenroth, J.; Xu, C.; Hermosilla, T. Effects of pre-processing methods on Landsat OLI-8 land cover classification using OBIA and random forests classifier. *Int. J. Appl. Earth Obs. Geoinf.* **2018**, *73*, 170–178. [CrossRef]
20. Valderrama-Landeros, L.; Flores-Verdugo, F.; Rodriguez-Sobreyra, R.; Kovacs, J.M.; Flores-de-Santiago, F. Extrapolating canopy phenology information using Sentinel-2 data and the Google Earth Engine platform to identify the optimal dates for remotely sensed image acquisition of semiarid mangroves. *J. Environ. Manag.* **2021**, *279*, 111617. [CrossRef]
21. Chen, N.; Yu, L.; Zhang, X.; Shen, Y.; Zeng, L.; Hu, Q.; Niyogi, D. Mapping Paddy Rice Fields by Combining Multi-Temporal Vegetation Index and Synthetic Aperture Radar Remote Sensing Data Using Google Earth Engine Machine Learning Platform. *Remote Sens.* **2020**, *12*, 2992. [CrossRef]
22. Gorelick, N.; Hancher, M.; Dixon, M.; Ilyushchenko, S.; Thau, D.; Moore, R. Google Earth Engine: Planetary-scale geospatial analysis for everyone. *Remote Sens. Environ.* **2017**, *202*, 18–27. [CrossRef]
23. Zhou, B.; Okin, G.S.; Zhang, J. Leveraging Google Earth Engine (GEE) and machine learning algorithms to incorporate in situ measurement from different times for rangelands monitoring. *Remote Sens. Environ.* **2020**, *236*, 111521. [CrossRef]
24. Nyamekye, C.; Kwofie, S.; Agyapong, E.; Ofosu, S.A.; Arthur, R.; Appiah, L.B. Integrating support vector machine and cellular automata for modelling land cover change in the tropical rainforest under equatorial climate in Ghana. *Curr. Res. Environ. Sustain.* **2021**, *3*, 100052. [CrossRef]

25. Mishra, V.N.; Rai, P.K.; Prasad, R.; Punia, M.; Nistor, M.-M. Prediction of spatio-temporal land use/land cover dynamics in rapidly developing Varanasi district of Uttar Pradesh, India, using geospatial approach: A comparison of hybrid models. *Appl. Geomat.* **2018**, *10*, 257–276. [CrossRef]
26. Mishra, V.N.; Rai, P.K. A remote sensing aided multi-layer perceptron-Markov chain analysis for land use and land cover change prediction in Patna district (Bihar), India. *Arab. J. Geosci.* **2016**, *9*, 249. [CrossRef]
27. Thapa, R.B.; Murayama, Y. Scenario based urban growth allocation in Kathmandu Valley, Nepal. *Landsc. Urban Plan.* **2012**, *105*, 140–148. [CrossRef]
28. Ludwig, M.; Morgenthal, T.; Detsch, F.; Higginbottom, T.P.; Lezama Valdes, M.; Nauß, T.; Meyer, H. Machine learning and multi-sensor based modelling of woody vegetation in the Molopo Area, South Africa. *Remote Sens. Environ.* **2019**, *222*, 195–203. [CrossRef]
29. De Torres, F.N.; Richter, R.; Vohland, M. A multisensoral approach for high-resolution land cover and pasture degradation mapping in the humid tropics: A case study of the fragmented landscape of Rio de Janeiro. *Int. J. Appl. Earth Obs. Geoinf.* **2019**, *78*, 189–201. [CrossRef]
30. Li, W.; Dong, R.; Fu, H.; Wang, J.; Yu, L.; Gong, P. Integrating Google Earth imagery with Landsat data to improve 30-m resolution land cover mapping. *Remote Sens. Environ.* **2020**, *237*, 111563. [CrossRef]
31. Sánchez-Ruiz, S.; Moreno-Martínez, Á.; Izquierdo-Verdiguier, E.; Chiesi, M.; Maselli, F.; Gilabert, M.A. Growing stock volume from multi-temporal landsat imagery through google earth engine. *Int. J. Appl. Earth Obs. Geoinf.* **2019**, *83*, 101913. [CrossRef]
32. Jensen, J.R. *Introductory Digital Image Processing: A Remote Sensing Perspective University of South Carolina*; Pearson: London, UK, 2015; ISBN 978-0-13-405816-0.
33. Dzieszko, P. Land-Cover Modelling Using Corine Land Cover Data and Multi-Layer Perceptron. *Quaest. Geogr.* **2014**, *33*, 5–22. [CrossRef]
34. Balogun, I.A.; Ishola, K.A. Projection of Future Changes in Landuse/Landcover Using Cellular Automata/Markov Model over Akure City, Nigeria. *J. Remote Sens. Technol.* **2017**, *5*, 22–31. [CrossRef]
35. Wang, Q.; Guan, Q.; Lin, J.; Luo, H.; Tan, Z.; Ma, Y. Simulating land use/land cover change in an arid region with the coupling models. *Ecol. Indic.* **2021**, *122*, 107231. [CrossRef]
36. Eastman, J.R. *IDRISI Taiga Guide to GIS and Image Processing*; Clark University: Worcester, MA, USA, 2009.
37. Xiong, J.; Thenkabail, P.S.; Gumma, M.K.; Teluguntla, P.; Poehnelt, J.; Congalton, R.G.; Yadav, K.; Thau, D. Automated cropland mapping of continental Africa using Google Earth Engine cloud computing. *ISPRS J. Photogramm. Remote Sens.* **2017**, *126*, 225–244. [CrossRef]
38. Roshan, S.A.; Heydari, M. Short Communication: Flora, life form and chorological study of Quercus brantii habitat in Emamzadeh Abdullah woodland, Iran. *Biodiversitas J. Biol. Divers.* **2014**, *15*. [CrossRef]

remote sensing

MDPI

Article

Developing Spatial and Temporal Continuous Fractional Vegetation Cover Based on Landsat and Sentinel-2 Data with a Deep Learning Approach

Zihao Wang [1,2], Dan-Xia Song [1,2,*], Tao He [3], Jun Lu [3], Caiqun Wang [3] and Dantong Zhong [1,2]

1 Hubei Provincial Key Laboratory for Geographical Process Analysis and Simulation, Central China Normal University, Wuhan 430079, China; wonka1025@126.com (Z.W.); winter@mails.ccnu.edu.cn (D.Z.)
2 College of Urban and Environmental Sciences, Central China Normal University, Wuhan 430079, China
3 School of Remote Sensing and Information Engineering, Wuhan University, Wuhan 430079, China; taohers@whu.edu.cn (T.H.); junlurs@whu.edu.cn (J.L.); cqwangrs@whu.edu.cn (C.W.)
* Correspondence: dxsong@ccnu.edu.cn

Abstract: Fractional vegetation cover (FVC) has a significant role in indicating changes in ecosystems and is useful for simulating growth processes and modeling land surfaces. The fine-resolution FVC products represent detailed vegetation cover information within fine grids. However, the long revisit cycle of satellites with fine-resolution sensors and cloud contamination has resulted in poor spatial and temporal continuity. In this study, we propose to derive a spatially and temporally continuous FVC dataset by comparing multiple methods, including the data-fusion method (STARFM), curve-fitting reconstruction (S-G filtering), and deep learning prediction (Bi-LSTM). By combining Landsat and Sentinel-2 data, the integrated FVC was used to construct the initial input of fine-resolution FVC with gaps. The results showed that the FVC of gaps were estimated and time-series FVC was reconstructed. The Bi-LSTM method was the most effective and achieved the highest accuracy (R^2 = 0.857), followed by the data-fusion method (R^2 = 0.709) and curve-fitting method (R^2 = 0.705), and the optimal time step was 3. The inclusion of relevant variables in the Bi-LSTM model, including LAI, albedo, and FAPAR derived from coarse-resolution products, further reduced the RMSE from 5.022 to 2.797. By applying the optimized Bi-LSTM model to Hubei Province, a time series 30 m FVC dataset was generated, characterized by a spatial and temporal continuity. In terms of the major vegetation types in Hubei (e.g., evergreen and deciduous forests, grass, and cropland), the seasonal trends as well as the spatial details were captured by the reconstructed 30 m FVC. It was concluded that the proposed method was applicable to reconstruct the time-series FVC over a large spatial scale, and the produced fine-resolution dataset can support the data needed by many Earth system science studies.

Keywords: fractional vegetation cover; spatio-temporal reconstruction; gap filling; LSTM; deep learning

1. Introduction

Vegetation is an essential component of Earth's terrestrial ecosystems, since it serves as a bridge connecting the soil, water, and atmosphere. Forest carbon sinks are an important potential solution for global climate change, and the absorption of carbon dioxide can be enhanced by protecting and restoring vegetation [1]. Therefore, it is vital to obtain high-resolution and real-time information on vegetation cover. When monitoring the dynamics of the Earth's vegetation, fractional vegetation cover (FVC) can be used to quantify the surface vegetation. The FVC is the percentage of the vertical projection area of green vegetation on the ground to the total statistical area [2]. The temporal dynamics of FVC are a useful indicator for environmental changes, such as drought and extreme wet conditions, and are an accurate indicator of the start of the growing season [3]. Additionally, this time series can be utilized to compare vegetation status in agriculture and forestry from year to year [4]. Therefore, FVC is an essential candidate for replacing classical vegetation indices for monitoring green vegetation [5]. To date, several methods have been proposed to estimate the FVC. Typically, these methods are divided into three major categories: empirical-based, physical-model-based, and spectral- unmixing-based [6].

With the rapid development and popularization of remote sensing technology, satellite data present continuous improvements in the detection range, temporal coverage, and spatial and temporal resolution [7]. Large-scale dynamic vegetation monitoring technology is becoming increasingly mature and is very suitable for the rapid estimation of vegetation coverage and understanding of the dynamic changes in global vegetation coverage [8]. Many institutions around the world have obtained FVC products using various retrieval methods based on satellite data (e.g., MERIS, SPOT, and POLDER) [9–11]. Current global-scale FVC products have been generated using remotely sensed data with medium spatial resolution, including VGT bioGEOphysical product Version 2 (GEOV2), Copernicus Global Land Operations (CGLOPS), and Global Land Surface Satellite (GLASS) FVC products [12–14]. However, there is an increasing demand for fine data applications. This means that FVC products based on low- or medium-spatial-resolution satellite data are not adequate to address the current demand for vegetation analysis. Meanwhile, owing to low spatial resolution, it is difficult to capture the precise spatial distribution of vegetation in highly heterogeneous and complex terrains. As a result, vegetation information extraction is less accurate.

With the continuous improvement of satellite sensors and computing capabilities, high spatial resolution satellites, such as Landsat, Sentinel, and GF, have been widely used for various quantitative remote sensing studies [15]. Several studies have used these satellites as data sources, combined with different retrieval methods, to obtain fine-resolution FVC products. The MuSyQ GF-series FVC product used Gaofen-1 satellite data to generate FVC with 16 m/10 d resolution in China from 2018 to 2020. However, this FVC product has only been available for 3 years in China, restricting time-series analysis [16]. To assess the usefulness of the hyperspectral imager (HSI) onboard the Chinese HJ-1-A small satellite for vegetation mapping, Zhang et al. [17] used dimidiate pixel models to retrieve 100 m FVC in the Shihezi Area of Xinjiang, China. The retrieved 100 m FVC from the HJ-1/HIS data was consistent with the in situ measurements with an R^2 of 0.86 and RMSE of 10.9%, indicating that HJ-1/HSI data can provide fine-scale monitoring of vegetation cover changes. Ludwig et al. [18] generated 10 m/7 d resolution of woody vegetation cover for the Molopo region in 2016 based on Sentinel-1 and Sentinel-2 data. Gill et al. [19] used Landsat 5 TM and Landsat 7 ETM+ data with a spectral unmixing approach to generate woody vegetation cover data for Australia from 2000 to 2010 at 30 m resolution. In addition, the data had a coefficient of determination of R^2 = 0.918, as verified by field measurements. In summary, despite the expanding number of fine-resolution FVC products from regional to national scales, most FVC models only estimate fine-resolution FVC for small or medium-sized regions, and the short period of the estimated FVC is still not conducive to large-scale, long-term vegetation analysis.

The Landsat-based FVC product is a global FVC product with 30 m resolution and long time series, using Landsat data as its data source and selecting training samples from Landsat 8 and Gao Fen 2 based on the Random Forest (RF) algorithm [20]. The Landsat FVC product outperforms other FVC products in terms of spatial resolution. This makes it more effective in investigating vegetation cover in small areas by providing finer details of the distribution of vegetation cover and allowing for the long-term monitoring of vegetation dynamics in recent decades. However, there are several issues with Landsat FVC products. Given that Landsat satellites revisit every 16 days, the Landsat FVC is frequently discontinuous in time series and is often contaminated by clouds or high concentrations of aerosols, resulting in fluctuations or gaps in the FVC product. In addition, certain areas are cloudy all year round, making it difficult for Landsat satellites to obtain accurate observations in these areas. As a result, they often have long periods of missing data, causing the vegetation cover data to be incomplete and discontinuous in both temporal and spatial aspects. Therefore, there is an urgent need to reconstruct the Landsat FVC to improve its spatial integrity and temporal continuity.

To fill in the gaps between the missing dates and cloud occlusion in the fine-resolution FVC, we attempt to reconstruct the time-series FVC spatially and temporally. Currently, the most effective method of achieving such objectives is through the integration of multi-source spatial resolution data, referring to other vegetation time-series parameters, and using deep learning methods. Additionally, there has been much research on the spatio-temporal reconstruction of satellite data products. For example, Chen et al. [21] obtained high-quality interpolated reconstructions of missing time in time-series images by fitting synthetic NDVI time series produced by combining MODIS NDVI data with Landsat cloud-free observations using a Savitzky–Golay (S-G) filter. Gao et al. first proposed the Spatial and Temporal Adaptive Reflectance Fusion Model (STARFM), which improves remote sensing data by fusing high-resolution Landsat data with low- and medium-resolution MODIS data [22]. The Flexible Spatio-temporal Data Fusion (FSDAF) method was proposed by Zhu et al. based on spectral unmixing analysis and a thin plate spline interpolator [23]. Morteza used the FSDAF algorithm to produce Landsat-like land surface temperature images with Landsat spatial and MODIS temporal resolutions [24]. Zhou fused MODIS and Landsat data to generate time-series NDVI and EVI with 30 m spatial resolution for dryland vegetation monitoring in Qinghai Province [25]. Walker fused MODIS and Landsat data to generate a 30 m spatial resolution time series of NDVI and EVI for dryland vegetation monitoring in Arizona [26]. Meng combined NDVI from HJ-1 CCD and MODIS for crop biomass estimation [27].

In recent years, deep learning methods have been widely applied in various remote sensing estimation applications and have outperformed traditional machine learning methods. Among the many deep learning architectures, recurrent neural networks (RNNs) can address the time dependence of time-series data [28] and handle time-series regression or prediction of satellite data. In vegetation index estimation based on time-series data, this approach has excellent noise tolerance and nonlinear processing ability, which can solve the limitation problem when data saturation exists. For instance, the reconstruction of high-quality NDVI time-series data was achieved using Long Short-Term Memory (LSTM) and the predictive reconstruction of MODIS NDVI time-series data, the results of which were well adapted to NDVI series trends [29]. Perceptual layer neural networks were used to extract vegetation cover from randomly selected training samples [30]. The LSTM deep learning framework was applied to spatio-temporal environmental dataset prediction based on experimental results from three marine datasets [31].

In this study, we aim to optimize the spatio-temporal integrity of FVC results and produce an accurate fine-scale vegetation dataset at a regional scale. All Sentinel-2 and Landsat 8 observations within a single year were integrated to derive the 30 m FVC, which contained gaps, at least twice a month. We compared three common reconstruction methods by applying them to the incomplete 30 m FVC. We selected the method with the highest accuracy at sample sites and optimized the method in terms of the model input and parameters to satisfy a regional application. Finally, the optimized method was selected and applied to produce the 30 m/16 d FVC of the study region.

2. Study Area and Materials

2.1. Study Area

Hubei Province is located in the middle Yangtze River basin of central China, and it includes the Jianghan Plain for cropland. The vegetation cover types in Hubei are mainly deciduous and evergreen forests and croplands. To compare the accuracy of different reconstruction methods, we selected three sample sites in Hubei Province representing different types of vegetation (Figure 1). The method with the highest accuracy was further applied to the entire province for regional analysis.

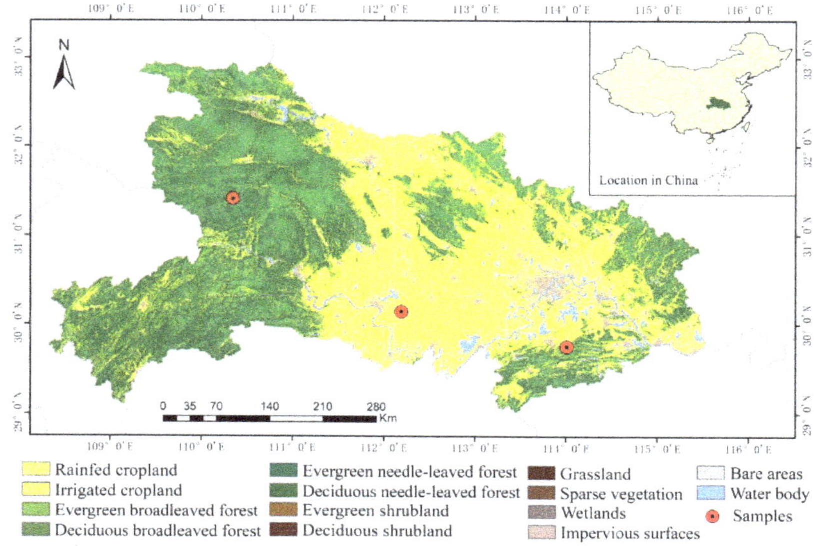

Figure 1. The study areas and the major land cover types of Hubei province.

2.2. *Fine-Resolution Satellite Data and Preprocessing*

2.2.1. Landsat 8 Data and Preprocessing

Launched in February 2013, the Operational Land Imager (OLI) sensor is a high spatial resolution multi-spectral imager onboard the Landsat 8 satellite in a sun-synchronous orbit (705 km altitude) with a 16-day repeat cycle. Landsat 8 data are suitable for pixel-level time-series analysis [32]. Thus, using Landsat surface reflectance data as the data source, the Landsat FVC product was developed based on the RF algorithm to select training samples. Landsat FVC has a spatial resolution of 30 m on a global scale and a temporal resolution of 16 d, covering the period from 2013 to the present [20]. In this study, we used 27 Landsat 8 images acquired in 2017 covering 3 sample sites to test reconstruction methods and 129 images covering the entire province for regional FVC retrieval.

2.2.2. Sentinel-2 Data and Preprocessing

The Sentinel-2A/B product available for users is Level-1C data (https://scihub.copernicus.eu, accessed on 1 October 2022), which refers to top-of-atmosphere reflectance in cartographic geometry in the UTM/WGS84 projection, with a size of 100 km × 100 km. The study area was covered by one Sentinel-2 tile (T49RGP). Sentinel-2 L1C data covering 2016 to 2018 in the study area with less than 10% cloud cover was used. Sentinel-based FVC data were generated using the RF model.

Sentinel-based FVC was generated in two steps: preprocessing and FVC estimation. The preprocessing of the Sentinel-2 data included atmospheric correction, spatial resampling, BRDF normalization, and band normalization. The European Space Agency (ESA) recommends free, open-source Sentinel Application Platform (SNAP) toolboxes developed by the ESA for the scientific exploitation of Sentinel-based information. The Sen2Cor algorithm in the SNAP toolbox, version 2.4.0, was utilized for atmospheric correction. It eliminated the effects of the atmosphere and delivered the Level-2A product, which is a bottom-of-atmosphere reflectance in cartographic geometry. Band-pass normalization was used to reduce the small difference between the spectral bands of MSI and OLI sensors [33]. Next, the preprocessed Sentinel-2 images were used as input to produce the FVC using the same RF model as the Landsat FVC [20]. In this case, we manually selected 61 Sentinel-2 tiles acquired in 2017, covering the 3 sample sites, as the data source to produce the original fine-resolution FVC. For regional analysis, 548 Sentinel-2 tiles were utilized.

2.3. Coarse-Resolution Satellite Products

CGLOPS produced a global 300 m resolution FVC product (FCover 300) by applying a dedicated neural network to instantaneous top-of-canopy reflectance from Sentinel-3 OLCI (v1.1 products) and daily top-of-aerosol input reflectance from PROBA-V (v1.0) from 2014 to the present (https://land.copernicus.eu/global/products/fcover, accessed on 1 October 2022). The GEOV2 (v2.0) FVC product was derived through the Geoland2/BioPar project from SPOT/VEGETATION (SPOT/VGT) and PROBA-V (GEOV2/PV) observations at 1 km and a 10 d step, extending the temporal coverage of global FVC from 1981 to the present. The FVC estimated from PROBA-V was based on neural networks [12].

GLASS products include Leaf area index (LAI), Fraction of Absorbed Photosynthetically Active Radiation (FAPAR), Broadband Albedo (albedo), Broadband Emissivity (BBE), Land Surface Temperature (LST), Daily Net Radiation (NR), Evapotranspiration (ET), Gross Primary Production (GPP), Net Primary Production (NPP), and Air Temperature (AT). GLASS products have been demonstrated to be of high quality and accuracy and are widely used [34]. This study used data from each GLASS product from 2016 to 2018 and selected GLASS products with similar (no more than three days) dates to the Landsat-based and Sentinel-based FVC data as training samples. Furthermore, we resampled the resolution of each GLASS product uniformly to 1 km as multivariate features were added during the deep learning training.

3. Methodology

3.1. Research Framework

In this study, the spatio-temporal reconstruction of fine-resolution FVC was based on data fusion, time-series fitting, and a deep learning approach. These methods were evaluated to identify the most efficient methods for spatio-temporal reconstruction. A schematic overview of the reconstruction approach is illustrated in Figure 2, which includes three key steps. First, the preprocessed Sentinel-2 data were used to generate the Sentinel-2-based FVC using the RF model. Then, Landsat and Sentinel FVC images with a 30 m spatial resolution were reconstructed by applying different spatial and temporal reconstruction methods. These included STARFM, S-G filtering, and different LSTM models with changing variables. Finally, the accuracy of the predicted FVC was estimated, and the advantages and disadvantages of these three methods were assessed. The predicted results were consolidated at a resolution of 30 m to reconstruct the 30 m FVC.

3.2. Reconstructing FVC Using STARFM

To optimize the temporal continuity of 30 m FVC, we used the Spatial and Temporal Adaptive Reflectance Fusion Model (STARFM) algorithm [22] to combine a pair of coarse- and fine-resolution images at moment t_1 and coarse-resolution image data at the moment t_2 and simulated and predicted the fine-resolution image data at the moment t_2 with different spatial weights. The formula is as follows:

$$L\left(x_{\frac{w}{2}}, y_{\frac{w}{2}}, t_2\right) = \sum_{i=1}^{w}\sum_{j=1}^{w}\sum_{k=1}^{n} w_{ijk} \times (M(x_i, y_i, t_2) + L(x_i, y_i, t_1) - M(x_i, y_i, t_1)) \quad (1)$$

where $(x_{w/2}, y_{w/2}, t_2)$ represents the central pixel to be predicted by the moving window at the moment t_2, w is the size of the moving window, w_{ijk} is the weight factor of pixels in the window similar to the central pixel, and L and M denote the pixel values of the fine-resolution image and the coarse-resolution image, respectively. The fusion method predicted the target pixel using the uniqueness of the spectrum, uniqueness of time, and relative distance between neighboring pixels and the target pixel. This was incorporated to give weight to the central pixel value at the moment of prediction. The FVC data in this study were input into the model with a window size set to 25. Additionally, pixels with spectral similarity to the central pixel FVC were sought in the 30 m FVC data based on spectral correlation and then combined with different weights and conversion coefficients to calculate the predicted value of the central pixel.

Here, using a subset of T49RGP as an example, the full year 2017 was used for this study, with 7 tiles of Landsat FVC, 14 tiles of Sentinel FVC, and 36 tiles of FCover 300 (1 tile every 10 d, from 1 January 2017, to 31 December 2017). After normalizing the DN values and ensuring that the two pairs of images cover the same spatial extent, we used Landsat FVC products based on Landsat and Sentinel-2 (30 m) as the fine-resolution image data and FCover (300 m) as the coarse-resolution image data to predict the FVC values in a particular region on a particular day.

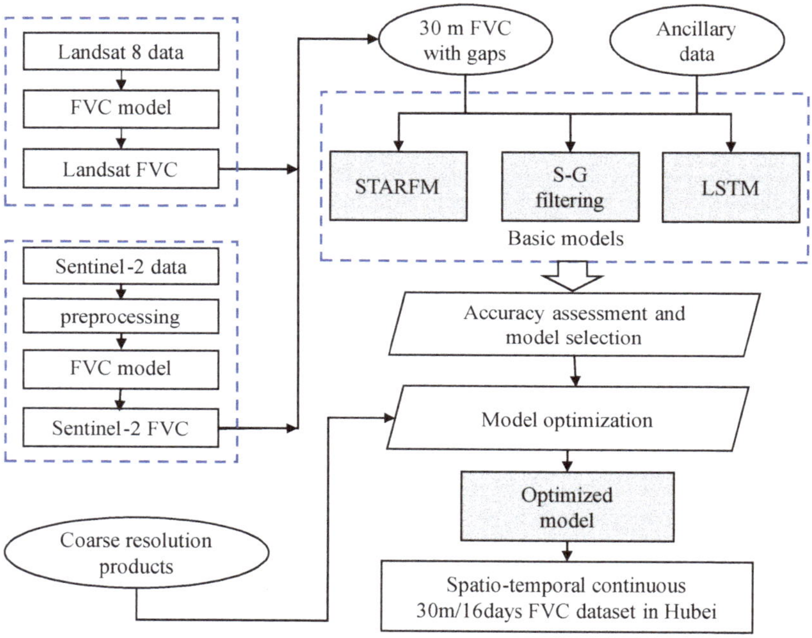

Figure 2. The framework of spatio-temporal reconstruction of 30 m FVC using different methods.

3.3. *Reconstructing FVC Using S-G Filtering Method*

The S-G filtering method, also known as the data-smoothing polynomial filtering method, is a convolution algorithm based on smoothed time-series data and the principle of least-squares [35]. The algorithm can be described as follows:

$$Y_j^* = \frac{\sum_{i=-m}^{m} C_i \times Y_{j+i}}{N} \quad (2)$$

where Y_j^* is the fitted value, Y_{j+i} is the original value of the image element, C_i is the coefficient when the first i value is filtered, m is the width of the filter window, and N is the filter length, which is equal to the width of the sliding array, $2m + 1$. In practice, S-G filtering requires two parameters to be set: the filter window and the order of polynomial operations. Based on the characteristics of the S-G filter, setting a filter window length that is too small may overfit the data points, making it difficult to capture the trend of the entire time series. However, setting a filter window length that is too large may miss some significant changes in the FVC time series. Different polynomial orders can also affect the fitting results of the S-G filter, and this parameter is generally set from two to four.

Considering the importance of temporal details in the FVC time series for phenology studies, we conducted a study with different combinations of w and d and finally chose a parameter combination of w = 7 and d = 2. This parameter choice provides an excellent trade-off between retaining temporal variation details and the overall time-series trend. Considering the edge effect seen in the S-G filter, the 0 m FVC during the period from December 2016 to January 2018 was employed to constitute a complete filter window for the first four points and the last four points of the NDVI time series. To evaluate the performance of the S-G filter, a subset of the T49RGP (113°37′–113°43′E, 29°43′–29°48′N) was used to construct a high-quality FVC time-series dataset.

Here, using Sentinel-based and Landsat-based FVC as input, we filled in the time series of FVC at 30 m resolution. A total of 31 scenes of the T49RGP from 2017 were selected as the time series. They were entered into the S-G filter model for spatio-temporal reconstruction. As the data were processed using Fmask4.0, the empty values caused by clouds and shadows were flagged as no values. Thus, we could reconstruct the data values in the image directly. Additionally, for scenes with missing dates, the entire scene was reconstructed as if it were all no-data values.

3.4. *Reconstructing FVC with LSTM and Optimized Parameters*

3.4.1. The LSTM Method

Time-series data prediction refers to learning patterns of change in past time series and predicting future trends. Traditional deep learning methods are unable to solve the problem of change over time, which has resulted in the development of RNNs. However, because RNNs perform poorly in solving long-term-dependent information, Hochreiter et al. proposed LSTM by adding the structure of gates to the RNN to selectively add or remove time-series information [36]. The LSTM cell states consist of two activation functions (sigmoid and tanh), which constitute the forgetting, input, and output gates. The sigmoid function outputs the input parameters as values between 0 and 1. These values are used to determine the impact on the model, with a maximum impact of 1. The tanh function outputs the input parameters as values between −1 and 1, which are used to normalize the results.

3.4.2. The Bi-LSTM Method

Bi-LSTM is another alternative RNN containing two separate recurrent net layers: forward and backward layers [37]. Bi-LSTM combines the strengths of RNNs and LSTM by applying LSTM-gated memory units to a bidirectional propagation model. The Bi-LSTM network has the advantage of using information from the entire input time-series sequence when predicting the current target, effectively solving the disadvantage that LSTM can use only information from the previous moment. Figure 3 illustrates LSTM and Bi-LSTM structures. Where *Ct* represents the long-term memory of the cell, *ht* represents the short-term memory of the unit, and *FVCt* represents the external input. It can be seen that *Ct* passes through each cell by deleting unimportant information and adding relevant information thereafter. In general, short-term memory, which is used as the final prediction, is the result of the interaction between *Ct*, *ht-1*, and *FVCt-1*.

To build the dataset for model training, the corresponding 2016 to 2018 time-series 30 m FVC data from the study area were extracted as training data. Owing to the small training set, the dropout method was used to randomly remove neurons during the learning process to prevent overfitting. The time interval of the training data for the input model was set to 16 d. Additionally, the masking layer incorporated the missing dates into the model to automatically skip them during training. This was to ensure that the dataset for the training model had the same time interval.

In this study, LSTM and Bi-LSTM were used to train the model. These two methods were designed as one input layer, one LSTM/Bi-LSTM layer, one masking layer, one dropout layer, and one output layer. The LSTM and Bi-LSTM models were trained using Landsat and Sentinel-based FVC time-series data. The ratio of the training set to the test set was 7:3.

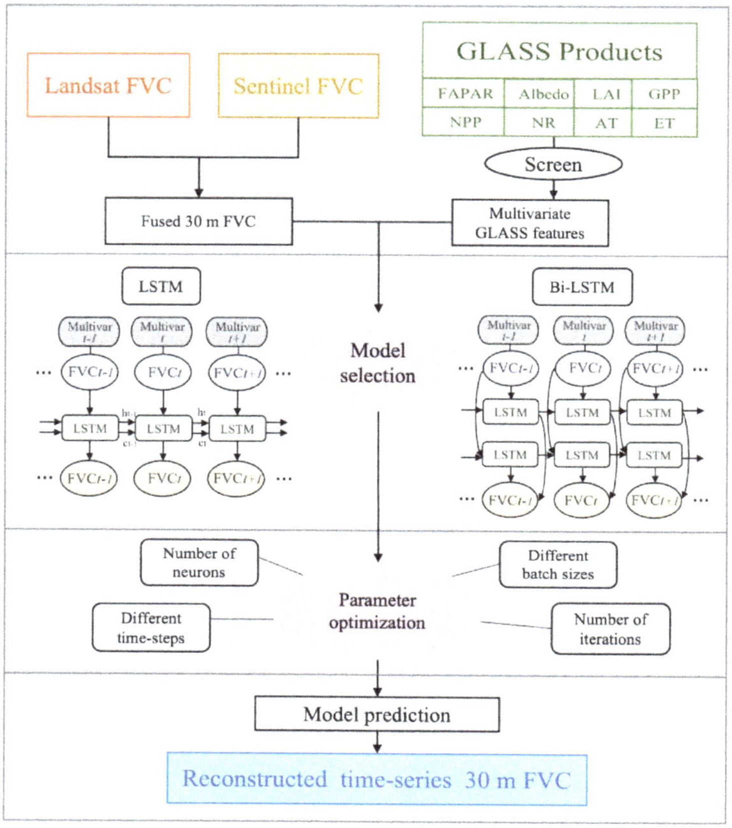

Figure 3. The structure of the LSTM/Bi-LSTM network.

3.4.3. Changing the Time Steps

To investigate the impact of the LSTM model's time step on reconstruction performance, we conducted separate experiments comparing single and multiple time steps. The single-step prediction (one-to-one) model takes the previous time step as input and predicts the next subsequent value in the time series, while the multi-step prediction (many-to-one) produces a predicted output using multiple inputs from the previous time steps. By comparing the estimated R^2 and RMSE performance of the one-step (1SP) and multi-step (3SP) models, we can determine which time step scheme yields the best performance.

3.4.4. The Inclusion of Various Input Variables

Furthermore, it was found that LSTM based on multiple feature inputs can improve prediction accuracy [38]; whereas, we currently only use FVC as a single variable to predict the model. In this study, we added GLASS products as multivariate data to the original model for multivariate LSTM prediction. We conducted statistical analyses of the data for the different factors for which GLASS products were selected as multiple variables. Based on the results of the analysis, we removed factors with low or insignificant trends from the time series. We retained factors with a high correlation with FVC. Univariate and multivariate models with different factors were trained separately, and their accuracies were compared. Here, we selected GLASS product data from 2016 to 2018, and the date of each product was kept consistent with the date of the FVC training data. Ideally, the difference in timepoints should not exceed three days. In addition, to simplify the calculation and training, we resampled all GLASS products at a spatial resolution of 1 km. To compare different GLASS variables, we employed the Phik (φk) coefficient, a new and practical correlation coefficient proposed by Baak et al. [39]. It is a practical correlation coefficient based on refinements to the Pearson correlation coefficient, which captures the nonlinear dependency, and is useful when measuring the correlation of different variables.

3.5. Validation of the Reconstructed FVC

Whether the FVC generated by spatio-temporal reconstruction can accurately and truly reflect the missing data of Landsat FVC needs to be confirmed by validity analysis. In this study, three approaches were adopted: (1) First, we used visual verification of the consistency of the spatial distribution at the image level. (2) Second, assuming the real date FVC as the missing date FVC to be predicted, the results predicted by the three methods were compared with the real FVC in a scatter plot. The closer the distribution of points in the scatter plot is to a 1:1 line, the less variability in the spatial distribution, and the more accurate the prediction. The validation of the results was further tested by comparing the fitting of the prediction results of the different methods with the true values of the time-series curves. (3) Third, we compared the reconstructed FVC with the available FVC product in the time-series curves to verify its consistency with the reference FVC.

To evaluate the accuracy of each method, we used the R^2 and RMSE values calculated based on the real FVC and predicted FVC. These values were calculated using the following equations:

$$R^2 = 1 - \frac{\sum_i \left(FVC_{i\_pred} - FVC_{i\_real}\right)^2}{\sum_i \left(FVC_{i\_pred} - FVC_{i\_aver}\right)^2} \tag{3}$$

$$RMSE = \sqrt{\frac{1}{num\_p} \sum_i \left(FVC_{i\_pred} - FVC_{i\_real}\right)^2} \tag{4}$$

where i is the number of testing samples; FVC_{i_pred} and FVC_{i_real} denote the predicted and real FVC for sample i, respectively; and FVC_{i_aver} is the average of all FVC predictions.

4. Results

4.1. Consistency between the Landsat and Sentinel-2 FVC

The comparison between the Sentinel-based and Landsat-based FVC is shown in Figure 4 in terms of image level and scatter plot over various vegetation cover types. The correlation was 0.83 for the summer woodland (Figure 4a) and 0.82 for grassland (Figure 4b). The high correlation between the two FVCs indicates that Landsat and Sentinel-2 FVCs were overall consistent after band normalizations. The Landsat FVC had slightly higher values than the Sentinel-2 FVC, likely due to terrain effects in mountainous regions. The overall FVC values were lower than for cropland. In Figure 4c, the window was mainly featured with rainfed and irrigated cropland, and the FVC values were concentrated within the lower ranges (10–30%). The overall correlation was 0.732, which is slightly lower than other vegetation types. The Sentinel-2 FVC was slightly lower than the Landsat FVC. The points that deviated from the 1:1 line were possibly owing to residues in the band normalization and uncertain FVC values for cropland. At the image level, the Landsat-based and Sentinel-based FVCs have similar spatial distributions across different vegetation types. The results demonstrated that the combination of FVC from Landsat and Sentinel-2 can be used as input to reconstruct the time-series 30 m FVC.

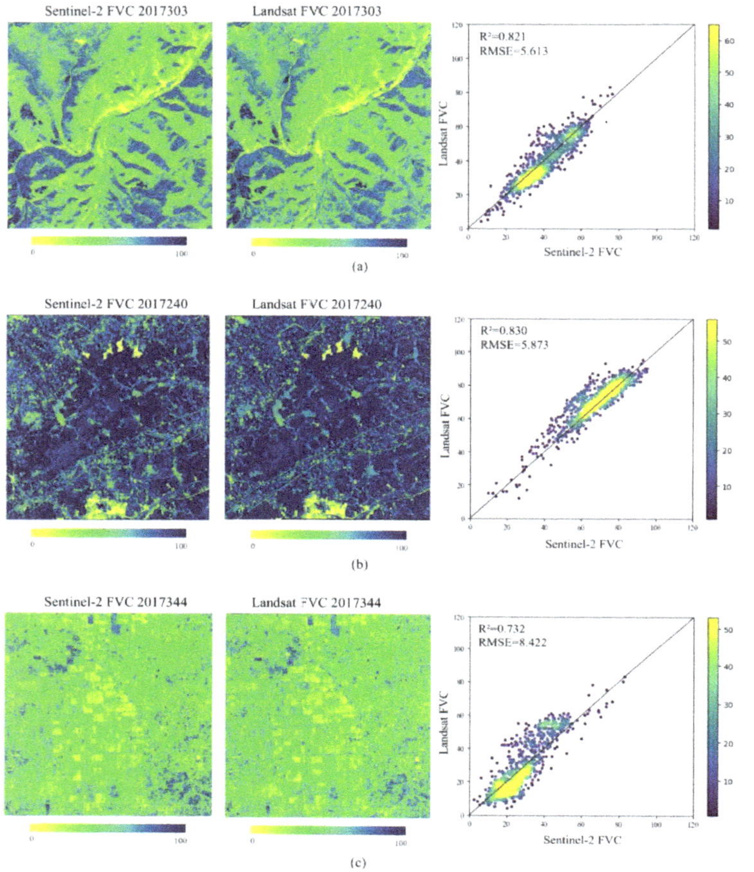

Figure 4. Comparison between the Landsat and Sentinel-2 FVC acquired for the same date over (**a**) forest, (**b**) grassland, and (**c**) cropland.

4.2. Comparison of Different Reconstruction Methods

4.2.1. Accuracy Comparison of Different FVC Reconstructions

The results of the comparison between the real and predicted FVC values are illustrated for a subset of the data located in T49RGP. The scene located in this subset was mainly composed of cropland and evergreen broad-leaved forests. Visually, the reconstruction results predicted by the three methods presented similar overall spatial distributions to the real FVC values, successfully capturing the details of the FVC variations (Figure 5). While there were unrealistic areas in all the strategies, the LSTM method generated the most reliable prediction. The distinction between low and high FVC values in the STARFM results was not as clear as that in the other methods. This was especially true in forested and non-forested areas, because STARFM uses coarse spatial resolution FVC as reference data and cannot accurately distinguish vegetation from non-vegetation. Both the S-G filter results and the STARFM results underestimated high FVC values to varying extents, while the LSTM results showed improved predictions for high FVC values, but slightly overestimated low FVC values, such as bare soil and cropland in the northwest of the subset and forest in the north.

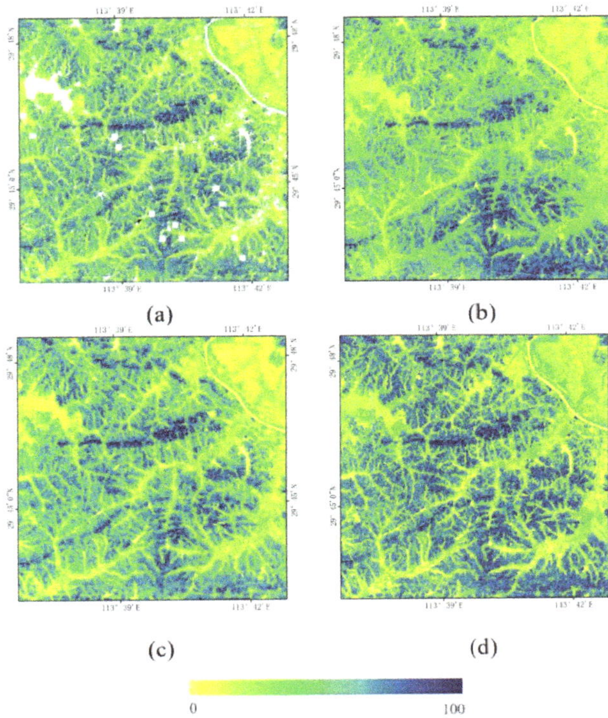

Figure 5. Comparisons between real FVC and predicted FVC on 24 December 2017: (**a**) is the real FVC with gaps; (**b**) is the FVC predicted by the STARFM method; (**c**) is the FVC predicted by the S-G filter; and (**d**) is the FVC predicted by LSTM.

To further evaluate the performances of these three strategies, we verified the prediction results by comparing the scatter plot with real data. Overall, the pixel-to-pixel validations showed significant consistency between the real and predicted FVC values (Figure 6). In terms of prediction accuracy, LSTM (R^2 = 0.857) performed better than STARFM (R^2 = 0.709) and S-G (R^2 = 0.705). Compared with the S-G filter and LSTM, STARFM had more points distributed far from the 1:1 reference line, and more points appeared to be biased. The dense distribution of points in the LSTM results in a low-value range of 10 to 30. Additionally, the high-value range of 50 to 70 indicates that the LSTM was able to effectively distinguish between high and low values. This characteristic was also found in the S-G filter results, where the points were more densely distributed in both intervals (Figure 6). However, the points in the STARFM results were less sensitive to the high-value range than the other two methods (Figure 6), probably because of the incorporation of coarse-resolution data.

4.2.2. Time-Series FVC Derived from Different Reconstruction Methods

We performed spatio-temporal reconstructions on different samples using three reconstruction strategies. In the results of the consistency analysis of the FVC time series of randomly selected pixels (Figure 7), the black line represents the time series of the real FVC data, and the other three colors represent the time series of the results obtained using different time reconstruction methods. The overall fitting results of the three methods reflected the normal growth conditions and seasonal change patterns of the vegetation. Among them, LSTM had the most accurate fit (RMSE = 8.175), followed by STARFM (RMSE = 19.628) and S-G filtering (RMSE = 22.857). The results of the small subset spatiotemporal reconstruction indicated that all three strategies were effective in reconstructing the FVC time series.

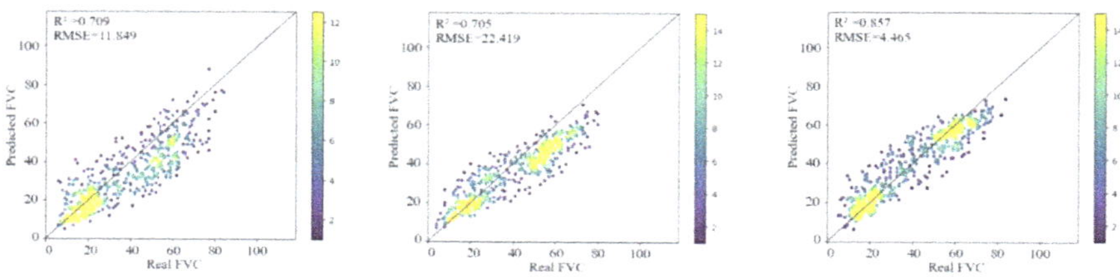

Figure 6. The scatter plot comparisons of real FVC and predicted FVC on 24 December 2017: the (**left**) plot is the real FVC compared with STARFM-predicted FVC; the (**middle**) plot is the real FVC compared with S-G-filter-predicted FVC; the (**right**) plot is the real FVC compared with LSTM-predicted FVC.

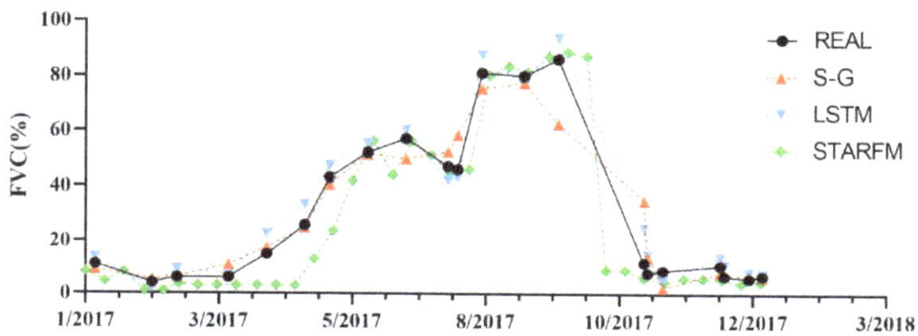

Figure 7. Time-series curves from real data and from other time reconstruction methods.

Additionally, FVCs reconstructed from three methods are capable of revealing the annual variation and phenological characteristics of various vegetation types. The temporal resolution of the reconstruction results was improved. In addition, the number of reconstructed images increased owing to the integration of the STARFM method with the 10 d resolution FCover FVC data. However, there were several abruptly changing values in the STARFM reconstructions (e.g., April, 15, 19 July and 10 October), which may be partially due to the quality of the referenced coarse-resolution data and may also be due to differences in satellite images before and after crop harvest. The reconstructed results of the S-G filtering method were solely based on Sentinel FVC and Landsat FVC as inputs, without other reference data. Therefore, the S-G filter is limited by the original time-series data, and the prediction results may be affected by the absence of the original data. The Bi-LSTM deep learning model runs more slowly than the other two algorithms but solves the problem of noise (contamination from clouds and snow) or missing values contained in the FVC data more efficiently than other methods. This is because Bi-LSTM can effectively extract information from the entire FVC time series, rather than a single point in time, which means that the Bi-LSTM model enables us to capture the dependencies of the FVC time sequence data and predict the missing or future FVC values with high accuracy.

4.3. *Accuracies of LSTM Models with Changing Parameters*

4.3.1. Accuracies of LSTM and Bi-LSTM

The performance of the deep learning model in reconstructing the 30 m time-series FVC was assessed using two metrics, the R^2 and RMSE. We trained the deep learning model using an I80V598 desktop computer (Intel i7, 4 CPU, RX550 GPU, 3.6 GHz, 24 GB RAM). To obtain the optimal parameters, we performed a tuning experiment. The results showed that the model predicted accurately when the number of neurons was 128 and the number of iterations was 2000. We trained both the LSTM and Bi-LSTM models with the training dataset separately, and the training results showed that Bi-LSTM performed more effectively, as assessed by both R^2 and RMSE (Figure 8). This is because the Bi-LSTM model can use information from the entire input time series when predicting the current target, effectively solving the drawback that traditional LSTM can only use information from the previous moment.

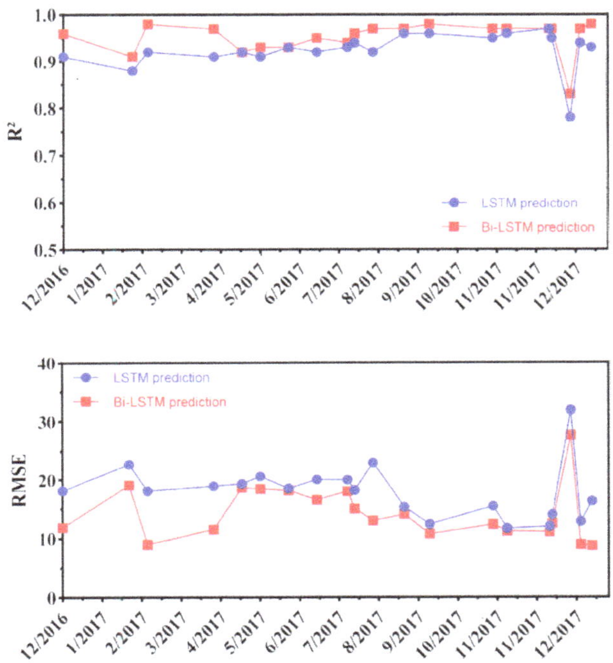

Figure 8. Comparison of the validation results of the LSTM model (blue) and Bi-LSTM model (red).

4.3.2. Accuracy of Changing the Time Step

A comparison of the prediction results with different time steps (Figure 9) shows that the prediction results using three time steps (3SP) were more accurate than the prediction results with a single time step (1SP). This is because single-step prediction is based on the most recent data and cannot be applied across input or output time steps, whereas multistep prediction is based on multiple data points and has more temporal dependency characteristics.

4.3.3. Accuracy of Changing Input Variables

According to the statistical results of the Phik (φk) correlation coefficient (Figure 10), FAPAR, Albedo, LAI, GPP, and NPP were highly correlated (here, GPP and NPP were removed because the same vegetation indices were used for GPP, NPP, and FVC), indicating that they can be used as training features for multivariate analysis. We trained the LSTM neural network based on different features. A comparison of the training results (Table 1) shows that the accuracy of the model with GLASS LAI data added is higher than that of the original model. However, this was not the case when there were more variables. When variables such as LAI, Albedo, FAPAR, AT, ET, NR, BBE, and LST were added to the model as multivariate features, model accuracy decreased. The model accuracy was highest when LAI, Albedo, and FAPAR were selected as multivariate features, in accordance with the correlation results.

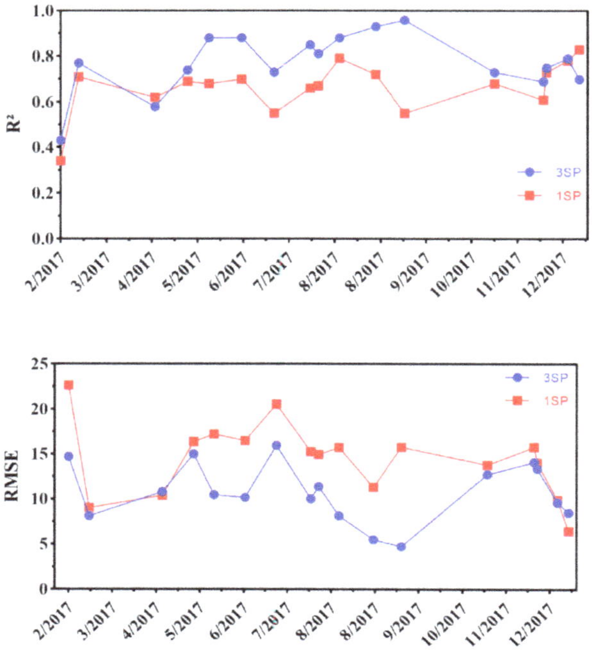

Figure 9. Model validation results with different time steps (blue line is the result of three steps, red line is the result of one step).

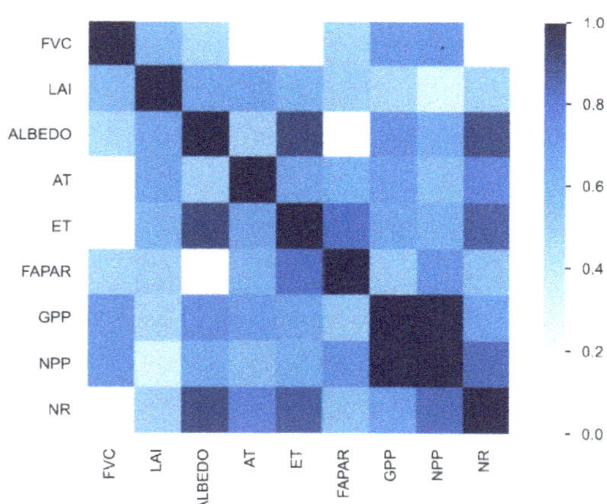

Figure 10. Phik (φk) correlation coefficients for different GLASS products and FVC.

Table 1. LSTM model accuracy results with different input feature variables.

Statistical Test	Feature			
	FVC	**FVC, LAI**	**FVC, LAI, Albedo, FAPAR**	**FVC, LAI, Albedo, AT FAPAR, ET, NR, BBE, LST**
R^2	0.94	0.97	**0.98**	0.94
RMSE	5.022	3.012	**2.797**	4.696

4.4. Reconstructed Time-Series FVC in Hubei Using the Optimized Model

4.4.1. Spatial Performance of the Reconstructed FVC Image

In the spatial domain, the proposed method successfully reconstructed FVC values of gap pixels over different landscapes (Figure 11). The optimized Bi-LSTM model rebuilt the spatial characteristics omitted in gaps by learning from fine-resolution images at neighboring dates. The spatial details, such as roads, terrain ridges, and field boundaries, were presented in the reconstructed images. By contrast, the original FVC values were substituted by the model estimation derived from FVCs at forward and backward dates, as well as the temporal trends of coarse-resolution products. Thus, the similarity between neighboring pixels of the reconstructed images was reduced in comparison to the original images.

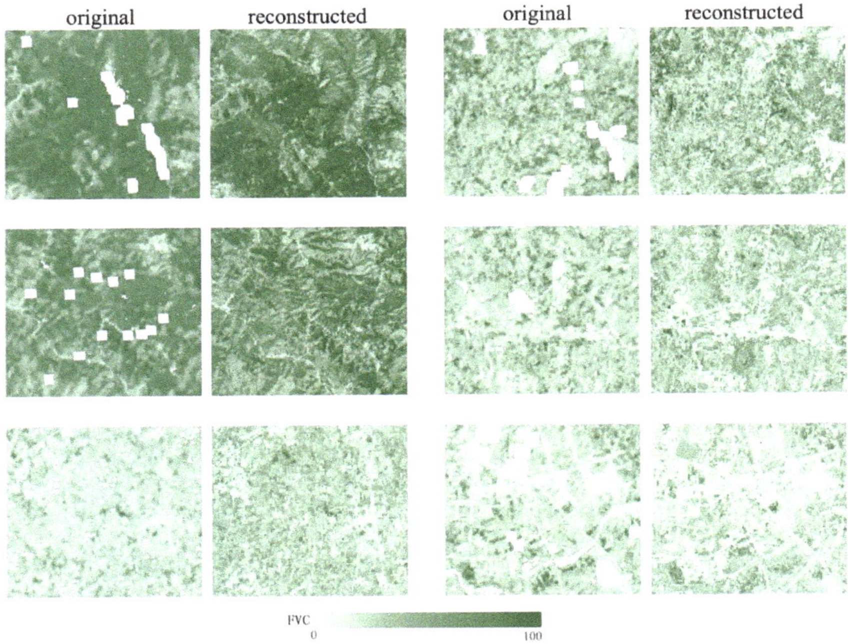

Figure 11. The image pairs of 30 m FVC before and after reconstructions using the optimized Bi-LSTM method with multiple variables.

4.4.2. Temporal Trend of the Reconstructed FVC Pixels

At the pixel level, the reconstructed 30 m FVC differentiated seasonal trends of different vegetation types (Figure 12). Although grassland and forest pixels corresponded to a single growing season per year, the peak FVC of forests was higher than that of other vegetation types. Double cropping and single cropping were both practiced in Hubei Province in 2017 [40], as captured by the reconstructed FVC (Figure 12). The seasonal trends showed that FVC at fine resolution was close to that captured by coarse-resolution FVC products and characterized by similar start- and end-of-season dates (Figure 12). The peak occurred in July, likely owing to grasses being restored between double cropping periods [40]. Regarding the three major vegetation types in Hubei, the 30 m FVC values were usually closer to GLASS FVC, which was lower than CGLOPS and GEOV2 FVC during the growing season and higher than CGLOPS and GEOV2 FVC from November to March. Furthermore, the reconstructed FVC values of grass and forest pixels fluctuated during the peak growing periods, which was likely caused by the difference between Landsat and Sentinel-2 FVC, as well as the fluctuating trends introduced by the coarse-resolution products used in the reconstruction model.

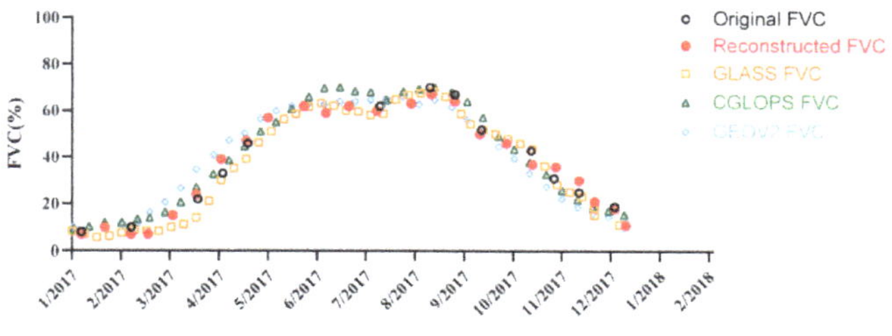

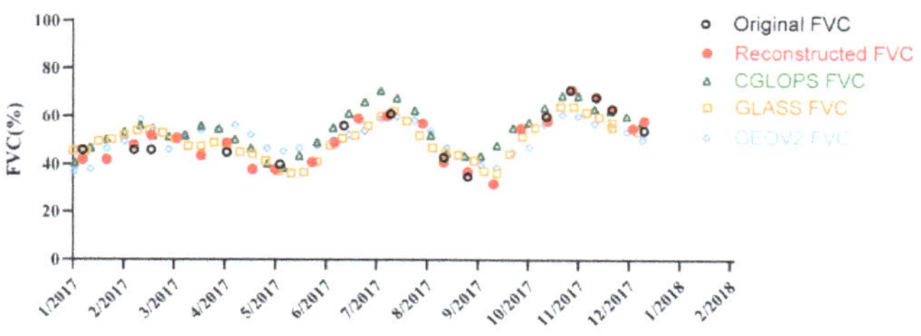

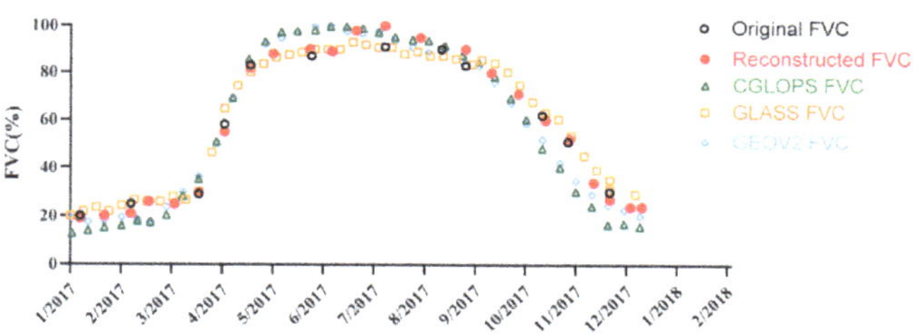

Figure 12. Comparison of the reconstructed FVC in 2017 using the optimized multivariate Bi-LSTM model to three reference FVC products for different vegetation types, including grassland, cropland, and forest. The 30 m FVC has been aggregated to 500 m for the purpose of comparison.

4.4.3. Regional Accuracy of the Reconstructed FVC

The 30 m/16 d FVC time series of Hubei Province in 2017 was reconstructed using the optimized Bi-LSTM model. From the reconstruction results, the FVC mosaics in Hubei illustrated spatial and seasonal variations of different vegetation covers (Figure 13). The forests located in the west of the province usually have much higher FVCs than croplands in central regions. The contrast shrank until the middle of the year, when crops, especially rice, cotton, maize, etc., were at their peak growing stages and had high FVCs. The reconstructed FVC and three reference FVC products, including GLASS, GEOV2, and CGLOPS FVC, showed a high degree of consistency across different vegetation types. Statistically, the R^2 ranged from 0.4 to 0.8 and the RMSEs were generally less than 10 (FVC values ranged from 0 to 100). The R^2 for cropland (0.670–0.709) and grassland (0.806–0.857) was higher than that for broadleaf forest (0.402–0.673), and needleleaf forest had the lowest R^2 (0.332–0.444). The

R^2 and RMSE were stable for forests; however, the R^2 decreased rapidly from 0.70 to 0.589 for crops from January (Figure 14) to July (Figure 15). This is likely because different types of crops were in different growing stages in July, causing uncertainties in FVC estimations for crops. The reconstructed FVC presented the highest consistency with the GLASS FVC product for both seasons and vegetation types (Figures 14 and 15). This suggests that the reconstructed FVC can effectively capture information on seasonal changes in various vegetation types. As a result, the time-series FVC reconstruction of the region demonstrates the feasibility of the proposed method for rebuilding a 30 m spatial resolution FVC at a larger scale.

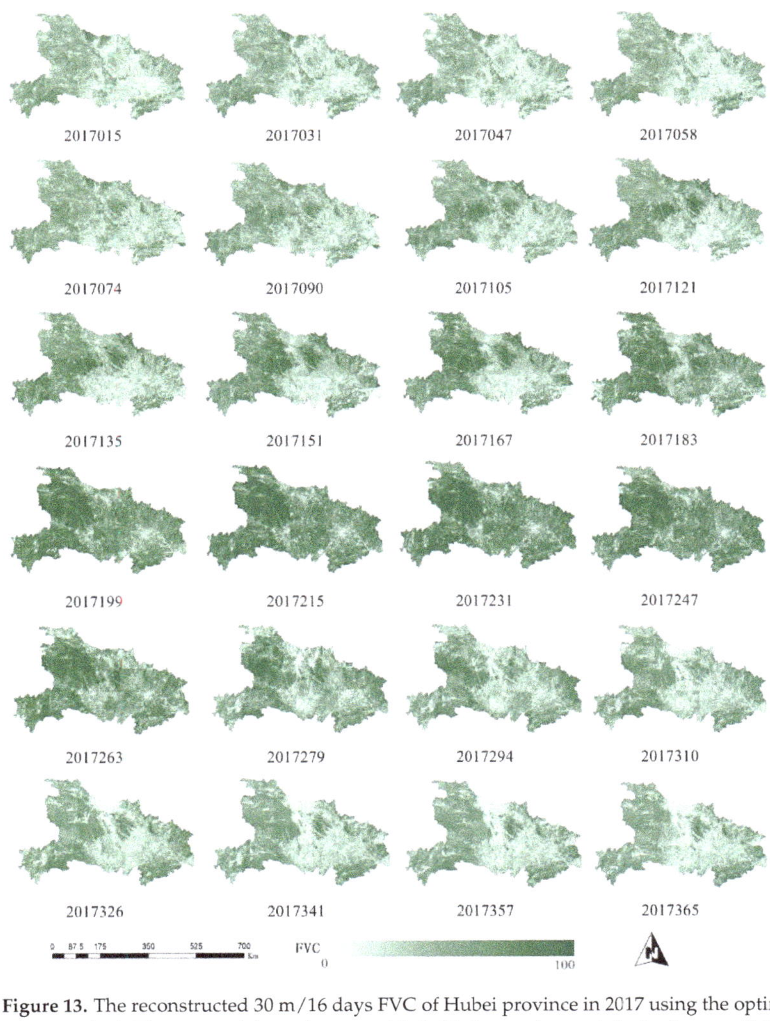

Figure 13. The reconstructed 30 m/16 days FVC of Hubei province in 2017 using the optimized Bi-LSTM method. The year and date-of-year are labeled below each mosaic in the format of YEARDOY.

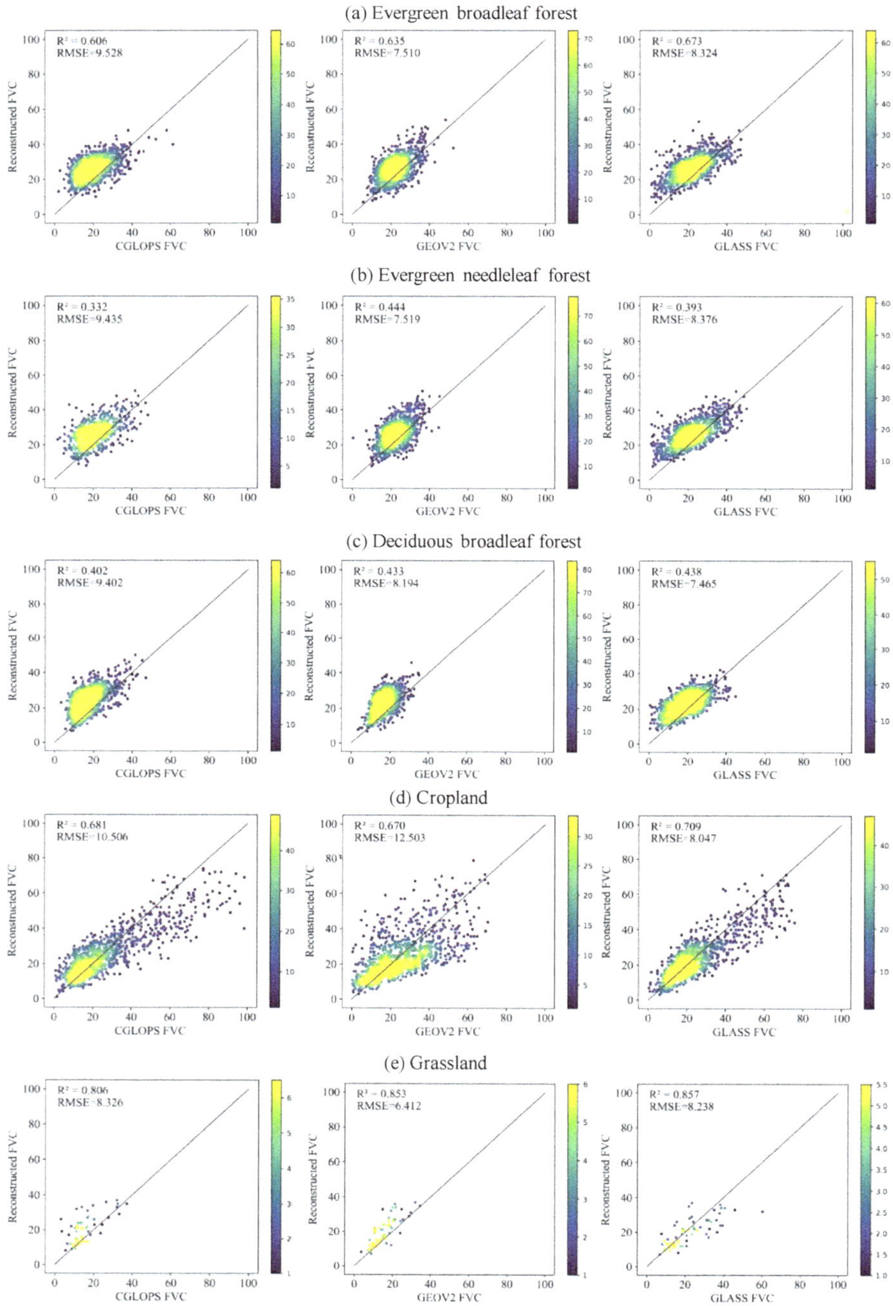

Figure 14. Comparison between the reconstructed FVC in January and coarse-resolution products for major vegetation types in Hubei. For illustration purpose, all FVC pixels have been aggregated to 1 km resolution.

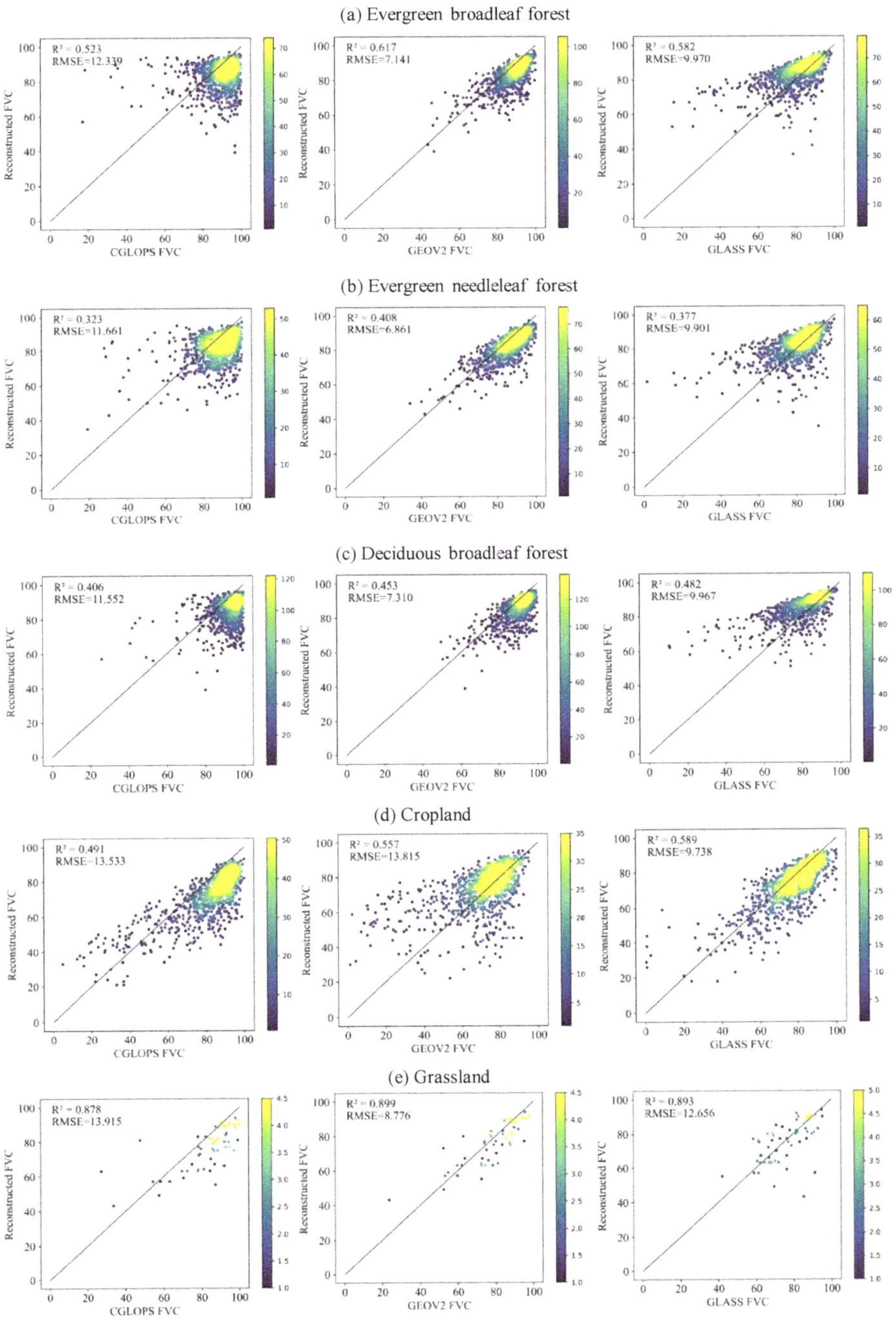

Figure 15. Comparison between the reconstructed FVC in July and coarse-resolution products for major vegetation types in Hubei. For illustration purposes, all FVC pixels have been aggregated to 1 km resolution.

5. Discussion

5.1. Implications of the Reconstructed 30 m FVC Dataset

Current FVC products, such as the GEOV2, GLASS, and CGLOPS, are usually generated at a spatial resolution of several hundred meters [12–14], enabling the large-scale analysis of vegetation dynamics while hindering the investigation in heterogenous landscapes. The Landsat FVC provides global long time-series FVC datasets at a 30 m spatial resolution [20], but it suffers from spatial incompleteness and temporal discontinuity caused by cloud cover and a long revisit cycle. To overcome these limitations and improve the efficiency of utilizing 30 m FVC imagery, this study compared different reconstruction methods and proposed a deep learning based method to develop 30 m FVC products. The reconstructed FVC dataset is complete in a spatial extent and continuous with a time interval of ~16 d. The reconstructed 30 m FVC in Hubei Province can greatly improve the current understanding of spatial and temporal dynamics of vegetation in vital ecological regions, which previously was derived from coarse-resolution products [41]. By applying the proposed reconstruction method to the entire Landsat 8 FVC, a time-series, 16 d FVC dataset can be derived for the past decade. Such a fine-resolution dataset is highly desired by researchers from the field of urban plant phenology, who address vegetation characteristics and changes caused by long-term urbanization processes in complex urban landscapes [42,43].

5.2. Uncertainties of Different Spatio-Temporal Reconstruction Methods

Each of the three reconstruction methods had advantages and limitations. When using the STARFM model for prediction, the combination of the spatial resolution of Landsat FVC and the temporal resolution of CGLOPS FVC allows the reconstruction results to reflect the dynamic changes in vegetation growth with high accuracy. Owing to the difference in temporal resolution between the reference data and the original 30 m FVC data, there may be a bias of several days in the selection of matched images, which is one of the reasons for the uncertainty in the reconstruction results. Coarse-resolution data cannot accurately distinguish vegetation and non-vegetation on the surface, causing the STARFM predictions to fail to distinguish between high and low FVC values. In addition, the reconstruction of the STARFM method is highly dependent on coarse-resolution reference data; without reference data, the STARFM method cannot make predictions. However, a more effective ESTARFM algorithm is available [44]. Given the small amount of raw data and large date interval, which does not effectively provide input data for both time pairs, the ESTARFM algorithm was not chosen for the reconstruction.

The uncertainty of the S-G filter reconstruction results is mainly embodied in the time interval of the input data, which can adversely affect reconstruction accuracy. In the case of long date intervals especially, S-G filters are unable to fit the time-series curve well since there are too few points in the raw time-series data input [35]. Therefore, the method is limited by the original time-series data; when the original data contains more gaps and poor time continuity, the S-G filtering method is no longer applicable. The overall accuracy of the first two models was slightly inferior compared to that of the deep learning methods. However, they have advantages such as being simple and easy to implement, having low computational complexity, high productivity, not requiring sample selection or training, and direct dependency of the accuracy of the model reconstruction on the quality of the original data. Other methods, such as the FSDAF [23] and GF-SG [21], can be utilized to test the accuracy of the FVC reconstruction.

Bi-LSTM reconstructions are the most accurate method for addressing noise and missing values in FVC data, such as contamination from clouds and shadows. This is because the Bi-LSTM network can efficiently extract information from the entire FVC time series and accurately predict missing or future FVC values with high precision [45]. Furthermore, the time efficiency of training is twice that of traditional LSTM. In this study, the training samples of the Bi-LSTM model were primarily pixels of various vegetation types (mainly evergreen broad-leaved forests and croplands) in tile T49RGP; therefore, the representativeness of the training samples and the transferability of the training model still need to be improved. To avoid the use of the Convolutional LSTM for image processing specifically, it was discovered that using the Convolutional LSTM would destroy the original spatial distribution structure of the surface vegetation because of the heterogeneity of the reconstructed regional surface. During future research, we hope to expand the sample area to provide a more representative sample for training LSTM models. In addition, we intend to use in situ weather station measurements (e.g., temperature and precipitation) as training features for multivariate LSTM models. Deep learning methods also have many advantages in large-scale and multi-temporal FVC mapping. These advantages can be summarized as follows: (1) the model can achieve spatio-temporal reconstruction with higher accuracy without using auxiliary data when making predictions; (2) the neural network model can ensure that the reconstruction interval is fixed, which means that the LSTM model can reconstruct high-accuracy FVC even during long periods of poor data quality or no observational data.

5.3. Further Improvements to the Proposed Method

In this study, the Sentinel-based FVC was produced using the RF model proposed by Song [20]. By comparing the FVC images under three typical vegetation covers, the consistency of the FVC produced by the two different sensors was verified. However, the model was trained using Landsat data, and despite the normalization operation, some differences were observed between the data obtained from the two sensors. In addition, when the GLASS product data were added to the Bi-LSTM model as multiple variables, we resampled each GLASS product to a spatial resolution of 1 km to simplify the computation, rather than one variable value per pixel, as should be the case in theory, which may be a source of uncertainty. In the future, we expect to use other parameters as multivariate training features to train the Bi-LSTM, aiming to ensure that it can capture more long-term and short-term time-dependent information and make the prediction model more robust. Furthermore, owing to differences in the temporal resolution of the sensors, all data cannot be obtained at the same time. This can lead to uncertainty during the model training and validation of the comparison. Owing to the lack of ground measurement data, the FVC in this study can only be compared with existing FVC products for accuracy verification. The reference FVC products (GLASS FVC, CGLOPS FVC, GEOV2 FVC) were selected based on their spatial coverage, time span, and spatio-temporal resolution, as well as the availability of data. However, these products have different spatial resolutions; therefore, the resolution was uniformly resampled to 300 m in the comparison, which may lead to uncertainty in the verification results [46]. While the Bi-LSTM model can produce smooth and continuous spatio-temporal predictions, the smoothing of noise in the time series may classify important ecological changes and disturbances as noise and filter them out, creating uncertainty [45]. Further research is needed to investigate how rapidly and abruptly changing FVC values may affect the model performance.

6. Conclusions

In this study, we compared three types of spatio-temporal reconstruction methods by applying them to the 30 m resolution FVC derived from integrated use of Landsat 8 and Sentinel 2 data. Among the three methods, the Bi-LSTM model was the most effective and had the highest accuracy, followed by the STARFM method and the S-G filtering method. In comparison to the data fusion and filtering method, the deep learning Bi-LSTM model captures the dependencies of the FVC time sequence data and can predict missing or future FVC values with high accuracy. Setting the proper time step in Bi-LSTM and inclusions of relevant variables, such as LAI and albedo, have further improved the prediction accuracy. Moreover, a 30 m FVC dataset of every 16 days in Hubei Province of China was produced using the optimized Bi-LSTM method. The derived time-series FVC dataset has shown the tremendous details of the spatial distribution of FVC values as well as the seasonal dynamics of different vegetation types in Hubei. It is concluded that the proposed model can effectively accomplish reconstruction of fine-resolution FVC in a province or larger spatial scale, and the generated 30 m FVC product can support many Earth system science studies.

Author Contributions: Conceptualization, D.-X.S. and T.H.; methodology, D.-X.S. and Z.W.; validation, Z.W., D.-X.S., J.L., C.W. and D.Z.; formal analysis, Z.W. and D.-X.S.; writing—original draft preparation, Z.W. and D.-X.S.; writing—review and editing, D.-X.S. and T.H. All authors have read and agreed to the published version of the manuscript.

Funding: This research was funded by the National Key Research and Development Program of China [2022YFB3903502], the National Natural Science Foundation of China [42271404], the Fundamental Research Funds for the Central Universities [CCNU22QN018], and the Natural Science Foundation of Hubei [2021CFA082].

Data Availability Statement: Not applicable.

Conflicts of Interest: The authors declare no conflict of interest. The funders had no role in the design of the study; in the collection, analyses, or interpretation of data; in the writing of the manuscript; or in the decision to publish the results.

References

1. Huang, L.; Chen, K.F.; Zhou, M. Climate change and carbon sink: A bibliometric analysis. *Environ. Sci. Pollut. Res.* **2020**, *27*, 8740–8758. [CrossRef] [PubMed]
2. Jia, K.; Liang, S.; Wei, X.; Yao, Y.; Yang, L.; Zhang, X.; Liu, D. Validation of Global LAnd Surface Satellite (GLASS) fractional vegetation cover product from MODIS data in an agricultural region. *Remote Sens. Lett.* **2018**, *9*, 847–856. [CrossRef]
3. Sha, Z.; Bai, Y.; Li, R.; Lan, H.; Zhang, X.; Li, J.; Liu, X.; Chang, S.; Xie, Y. The global carbon sink potential of terrestrial vegetation can be increased substantially by optimal land management. *Commun. Earth Environ.* **2022**, *3*, 8. [CrossRef]
4. Gao, L.; Wang, X.; Johnson, B.A.; Tian, Q.; Wang, Y.; Verrelst, J.; Mu, X.; Gu, X. Remote sensing algorithms for estimation of fractional vegetation cover using pure vegetation index values: A review. *ISPRS J. Photogramm. Remote Sens.* **2020**, *159*, 364–377. [CrossRef] [PubMed]
5. Jia, K.; Yang, L.; Liang, S.; Xiao, Z.; Zhao, X.; Yao, Y.; Zhang, X.; Jiang, B.; Liu, D. Long-Term Global Land Surface Satellite (GLASS) Fractional Vegetation Cover Product Derived from MODIS and AVHRR Data. *IEEE Trans. Geosci. Remote Sens.* **2019**, *12*, 508–518.
6. Jia, K.; Liang, S.; Gu, X.; Baret, F.; Wei, X.; Wang, X.; Yao, Y.; Yang, L.; Li, Y. Fractional vegetation cover estimation algorithm for Chinese GF-1 wide field view data. *Remote Sens. Environ.* **2016**, *177*, 184–191. [CrossRef]
7. North, P.R.J. Estimation of f(APAR), LAI, and vegetation fractional cover from ATSR-2 imagery. *Remote Sens. Environ.* **2002**, *80*, 114–121. [CrossRef]
8. Baret, F.; Weiss, M.; Lacaze, R.; Camacho, F.; Makhmara, H.; Pacholcyzk, P.; Smets, B. GEOV1: LAI and FAPAR essential climate variables and FCOVER global time series capitalizing over existing products. Part1: Principles of development and production. *Remote Sens. Environ.* **2013**, *137*, 299–309. [CrossRef]
9. García-Haro, F.J.; Camacho, F.; Meliá, J. Inter-comparison of SEVIRI/MSG and MERIS/ENVISAT biophysical products over Europe and Africa. In *MERIS/(A)ATSR User Workshop*, 2nd ed.; ESA SP-666: Frascati, Italy, 2008.
10. Roujean, J.L.; Lacaze, R. Global mapping of vegetation parameters from POLDER multiangular measurements for studies of surface-atmosphere interactions: A pragmatic method and its validation. *J. Geophys. Res.-Atmos.* **2002**, *107*, ACL-6. [CrossRef]
11. Fillol, E.; Baret, F.; Weiss, M.; Dedieu, G.; Demarez, V.; Gouaux, P.; Ducrot, D. Cover fraction estimation from high resolution SPOT HRV&HRG and medium resolution SPOTVEGETATION sensors, Validation and comparison over South-West France. In *Second Recent Advances in Quantitative Remote Sensing Symposium*; Publicacions de la Universitat de València: Valencia, Spain, 2006; pp. 659–663.
12. Verger, A.; Baret, F.; Weiss, M. Algorithm Theoretical Basis Document, Leaf Area Index (LAI), Fraction of Absorbed Photosynthetically Active Radiation (FAPAR), Fraction of Green Vegetation Cover (FCover), Collection 1 km, Version 20. Operations, C.G.L., Ed.; 2019. Available online: https://land.copernicus.eu/global/sites/cgls.vito.be/files/products/CGLOPS1_ATBD_LAI1km-V2_I1.41.pdf (accessed on 1 October 2022).
13. Verger, A.; Descals, A. Algorithm Theoretical Basis Document, Leaf Area Index (LAI), Fraction of Absorbed Photosynthetically Active Radiation (FAPAR), Fraction of Green Vegetation Cover (FCover), Collection 300m, Version 1.1. Copernicus Global Land Operations. Pubmed Partial Author Stitle Stitle. 2021. Available online: https://land.copernicus.eu/global/sites/cgls.vito.be/files/products/CGLOPS1_ATBD_FCOVER300m-V1.1_I1.02.pdf (accessed on 1 October 2022).
14. Jia, K.; Liang, S.; Liu, S.; Li, Y.; Xiao, Z.; Yao, Y.; Jiang, B.; Zhao, X.; Wang, X.; Xu, S.; et al. Global land surface fractional vegetation cover estimation using general regression neural networks from MODIS surface reflectance. *IEEE Trans. Geosci. Remote Sens.* **2015**, *53*, 4787–4796. [CrossRef]
15. Claverie, M.; Ju, J.; Masek, J.G.; Dungan, J.L.; Vermote, E.F.; Roger, J.-C.; Skakun, S.V.; Justice, C. The Harmonized Landsat and Sentinel-2 surface reflectance data set. *Remote Sens. Environ.* **2018**, *219*, 145–161. [CrossRef]

16. Zhao, J.; Li, J.; Zhang, Z.; Wu, S.; Zhong, B.; Liu, Q. MuSyQ GF-Series 16m/10days Fractional Vegetation Cover Product (from 2018 to 2020 across China Version 01). Science Data Bank. 2021. Available online: http://cstr.cn/31253.11.sciencedb.j00001.00266 (accessed on 4 September 2022).
17. Zhang, X.; Liao, C.; Li, J.; Sun, Q. Fractional vegetation cover estimation in arid and semi-arid environments using HJ-1 satellite hyperspectral data. *Int. J. Appl. Earth Obs. Geoinf.* **2013**, *21*, 506–512. [CrossRef]
18. Ludwig, M.; Morgenthal, T.; Detsch, F.; Higginbottom, T.P.; Valdes, M.L.; Nauss, T.; Meyer, H. Machine learning and multi-sensor based modelling of woody vegetation in the Molopo Area, South Africa. *Remote Sens. Environ.* **2019**, *222*, 195–203. [CrossRef]
19. Gill, T.; Johansen, K.; Phinn, S.; Trevithick, R.; Scarth, P.; Armston, J. A method for mapping Australian woody vegetation cover by linking continental-scale field data and long-term Landsat time series. *Int. J. Remote Sens.* **2017**, *38*, 679–705. [CrossRef]
20. Song, D.X.; Wang, Z.; He, T.; Wang, H.; Liang, S. Estimation and Validation of 30 m Fractional Vegetation Cover over China Through Integrated Use of Landsat 8 and Gaofen 2 Data. *Sci. Remote Sens.* **2022**, *6*, 100058. [CrossRef]
21. Chen, Y.; Cao, R.; Chen, J.; Liu, L.; Matsushita, B. A practical approach to reconstruct high-quality Landsat NDVI time-series data by gap filling and the Savitzky–Golay filter. *ISPRS J. Photogramm. Remote Sens.* **2021**, *180*, 174–190. [CrossRef]
22. Gao, F.; Masek, J.; Schwaller, M.; Hall, F. On the blending of the Landsat and modis surface reflectance: Predicting daily Landsat surface reflectance. *IEEE Trans. Geosci. Remote Sens.* **2006**, *44*, 2207–2218.
23. Zhu, X.; Helmer, E.H.; Gao, F.; Liu, D.; Chen, J.; Lefsky, M.A. A flexible spatiotemporal method for fusing satellite images with different resolutions. *Remote Sens. Environ.* **2016**, *172*, 165–177. [CrossRef]
24. Kaffash, M.; Nejad, H.S. Spatio-temporal fusion of Landsat and MODIS land surface temperature data using FSDAF algorithm. *J. Water Soil Sci.* **2021**, *25*, 45–62.
25. Zhou, L.; Lyu, A. Investigating natural drivers of vegetation coverage variation using MODIS imagery in Qinghai, China. *J. Arid. Land* **2016**, *8*, 109–124. [CrossRef]
26. Walker, J.J.; Beurs, K.M.; Wynne, R.H.; Gao, F. Evaluation of Landsat and MODIS data fusion products for analysis of dryland forest phenology. *Remote Sens. Environ.* **2012**, *117*, 381–393. [CrossRef]
27. Meng, J.; Du, X.; Wu, B. Generation of high spatial and temporal resolution NDVI and its application in crop biomass estimation. *Int. J. Digit. Earth* **2013**, *6*, 203–218. [CrossRef]
28. Sherstinsky, A. Fundamentals of Recurrent Neural Network (RNN) and Long Short-Term Memory (LSTM) network. *Phys. D Nonlinear Phenom.* **2020**, *404*, 132306. [CrossRef]
29. Reddy, D.S.; Prasad, P.R. Prediction of vegetation dynamics using NDVI time series data and LSTM. *Model. Earth Syst. Environ.* **2018**, *4*, 409–419. [CrossRef]
30. Van de Voorde, T.; Vlaeminck, J.; Canters, F. Comparing Different Approaches for Mapping Urban Vegetation Cover from Landsat ETM + Data: A Case Study on Brussels. *Sensors* **2008**, *8*, 3880–3902. [CrossRef]
31. O'Donncha, F.; Hu, Y.; Palmes, P.; Burke, M.; Filgueira, R.; Grant, J. A spatio-temporal LSTM model to forecast across multiple temporal and spatial scales. *Ecol. Inform.* **2022**, *69*, 101687. [CrossRef]
32. Markham, B.L.; Arvidson, T.; Barsi, J.A.; Choate, M.; Kaita, E.; Levy, R.; Lubke, M.; Maesk, J.G. Landsat program. In *Comprehensive Remote Sensing*; Liang, S., Ed.; Elsevier: Oxford, UK, 2018; pp. 27–90.
33. Xu, G.; Xu, H. Cross-comparison of Sentinel-2A MSI and Landsat 8 OLI Multispectral Information. *Remote Sens. Technol. Appl.* **2021**, *36*, 165–175.
34. Liang, S.; Zhao, X.; Yuan, W.; Liu, S.; Cheng, X.; Xiao, Z.; Zhang, X.; Liu, Q.; Cheng, J.; Tang, H.; et al. A Long-term Global LAnd Surface Satellite (GLASS) Dataset for Environmental Studies. *Int. J. Digit. Earth* **2013**, *6*, 5–33. [CrossRef]
35. Chen, J.; Jönsson, P.; Tamura, M.; Gu, Z.; Matsushita, B.; Eklundh, L. A simple method for reconstructing a high-quality NDVI time-series data set based on the Savitzky–Golay filter. *Remote Sens. Environ.* **2004**, *91*, 332–344. [CrossRef]
36. Hochreiter, S.; Schmidhuber, J. Long Short-term Memory. *Neural Comput.* **1997**, *9*, 1735–1780. [CrossRef]
37. Graves, A.; Schmidhuber, J. Framewise phoneme classification with bidirectional LSTM and other neural network architectures. *Neural Netw.* **2005**, *18*, 602–610. [CrossRef] [PubMed]
38. Tian, H.; Wang, P.; Tansey, K.; Zhang, J.; Zhang, S.; Li, H. An LSTM neural network for improving wheat yield estimates by integrating remote sensing data and meteorological data in the Guanzhong Plain, PR China. *Agric. For. Meteorol.* **2021**, *310*, 108629. [CrossRef]
39. Baak, M.; Koopman, R.; Snoek, H.; Klous, S. A new correlation coefficient between categorical, ordinal and interval variables with Pearson characteristics. *Comput. Stat. Data Anal.* **2020**, *152*, 107043. [CrossRef]
40. Tao, J.; Wang, Y.; Qiu, B.; Wu, W. Exploring cropping intensity dynamics by integrating crop phenology information using Bayesian networks. *Comput. Electron. Agric.* **2022**, *193*, 106667. [CrossRef]
41. Yan, Y.; Liu, H.; Bai, X.; Yan, Y.; Liu, H.; Bai, X.; Zhang, W.; Wang, S.; Luo, J.; Cao, Y. Exploring and attributing change to fractional vegetation coverage in the middle and lower reaches of Hanjiang River Basin, China. *Environ. Monit. Assess* **2023**, *195*, 131. [CrossRef] [PubMed]
42. Zhou, Y. Understanding urban plant phenology for sustainable cities and planet. *Nat. Clim. Chang.* **2022**, *12*, 302–304. [CrossRef]
43. Zhang, M.; Du, H.; Zhou, G.; Mao, F.; Li, X.; Zhou, L.; Zhu, D.; Xu, Y.; Huang, Z. Spatiotemporal Patterns and Driving Force of Urbanization and Its Impact on Urban Ecology. *Remote Sens.* **2022**, *14*, 1160. [CrossRef]
44. Zhu, X.; Chen, J.; Gao, F.; Chen, X.; Masek, J.G. An enhanced spatial and temporal adaptive reflectance fusion model for complex heterogeneous regions. *Remote Sens. Environ.* **2010**, *114*, 2610–2623. [CrossRef]

45. Ma, H.; Liang, S. Development of the GLASS 250-m leaf area index product (version 6) from MODIS data using the bidirectional LSTM deep learning model. *Remote Sens. Environ.* **2022**, *273*, 112985. [CrossRef]
46. Liu, D.; Jia, K.; Wei, X.; Xia, M.; Zhang, X.; Yao, Y.; Zhang, X.; Wang, B. Spatiotemporal comparison and validation of three global-scale fractional vegetation cover products. *Remote Sens.* **2019**, *11*, 2524. [CrossRef]

remote sensing

MDPI

Article

Examining Spatiotemporal Photosynthetic Vegetation Trends in Djibouti Using Fractional Cover Metrics in the Digital Earth Africa Open Data Cube

Julee Wardle * and Zachary Phillips

Earth and Atmospheric Science Department, Saint Louis University, Saint Louis, MO 63103, USA; zachary.phillips@slu.edu
* Correspondence: julee.wardle@slu.edu

Abstract: The Horn of Africa has sensitive, arid ecosystems, with its vegetation commonly distressed by factors such as climate change, population increase, unstable water resources, and rarely enforced land use management practices. These factors make countries such as Djibouti highly variable locations for the growth of vegetation and agricultural products, and these countries are becoming more vulnerable to food insecurity as the climate warms. The rapid growth of satellite and digital image processing technology over the last five decades has improved our ability to track long-term agricultural and vegetation changes. Data cubes are a newer approach to managing satellite imagery and studying temporal patterns. Here, we use the cloud-based Digital Earth Africa, Open Data Cube to analyze 30 years of Landsat imagery and orthomosaics. We analyze long-term trends in vegetation dynamics by comparing annual fractional cover metrics (photosynthetic vegetation, non-photosynthetic vegetation, and bare ground) to the Normalized Difference Vegetation Index. Investigating Djibouti-wide and regional vegetation trends, we provide a comparison of trends between districts and highlight a primary agricultural region in the southeast as a detailed example of vegetation change. The results of the Sen's slope and Mann–Kendall regression analyses of the data cube suggest a significant decline in vegetation ($p = 0.00002$), equating to a loss of ~0.09 km^2 of arable land per year (roughly 2.7 km^2 over the 30-year period). Overall, decreases in photosynthetic vegetation and increases in both non-photosynthetic vegetation and bare soil areas indicate that the region is becoming more arid and that land cover is responding to this trend.

Keywords: Open Data Cube; spatiotemporal trends; fractional cover; photosynthetic vegetation; Djibouti; food security; land cover trends

1. Introduction

Globally, vegetation in arid regions is often susceptible to the small shifts in temperature and precipitation that arise due to climate change [1–4]. Current research shows that arid regions are a challenging environment for vegetation to grow, and that agriculture and food security are closely tied to trends in climate change [5,6]. Global climate models predict increased variability in precipitation, higher temperatures, and more intense precipitation events [7]. These short-term variations drive the need for new analytical tools that can efficiently examine the long-term temporal trends of vegetation in response to climate change. The Digital Earth Africa (DEA) Open Data Cube (ODC) is a relatively new tool that efficiently addresses the existing problem of analyzing large quantities of remote sensing data for wide-ranging spatial scales across multiple decades.

The growth of satellite technology over the last five decades has improved the ability of scientists to track long-term vegetation change in arid regions where vegetation is sensitive to shifting climates. Satellite remote sensing is a leading technology in providing comprehensive information about numerous earth systems, particularly in monitoring global vegetation health and trends [8–14]. In handling remote sensing data, data cubes are a newer approach to managing imagery, with hundreds or thousands of revisits to the same swath of land [15–17]. They are designed to efficiently store extensive sets of repeat imagery and are tailored for rapid access, processing, and analyses using a Python Application Programming Interface (API) [15,17,18]. The primary dimensions of a data cube for remote sensing purposes usually consist of the latitude, longitude, spectral data, and time [15]. The ODC is a non-profit, open-source data cube project driven by the need to uniformly manage and study temporally repetitive satellite data, and is freely available to the public [15–19].

DEA is an ODC initiative supported by the Australian Department of Foreign Affairs and Trade, funded in part by the Helmsley Charitable Trust and derived from Digital Earth Australia [20,21]. It maintains robust imagery libraries for the United States Geological Survey (USGS) Landsat mission, the European Commission (EC) and the European Space Agency (ESA) Copernicus Sentinel-2 mission. DEA also provides a cloud-based user computational platform in the form of a sandbox, operating in a Jupyter Notebook environment [20]. The ODC is freely available through DEA for users working on any type of African geospatial challenges [20–22].

The challenge being addressed here is analyzing the long-term temporal vegetation trends in arid Africa, with a focus on Djibouti. In the Horn of Africa, vegetation systems and agricultural areas are fragile ecosystems that are regularly distressed by extreme heat and drought, increasing populations, unstable water resources with limited irrigation options, and poor land use management practices such as a lack of crop rotation or a lack of controls for disease and pests [9,23–31]. Over half of the global population growth in the next 25 years is expected to occur in Africa. With this, the population of sub-Saharan Africa is projected to double by 2050 [8,32–34]. With future projections of climate warming and population growth, food systems and security will continue to be vulnerable in arid nations such as Djibouti [8,35–40].

Analyzing annual vegetation dynamics provides key information for determining the overall trends in natural vegetation and agricultural areas that may be subjected to abnormal drought, poor cropland management practices, or abandonment [41,42]. In arid regions, vegetation can be categorized either as photosynthetic (i.e., green leaves, healthy growing crops) or non-photosynthetic (dead/decaying/brown vegetation) [42–49]. Fractional cover algorithms use spectral unmixing [43,46–48] to determine the proportion of photosynthetic vegetation (pv), non-photosynthetic vegetation (npv), and bare soil (bs) (i.e., bare soil or rock) contained within a single pixel of imagery [44,46,49]. Comparing the ratios over months in a particular year and over multiple years indicates trends in a nation's land use management [47] and, more broadly, land cover.

This research (1) analyzed 30 years of Landsat fractional cover data products for Djibouti using DEA's ODC to quantify the fractional cover vegetation type (pv, npv, or bs), (2) compared the fractional cover results to the commonly used Normalized Difference Vegetation Index (NDVI) metric, and (3) tested for significant temporal trends using Mann–Kendall regressions and Sen's slope tests. For these statistical tests, we present a null hypothesis (H_0) of no trend in vegetation changes between 1990 and 2020, and an alternate hypothesis (H_a) of a significant trend in vegetation changes between 1990 and 2020. In the alternate hypothesis, the direction of the trend (positive or negative) is noted by the Mann–Kendall statistic and Sen's slope tests. For further clarification, in this study, the terms "photosynthetic cover" and "green cover" are used interchangeably to describe photosynthetic vegetation.

2. Materials and Methods

2.1. Area of Interest

Djibouti is a small country in the Horn of Africa (Figure 1), bordered by Eritrea to the north, Ethiopia to the west and south, and Somalia to the southeast. Djibouti has a population of approximately one million people and a land area of 23,200 km^2 [35,37,40,50–60]. However, only a small percentage of this land is considered cultivable due to factors such as aridity and soil quality [52]. Only about 0.04% of Djibouti's land is classified as arable land, which is equivalent to around 9200 hectares or 92 square kilometers [40]. It receives an average of 147 mm/year of precipitation and had a water supply that could serve only half of the urban population and 21% of the rural population as of 1990 [40]. Its economy relies heavily on services, particularly those related to its status as a regional trade and logistics hub [52,53,61,62]. As a result, agriculture represents only a small fraction of its GDP, and it experiences frequent droughts that severely impact food security [50,51]. Djibouti is classified by the United Nations as a "chronic food-deficit country" due to its limited agricultural resources and recurrent droughts [40,52,53].

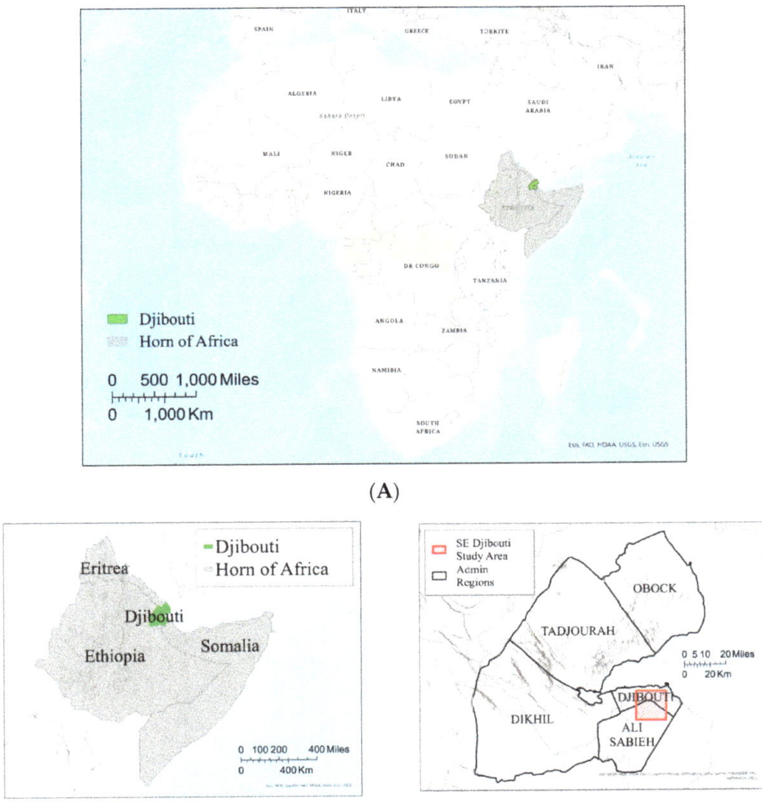

Figure 1. Location of Djibouti and study area in the arid Horn of Africa of the Eastern African Continent. Djibouti is one of ~20 African countries in arid regions. (**A**) Location of Djibouti and the Horn of Africa on the African Continent. (**B**) Djibouti location in the Horn of Africa. (**C**) SE Djibouti study area location and sub-regions in Djibouti.

For a more intent focus, a smaller, 23 km^2 study area (Figure 1C) was selected to exemplify the study results in raster format (Figure 2) and was selected in consultation with previous studies [63]. This smaller, exemplary region is the most visible cropland area in the region; it is prominently observable on satellite imagery and highlights an agriculturally important region in Djibouti District, just south of Djibouti City (Figure 1C). The region, while not inclusive of the entire country, served as a representation of the common regional geographic features and highlights one of the primary productive agricultural areas in

Djibouti. Agricultural production in Djibouti is concentrated in small pockets of fertile land located mainly in the southeast and northwest regions of the country [40,53].

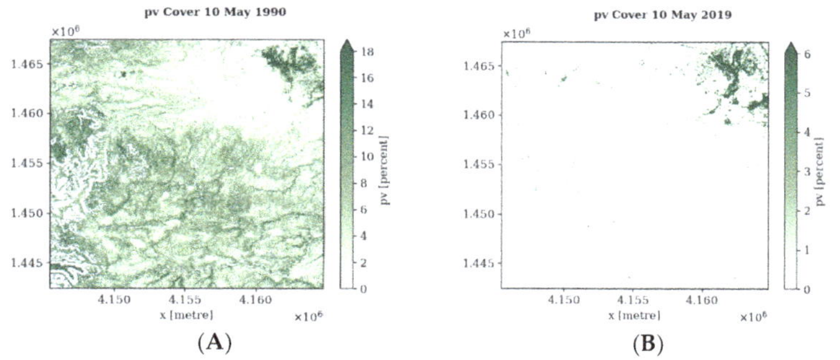

Figure 2. Example of photosynthetic vegetation (pv) cover in the SE Djibouti study area (Figure 1C) for the first and final portion of the 30-year period, May 1990 (**A**) and May 2019 (**B**).

2.2. Data Utilized

This study utilized several Landsat datasets, accessed for this study through the DEA platform. USGS Landsat Collection 2 Surface Reflectance imagery from Landsat 5, 7, 8, and 9 [20] was used by DEA to create the fractional cover (FC) products [41,64] for this study's 30-year study period (1990–2020). A Water Observations from Space product, also created by DEA [65], was used to mask the water and cloud pixels. An FC data cube was compiled using DEA's FC product to provide FC percentages at timesteps matching the Landsat satellite's 16-day temporal resolution. The FC products utilize each pixel in the study area at a 30 m cell size and use the WGS 1984 UTM Zone 38N Coordinate Reference System (CRS) (EPSG: 32638). Temporal data coverage is consistently available biweekly, aside from a two-year gap from 2006 to 2007 due to a Scan Line Corrector (SLC) failure with the Landsat 7 Enhanced Thematic Mapper plus (ETM+) sensor [66].

There are four bands in the FC dataset that express the percentage of fractional cover as an integer ranging between 0 and 100 for bare soil, photosynthetic (green) vegetation, and non-photosynthetic (dead/decaying) vegetation [12]. Also included is a fourth spectral unmixing error band [22] (Table 1).

Table 1. Data products and band IDs of fractional cover and water observations from space products derived from Landsat collection 2 surface reflectance.

Product	Description	Band ID	Name	Value Range	Units	No Data
fc_ls	Fractional Cover from Landsat	bs	bare soil	1–100	percent	255
		pv	photosynthetic vegetation	1–100	percent	255
		npv	non-photosynthetic vegetation	1–100	percent	255
		ue	unmixing error		1	255
wofs_ls	Water Observations from Space using Landsat	water	Water Observation Feature Layer	0–255	na	1

2.3. Processing in ODC

This study comprised three main processing phases: (1) computation structuring, (2) data compilation, and (3) analysis (Figure 3). We completed all three phases using DE Africa's online ODC python-based scripting environment. Parallel processing was used to process the large volume of time slices in the data cube covering the thirty years and was accomplished using a four-core computer cluster and 28 gigabytes of memory. DEA utilizes Dask for dynamic task scheduling, which is a flexible library for parallel computing in Python [67]. It uses parallel arrays and data frames that run on top of dynamic task schedulers. Here, we assembled multiple data cubes to compare the FC and NDVI at both national and regional levels. A water and cloud mask were used to mask the presence of water and cloud pixels from the data cube, providing a cloud-free analysis.

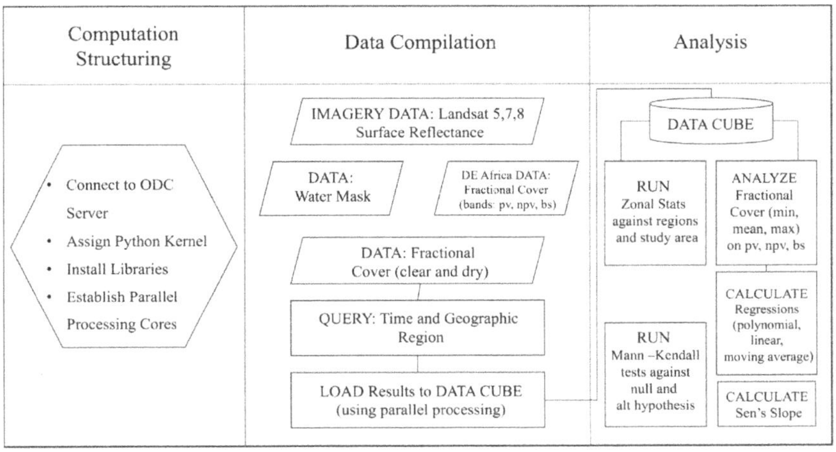

Figure 3. Methodology workflow encompassing the three processing phases of the study: computation structuring, data compilation, and analysis; these were all within the ODC environment.

2.4. Fractional Cover Calculations

The Joint Remote Sensing Research Program (JRSRP) developed the FC algorithms [41] used to build the DEA's bare soil (bs), photosynthetic vegetation (pv), and non-photosynthetic vegetation (npv) products, and DEA provided access to the algorithms and products as part of its analytic data sets [20]. The FC algorithm (Equation (1)), developed by Scarth et al. for Digital Earth Australia [41], is defined as

$$\begin{bmatrix} x \\ \delta \end{bmatrix} = \begin{bmatrix} M \\ \delta 1^T \end{bmatrix} f + e \quad (1)$$

where δ is a weighting for the sum to one constraint and $1^T = \begin{bmatrix} 1 & 1 & \ldots 1 \end{bmatrix}$ is a $c+1$ vector of ones. The optimal value of δ is determined during the 10-fold cross validation process [41].

A comparison between the fractional cover and the more common NDVI provides insight into general relationships with a well-studied metric [42,44,45,47,49,68–70].

A spectral unmixing approach using the NDVI and the Cellulose Absorption Index (CAI) was applied in the creation of the FC product for this project [41,44,47]. In total, 675 images were filtered for clouds, leading to 119 scenes with less than 10% cloud cover being processed for the FC data products covering the period from 1 January 1990 to 1 January 2020. Each pixel in the Djibouti study area was input to the fractional cover algorithm after masking out cloud and water pixels.

We also conducted a district-level (Figure 1C) statistical comparison to summarize the changes over the 30-year period, including the minimum, maximum, range, mean, and standard deviation of pv values. The maximum values for pv signify the maximum cover level of green vegetation throughout the study period, the mean pv provides a general measure, and the minimum pv signifies the lowest level of green vegetation throughout the study period. Variability/stability was measured using the range and standard deviation. Here, we present our results in the maximum, minimum, quartiles, and mean pv values by year. The maximum and minimum pv are the maximum or minimum pv value for any single pixel in any of the scenes contained within a given year, signifying both seasonal and spatial variations. The mean pv is the mean value in every pixel over the entire year. The presentation of results in this manner enables the average trends through the mean to be observed and highlights the locations and periods with the healthiest vegetation and most arid conditions.

Next, we summarized the FC results for each of the 119 scenes using the yearly FC maximum (wet season), minimum (dry season), and average to analyze consistent seasonal trends rather than include seasonal noise. We aggregated all scenes by year in the three categories (bs, pv, npv) for minimum, mean, and maximum levels, and calculated trendlines using linear and moving-average regressions for each.

2.5. Statistical Testing

Finally, the FC dataset was subjected to spatiotemporal statistical testing for photosynthetic vegetation, as photosynthetic vegetation is most frequently associated with healthy vegetation [47]. To test the direction of trends, we performed Sen's slope and Mann–Kendall trend tests (alpha = 0.05) to test the null (H_0) and alternate hypotheses (H_a):

- H_0: No significant temporal trend in FC types, by percent cover.
- H_a: Significant temporal trend in FC types, by percent cover.

Sen's slope and Mann–Kendall regressions are both methods of testing for trends in univariate, non-normally distributed temporal data. The direction of the trend (positive or negative) is noted by the Mann–Kendall (MK) test, where positive and negative values are, respectively, indicative of positive and negative temporal trends in photosynthetic vegetation over the entire 30-year study period. Mann–Kendall tests require computation of the z-statistic (Equation (2)), which is calculated as follows:

$$Z = (\tau - 1)/\left(\sqrt{Var(\tau)}\right) \tag{2}$$

where τ is Kendall's Tau, $Var(\tau)$ is the variance of Kendall's Tau, and Z is the test statistic. Kendall's Tau (Equation (3)) is the non-parametric correlation coefficient (or measure of association) for the sample and is defined as

$$\tau_n = \frac{c-d}{c+d} = \frac{S}{\binom{n}{2}} = \frac{2S}{n(n-1)} \tag{3}$$

where S is the number c of concordant pairs minus the number d of discordant pairs [71,72]. The Z-statistic is compared with the critical value to determine the significance of the trend in the time series data [71,73,74]. A significant result indicates the presence of a trend, either positive or negative, while a non-significant result suggests the absence of a significant trend [71,74].

Sen's slope was then used to provide an estimate of the 95% confidence interval around the mean trend. The Sen's slope test is a non-parametric method used to estimate the slope or trend in a time series by computing the median of all possible slopes between pairs of data points. It is robust to outliers and does not assume any specific distribution of the data. The calculation contains two steps:

1. For each pair of data points (x_i, x_j) in the time series, calculate the slope as $(x_j - x_i)/(j - i)$, where i and j are the indices of the data points.
2. Compute the median of all calculated slopes. This median slope is considered the Sen's slope [73,75–77].

3. Results

3.1. Fractional Cover Regression

Analysis of the temporal trends in total green vegetation showed both periods of increase and decrease over the thirty-year period, but the overall pv fractional cover decreased between 1990 and 2020 with a −0.00025 slope (Figure 4). Non-photosynthetic vegetation and bare soil comprised 35–60% of the study area, and photosynthetic vegetation accounted for only 10% of the fractional cover (Figure 4). The non-photosynthetic vegetation and bare soil had negative correlations with photosynthetic vegetation, and both displayed increasing trends over the study period. Each point on the graph (Figure 4) represents the average value for pv, npv, and bs for each of the 119 images in the data cube.

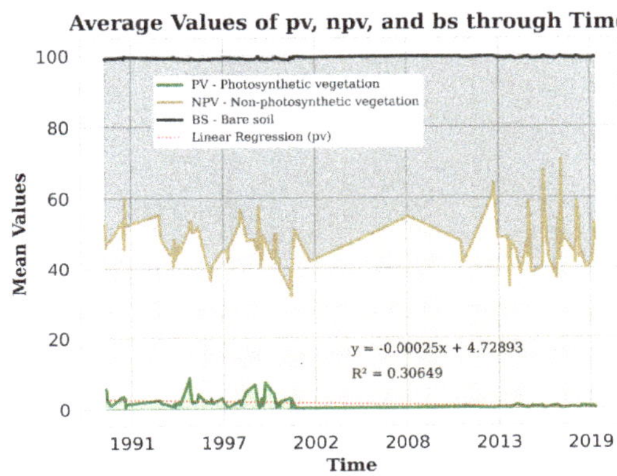

Figure 4. Image-averaged FC values (in percent) of the study area for the 30-year study period (119 cloud-free scenes total). The pv (photosynthetic vegetation) makes up less than 10% of cover annually and shows an overall decrease, whereas npv and bs show high variability and a slight increase throughout the study period.

The change in photosynthetic (green) vegetation for all 119 scenes of the study area showed a clear decreasing trend in the linear average (Figure 4). The linear regression slope was −0.00025 per year and showed a weak association with the input data (R^2 = 0.30649). Considering the area of Djibouti (23,200 km^2), this means that Djibouti has lost the equivalent of 5.8 hectares of pv per year, which equates to 174 hectares over the 30-year study period.

To account for seasonal variability, we aggregated the pv data to the yearly level, reducing the time cube from 119 images (Figure 5A) to 30 annual composites, one for each year, and re-analyzed the regressions (Figure 5B–D). The decreasing trend in green vegetation was more readily observable when the data were aggregated annually to show the mean (Figure 5B), minimum (Figure 5C), and maximum (Figure 5D) photosynthetic vegetation levels for each of the 30 years.

The aggregated mean pv regressions showed similar results to the pre-aggregated dataset with a regression coefficient of −0.00028 and an R-squared of 0.73902 (Figure 5B). The aggregated minimum pv showed a slightly lower regression coefficient of −0.00011 and an R-square of 0.67484 (Figure 5C). The maximum green FC showed a linear regression coefficient of −0.00051 and an R-squared of 0.61761 (Figure 5D). The mean, min and max regression values represent various stages of photosynthetic vegetative growth, whether early spring-like growth (min) or full peak-season growth (max).

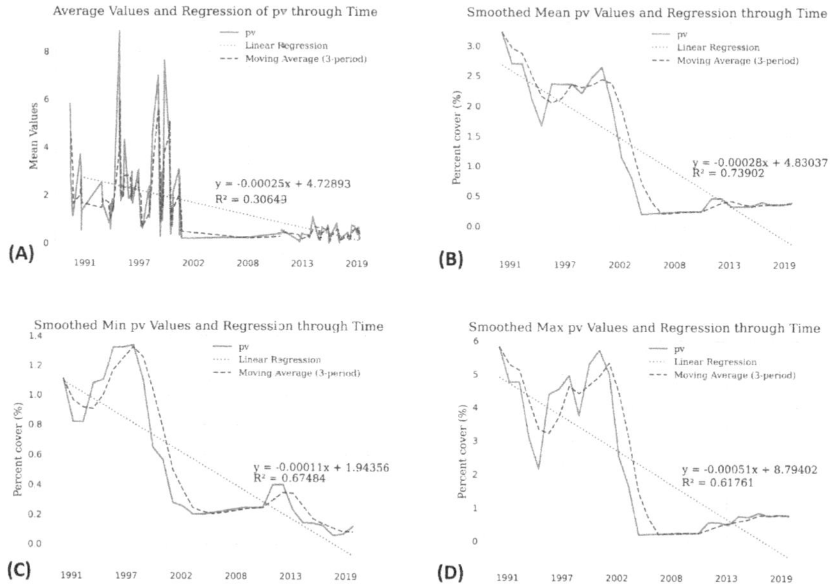

Figure 5. Summary trends of pv, 30-year linear trend, and three-year averages for all imagery in SE Djibouti study area (Figure 1C) (**A**); annual mean composite pv, 30-year linear trend, and three-year averages for mean pv (**B**); annual minimum composite pv, 30-year linear trend, and three-year averages for mean pv (**C**); and annual maximum composite pv, 30-year linear trend, and three-year averages for mean pv (**D**).

3.2. Mann–Kendall and Sen's Slope Tests

All trends show general decreases in green fractional cover throughout the 30 years. The Mann–Kendall test (Table 2) rejected the null hypothesis (H_0 = no change in pv cover) and showed decreasing trends in the minimum, maximum, and mean photosynthetic vegetation covers, with MK-stat values of −242, −134, and −190, respectively. The pv covers also showed negative Sen's slopes (Table 2) over the entire study period of −0.034, −0.156, and −0.091, respectively.

Table 2. Mann–Kendall and Sen's slope results for smoothed photosynthetic cover. We computed all statistics at 95% confidence (alpha = 0.05) [78,79].

	Mann-Kendall					Sen's Slope		
Photosynthetic (Smoothed)	**Alpha**	**MK-Stat**	**z-Stat**	***p*-Value**	**Trend**	**Alpha**	**Slope**	**Intercept**
MEAN	0.05	−190	−3.945	0.00008	yes	0.05	−0.091	1.984
MAXIMUM	0.05	−134	−2.776	0.00550	yes	0.05	−0.156	3.669
MINIMUM	0.05	−242	−5.030	0.00000	yes	0.05	−0.034	0.718

The rasterized change detection results further support the statistical outcomes, showing a distinct decrease in photosynthetic fractional vegetative cover. Comparing the May 1990 image (Figure 2A) and the May 2019 image (Figure 2B) shows a distinct decrease in pv cover in the study area (Figure 6). Most pixels in the study area, and across Djibouti, decreased in pv between 1990 and 2019, while only small areas had increasing pv levels.

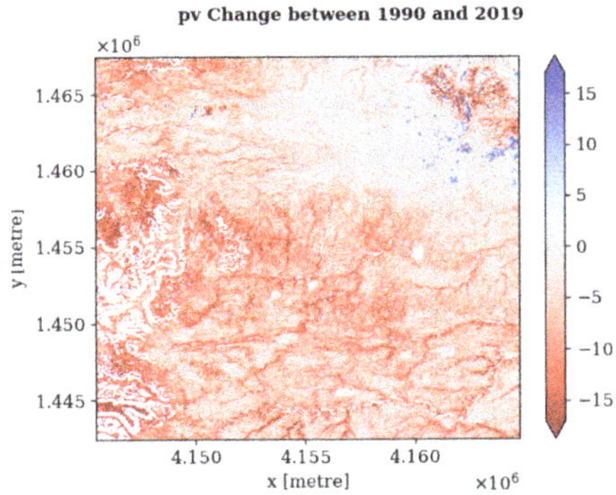

Figure 6. Change detection between 1990 pv and 2019 pv shows loss of pv in red, gain of pv in blue, and no change in grey. Most areas decreased in pv levels, while only small areas had increasing pv levels.

3.3. Zonal Summary Statistics

The pv and NDVI summary statistics over the 30-year study period (Figure 7) for the five districts of Djibouti (Figure 1C) show a positive relationship between pv and the NDVI. Ali Sabieh shows the greatest increase in pv between 1990 and 2005 (Figure 7C) and 2006 and 2020 (Figure 7D), and aligns with the NDVI results for the same region (Figure 7A,B). Dikhil and Tadjourah districts show maximum pv values at 100% in both the 1990–2005 (Figure 7C) and 2006–2020 (Figure 7D) periods. The statistical ranges for the Dikhil and Tadjourah districts were the greatest of all the districts for the full 30 years (showing the greatest variation in both pv and NDVI). Djibouti, Obock, and Tadjourah had the highest maximum NDVI values for both the 1990–2005 and 2006–20202 periods at 0.53, 0.59, and 0.65 (Figure 7A) and 0.61, 0.67, and 0.64 (Figure 7B), respectively.

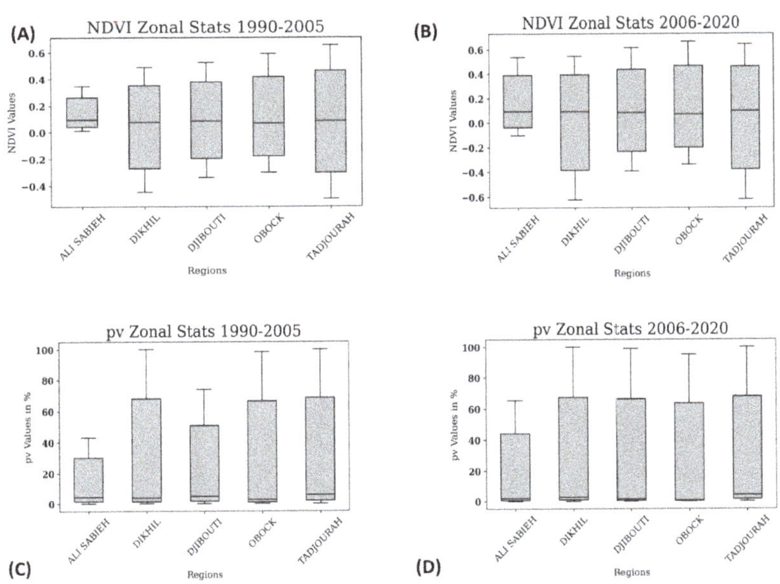

Figure 7. NDVI and pv zonal statistics by region for 15-year periods. (**A**) NDVI statistics for 1990–2005 period. (**B**) NDVI statistics for 2006–2020 period. (**C**) pv statistics for 1990–2005 period. (**D**) pv statistics for 2006–2020 period.

4. Discussion

Here, we discuss the above rasterized results for a study area in SE Djibouti, the relationship between the FC and NDVI, and their general decreases over time. We also highlight the capabilities of DE Africa's ODC for processing and analyzing high volumes of satellite imagery and the results displaying decreasing trends in overall vegetation abundance and health. We discuss their implications toward heightened vulnerability in terms of agriculture and food production.

4.1. Temporal Vegetation Trends in Djibouti

The results suggest that throughout the study period, the Djibouti study area was covered by 35–60% of both non-photosynthetic vegetation and bare soil, and less than 10% of photosynthetic vegetation (Figure 5). The results also show a negative correlation between pv versus npv and bs [80]. This suggests the death of vegetation and the transition of those areas into bare soil, or the growth of vegetation into photosynthetic vegetation. Consequently, both the npv and bs covers demonstrated increasing trends over the study period as the pv cover decreased.

The district-level zonal statistics revealed a generally positive relationship between pv and the NDVI. The Tadjourah and Dikhil districts had the highest pv FC and NDVI (Figure 4) estimations, as well as the widest ranges throughout the study period. This suggests a greater presence of photosynthetic vegetation in these districts over a longer period, and a higher variability in the vegetation in Tadjourah and Dikhil. This finding spatially correlates with agricultural production in Djibouti being mostly concentrated in the more fertile land located primarily in the SE Djibouti study area and northwest regions (Figure 1C) of the country [81,82].

Focusing on the SE Djibouti study area (Figure 1C), a decreasing trend in total green vegetation was observed between 1990 and 2020 (Figure 2), with some periods of smaller positive trends from 1990 to 1995 and 1997 to 1999 and negative trends from 1995 to 1997 and 1999 to 2001. Between the years of 1997 and 2003, Djibouti underwent its most severe decline in pv over the study period.

Analyzing the change in photosynthetic (green) vegetation across all 119 scenes (Figure 5) revealed a clear decreasing trend when examining the and three-year moving averages (Figure 5A). The linear regression coefficient for the entire period from 1990 to 2020 indicated a consistent decline. However, the R-squared value was relatively low, indicating that the linear regression model did not fit the data well. To account for seasonal variability, we performed a yearly aggregation of the data (Figure 5B–D), which further accentuated the decreasing trend in green vegetation.

More specifically, the yearly smoothed minimum green fractional cover (Figure 5B), while exhibiting a slightly smaller regression, also confirmed the decreasing trend in the lowest pv values. The smoothed mean (Figure 5C) and maximum (Figure 5D) green fractional covers displayed decreasing trends as well, indicating that the average and highest levels of pv cover also decreased throughout the 30 years. These quantitative results aligned with the qualitative imagery observation of decreasing green fractional cover over the 30 years (Figure 2). The Mann–Kendall and Sen's slope statistical tests (Table 2) supported the rejection of the null hypothesis (of no change), indicating a significant decline in photosynthetic vegetation cover.

The raster results further substantiate the drastic change in photosynthetic fractional vegetative cover over the thirty years (Figure 7), and between the 10 May 1990 image (Figure 2A) and 10 May 2019 image (Figure 2B). These results demonstrate a distinct decrease in photosynthetic vegetation within the study area, highlighting the increasingly challenging environment in which plants grow and their inability to quickly adapt to a warming climate by relocating to new locales.

While there are minimal studies in the published literature using the NDVI and FC to specifically assess Djibouti's vegetative trends [83], the findings of this study align with the concerns about Djibouti published by the United Nations' Food and Agriculture

Organization [40] and Development Program [52], the World Bank [56], and the U.S. Agency for International Development's Famine Early Warning System [35], who all report on Djibouti's food insecurity and need to strengthen resiliency over the long term.

4.2. Connections between FC and NDVI

In other published works, the relationship between the FC and NDVI have been found to correlate positively with one another [69,84–88]. The general positive relationship between the FC and NDVI in this study and others is not surprising given that each is a measure of vegetation existence, abundance, and health. Carlson and Ripley (1997) [87] found the NDVI and fractional vegetation cover to be dependent, particularly when the NDVI is scaled between the limits of the minimum (bare soil) and maximum fractional vegetative cover [87]. Gao et al. (2020) reviewed algorithms for estimating the fractional cover using pure vegetation index values [69]. Their review of 173 selected scientific publications found that relative vegetation abundance (RA) algorithms for determining FC based on scaled maximum/minimum vegetation index values (such as the NDVI) are those most widely used, as opposed to other methods such as spectral mixture analysis (SMA), spectral-based supervised classification algorithms or machine learning algorithms [83]. They also noted that there were few regional studies of the fractional vegetative cover over Africa, Oceania and South/Central America presented in the reviewed literature [83].

The comparison between the FC and NDVI in these results offers valuable insights into their respective implications for assessing vegetation dynamics [69,84–87]. Firstly, the positive relationship observed between the FC and NDVI throughout the 30-year study period underscores their correlation. The peaks in the FC and NDVI in different years suggest temporal variations in vegetation health, with the FC peaking in one year and the NDVI in another. This implies that while both the FC and NDVI are related, they respond differently to changing environmental conditions or land management practices.

One notable finding is the maximum pv values at 100% observed in the Dikhil and Tadjourah districts in both the first and second halves of the study period. These districts also consistently exhibit the highest NDVI values, indicating a more robust photosynthetic vegetation cover than the other districts. This suggests that, in these districts, the FC and NDVI align closely [69,87,88], confirming the utility of both metrics in assessing vegetation health [69,87,88]. However, the variations in the range of NDVI values in these districts over the 30-year study period indicate the sensitivity of the NDVI to environmental changes that the FC may not fully capture [69].

The raster results further substantiate the correlation between the FC and NDVI. The distinct drop in green vegetative cover shown in the raster results between 1990 and 2019 (Figure 2) aligns with the quantified trends in the NDVI. This evidence reinforces the importance of FC as a valuable tool for monitoring and assessing changes in green vegetation cover, especially in under-studied regions like Djibouti [83], where such changes can have significant ecological, environmental, and socio-economic impacts [42,47,69,89,90].

4.3. Advantages of Data Processing in Datacubes

Utilizing DE Africa's ODC to access and analyze the study data resulted in a substantial reduction in both the processing time and the need for high-power computing resources.

Data cubes store spatial, spectral, and temporal data in multidimensional arrays [91] as opposed to traditional methods that involve downloading and maintaining individual images referenced as single files, requiring separate analysis steps for each file [15]. DE Africa's ODC provided analysis-ready data (ARD) and a processing platform for analysis and sharing [20] that saved time, effort and computing resources, allowing the analyst to spend more time on research and minimal time on data preparation and management [15,18,91,92]. Conducting the same study on a local PC, requiring the need to download, preprocess, and analyze data, would have taken an estimated 500% longer than the same study conducted on the DE Africa servers using the ODC. The parallel computing of the DE Africa servers supports the complex workflows necessary for the sophisticated algorithms used in machine learning, image processing, and statistics [93].

4.4. Implications for Agriculture and Food Security

The results provide insights into the implications for agricultural vegetation in the 30-year period. Considering that agricultural vegetation is photosynthetic and that photosynthetic vegetation has declined over the study period, our results suggest that it has become more challenging to grow agricultural plant products in Djibouti. These results generally align with the findings of other researchers [9,94–96]. These implications are essential for understanding the dynamics and sustainability of agriculture and its larger implications for food security in Djibouti.

The negative trend of a mean pv of 0.00028% throughout the study period points to the loss of ~0.09 km^2 of arable land per year, and ~2.68 km^2 of arable land throughout the 30-year period between 1990 and 2020. Contrarily to the results of this study, the World Bank reports a 0.1% increase since 2000, and states that, in 2023, 0.15% of Djibouti's 23,200 km^2 was arable, which equates to 34.5 km^2 [50,56]. Considering these values, our study suggests that Djibouti loses about 0.2% of its arable land per year on average. Over the thirty-year period, Djibouti lost the equivalent of 7.7% of its currently arable land.

Unfortunately, in the absence of in situ data confirming the areas of true agricultural vegetation, it is difficult to precisely determine agricultural abundance and health over the study period. High-resolution remote sensing of the study area and hyperspectral imaging for the region are beyond the scope of this study but warrant further investigation to validate the results and confirm the agricultural trends independent of fractional cover. The trend toward decreasing photosynthetic vegetation is concerning enough to call for further investigations into why photosynthetic vegetation is decreasing, and possible mitigation strategies that could include increased political investments, adoption of climate-resilient farming styles, or increased planting of climate-resilient crops.

5. Conclusions

The major conclusion of this study is that, based on the FC and NDVI measures, Djibouti's vegetation has declined in abundance and health between the beginning of 1990 and the beginning of 2020. The ODC platform made processing quicker and easier than traditional GUI- or Python-based iterative methods. A lack of in situ data makes these conclusions less certain, but the comparison of FC and NDVI metrics suggests good agreement and provides supporting evidence for these further conclusions:

- The FC values for pv suggest an overall decline of vegetation abundance and health, in alignment with an increasingly arid environment and less photosynthetic vegetation, equating to a loss that is equivalent to ~7.7% of Djibouti's arable area.
- Districts with the greatest levels of pv and NDVI showed the greatest variability in those measures, pointing to the sensitive nature of photosynthetic vegetation in arid regions.
- Climate change, although not studied here, poses an imminent threat to photosynthetic vegetation, agriculture, and food security in Djibouti and other arid nations.

Overall, further investment in climate-resilient strategies would surely benefit Djibouti and the region, while improving geopolitical stability in the face of climate change and the connected challenges facing vegetation and agriculture in the Horn of Africa.

Author Contributions: Methodology, J.W.; Software, J.W.; Formal analysis, J.W.; Investigation, J.W.; Data curation, J.W.; Writing—original draft, J.W.; Writing—review & editing, Z.P.; Visualization, J.W.; Supervision, Z.P. All authors have read and agreed to the published version of the manuscript.

Funding: This research received no external funding.

Data Availability Statement: We analyzed publicly available datasets in this study. Data are freely available here: https://explorer.digitalearth.africa/products/fc_ls (accessed on 22 February 2023). Other data presented in this study are openly available at https://explorer.digitalearth.africa/products/wofs_ls (accessed on 22 February 2023).

Conflicts of Interest: The authors declare no conflicts of interest.

References

1. Sun, W.Y.; Song, X.Y.; Mu, X.M.; Gao, P.; Wang, F.; Zhao, G.J. Spatiotemporal vegetation cover variations associated with climate change and ecological restoration in the Loess Plateau. *Agric. For. Meteorol.* **2015**, *209*, 87–99. [CrossRef]
2. Xu, X.K.; Chen, H.; Levy, J.K. Spatiotemporal vegetation cover variations in the Qinghai-Tibet Plateau under global climate change. *Chin. Sci. Bull.* **2008**, *53*, 915–922. [CrossRef]
3. Dore, M.H.I. Climate change and changes in global precipitation patterns: What do we know? *Environ. Int.* **2005**, *31*, 1167–1181. [CrossRef] [PubMed]
4. Huang, J.; Ji, M.; Xie, Y.; Wang, S.; He, Y.; Ran, J. Global semi-arid climate change over last 60 years. *Clim. Dyn.* **2016**, *46*, 1131–1150. [CrossRef]
5. El-Beltagy, A.; Madkour, M. Impact of climate change on arid lands agriculture. *Agric. Food Secur.* **2012**, *1*, 3. [CrossRef]
6. Aldababseh, A.; Temimi, M.; Maghelal, P.; Branch, O.; Wulfmeyer, V. Multi-Criteria Evaluation of Irrigated Agriculture Suitability to Achieve Food Security in an Arid Environment. *Sustainability* **2018**, *10*, 803. [CrossRef]
7. Rishmawi, K.; Prince, S.D.; Xue, Y. Vegetation Responses to Climate Variability in the Northern Arid to Sub-Humid Zones of Sub-Saharan Africa. *Remote Sens.* **2016**, *8*, 910. [CrossRef]
8. Grote, U.; Fasse, A.; Nguyen, T.T.; Erenstein, O. Food security and the dynamics of wheat and maize value chains in Africa and Asia. *Front. Sustain. Food Syst.* **2021**, *4*, 617009. [CrossRef]
9. Qu, C.; Hao, X. Agriculture drought and food security monitoring over the horn of Africa (HOA) from space. In Proceedings of the 2018 7th International Conference on Agro-geoinformatics (Agro-geoinformatics), Hangzhou, China, 6–9 August 2018; pp. 1–4.
10. USDA. *CropScape and Cropland Data Layers Information*; USDA: Washington, DC, USA, 2022.
11. Yin, H.; Brandão, A.; Buchner, J.; Helmers, D.; Iuliano, B.G.; Kimambo, N.E.; Lewińska, K.E.; Razenkova, E.; Rizayeva, A.; Rogova, N.; et al. Monitoring cropland abandonment with Landsat time series. *Remote Sens. Environ.* **2020**, *246*, 111873. [CrossRef]
12. Weier, J. *Measuring Vegetation (NDVI and EVI)*; Herring, D., Ed.; NASA: Washington, DC, USA, 2000.
13. Hazaymeh, K.; Hassan, Q.K. Remote sensing of agricultural drought monitoring: A state of art review. *AIMS Environ. Sci.* **2016**, *3*, 604–630. [CrossRef]
14. Bégué, A.; Madec, S.; Lemettais, L.; Leroux, L.; Interdonato, R.; Becker-Reshef, I.; Barker, B.; Justice, C.; Kerdilés, H.; Meroni, M. How well do EO-based food security warning systems for food security agree? Comparison of NDVI-based vegetation anomaly maps in West Africa. *IEEE J. Sel. Top. Appl. Earth Obs. Remote Sens.* **2023**, *16*, 1641–1653. [CrossRef]
15. Kopp, S.; Becker, P.; Doshi, A.; Wright, D.J.; Zhang, K.; Xu, H. Achieving the Full Vision of Earth Observation Data Cubes. *Data* **2019**, *4*, 94. [CrossRef]
16. ODC. Open Data Cube. Available online: https://www.opendatacube.org (accessed on 14 December 2022).
17. Wagner, W.; Bauer-Marschallinger, B.; Navacchi, C.; Reuß, F.; Cao, S.; Reimer, C.; Schramm, M.; Briese, C. A Sentinel-1 Backscatter Datacube for Global Land Monitoring Applications. *Remote Sens.* **2021**, *13*, 4622. [CrossRef]
18. ODC. Open Data Cube Overview. Available online: https://www.opendatacube.org/overview (accessed on 14 December 2022).
19. Wellington, M.J.; Renzullo, L.J.; Hosseini, M. High-Dimensional Satellite Image Compositing and Statistics for Enhanced Irrigated Crop Mapping. *Remote Sens.* **2021**, *13*, 1300. [CrossRef]
20. Africa, D. Digital Earth Africa. Available online: https://www.digitalearthafrica.org (accessed on 14 October 2022).
21. Digital Earth Africa. GeoMAD. 2022. Available online: https://docs.digitalearthafrica.org/en/latest/data_specs/GeoMAD_specs.html#Technical-information (accessed on 14 October 2022).
22. Digital Earth Africa. Fractional Cover. 2021 ed. 2021. Available online: https://docs.digitalearthafrica.org/en/latest/data_specs/Fractional_Cover_specs.html?highlight=fractional%20cover (accessed on 14 October 2022).
23. Govind, A. Towards Climate Change Preparedness in the MENA's Agricultural Sector. *Agronomy* **2022**, *12*, 279. [CrossRef]
24. Brown, M.E.; Funk, C.C.; Galu, G.; Choularton, R. Earlier famine warning possible using remote sensing and models. *Eos Trans. Am. Geophys. Union* **2007**, *88*, 381–382. [CrossRef]
25. Fre, Z.; Tsegay, B. Economic contribution of pastoral and agro pastoral production to food security and livelihoods systems in Africa: The case of Eastern Sudan, Eritrea and Western Ethiopia in the Horn of Africa. *Ethiop. E—J. Res. Innov. Foresight* **2013**, *5*, 14–31.
26. Oroda, A. Application of remote sensing to early warning for food security and environmental monitoring in the Horn of Africa. *Int. Arch. Photogramm. Remote Sens. Spat. Inf. Sci.* **2001**, *34*, W6.
27. Pavanello, S. Working across Borders—Harnessing the Potential of Cross-Border Activities to Improve Livelihood Security in the Horn of Africa Drylands. Oversees Development Institute-Humanitarian Policy Group. 2010. Available online: https://odi.org/en/publications/working-across-borders-harnessing-the-potential-of-cross-border-activities-to-improve-livelihood-security-in-the-horn-of-africa-drylands/ (accessed on 14 October 2022).
28. Sasson, A. Food security for Africa: An urgent global challenge. *Agric. Food Secur.* **2012**, *1*, 1–16. [CrossRef]
29. Thrupp, L.A.; Megateli, N. *Critical Links: Food Security and the Environment in the Greater Horn of Africa*; World Resources Institute: Washington, DC, USA, 1999.

30. Vogels, M.; de Jong, S.; Sterk, G.; Douma, H.; Addink, E. Spatio-Temporal Patterns of Smallholder Irrigated Agriculture in the Horn of Africa Using GEOBIA and Sentinel-2 Imagery. *Remote Sens.* **2019**, *11*, 143. [CrossRef]
31. Vrieling, A.; de Leeuw, J.; Said, M. Length of Growing Period over Africa: Variability and Trends from 30 Years of NDVI Time Series. *Remote Sens.* **2013**, *5*, 982–1000. [CrossRef]
32. UN. United Nations Global Issues Population. Available online: https://www.un.org/en/global-issues/population#:~:text=Africa:%20fastest%20growing%20continent,projected%20to%20double%20by%202050 (accessed on 14 December 2022).
33. Devereux, S.; Edwards, J. *Climate Change and Food Security*; John Wiley & Sons: Hoboken, NJ, USA, 2004.
34. Shiferaw, B.; Prasanna, B.M.; Hellin, J.; Bänziger, M. Crops that feed the world 6. Past successes and future challenges to the role played by maize in global food security. *Food Secur.* **2011**, *3*, 307–327. [CrossRef]
35. USAID. FEWS NET Djibouti Monitoring Report. Available online: https://fews.net/east-africa/djibouti (accessed on 12 December 2022).
36. United Nations. No. 46495. International Fund for Agricultural Development and Djibouti. In *United Nations Treaty Series*; UN: New York, NY, USA, 2018; 380p.
37. Magin, G. Djibouti. BirdLife International. 1999. Available online: http://www.datazone.birdlife.org/userfiles/file/IBAs/AfricaCntryPDFs/Djibouti.pdf (accessed on 15 September 2022).
38. Hassen, T.B.; El Bilali, H. Food security in the gulf cooperation council countries: Challenges and prospects. *J. Food Secur.* **2019**, *7*, 159–169. [CrossRef]
39. Mutopo, P.; Beyene, A.; Haaland, H.; Boamah, F.; Widengård, M.; Skarstein, R. *Biofuels, Land Grabbing and Food Security in Africa*; Bloomsbury Publishing: London, UK, 2011.
40. Food and Agricultural Organization of the United Nations. Djibouti Country Profile. Available online: https://www.fao.org/countryprofiles/index/en/?iso3=DJI (accessed on 14 October 2022).
41. Scarth, P.; Roder, A.; Schmidt, M. Tracking grazing pressure and climate interaction—The role of Landsat fractional cover in time series analysis. In Proceedings of the 15th Australasian Remote Sensing & Photogrammetry Conference (ARSPC), Alice Springs, Australia, 13–17 September 2010.
42. Hill, M.J.; Guerschman, J.P. Global trends in vegetation fractional cover: Hotspots for change in bare soil and non-photosynthetic vegetation. *Agric. Ecosyst. Environ.* **2022**, *324*, 107719. [CrossRef]
43. Ustin, S.L.; Roberts, D.A.; Gamon, J.A.; Asner, G.P.; Green, R.O. Using Imaging Spectroscopy to Study Ecosystem Processes and Properties. *BioScience* **2004**, *54*, 523–534. [CrossRef]
44. Dennison, P.; Roberts, D.A.; Chambers, J.Q.; Daughtry, C.S.; Guerschman, J.P.; Kokaly, R.F.; Okin, C.G.S.; Scarth, P.F. *Global Measurement of Non-Photosynthetic Vegetation*; NASA: Washington, DC, USA, 2016.
45. Liu, J.; Fan, J.; Yang, C.; Xu, F.; Zhang, X. Novel vegetation indices for estimating photosynthetic and non-photosynthetic fractional vegetation cover from Sentinel data. *Int. J. Appl. Earth Obs. Geoinf.* **2022**, *109*, 102793. [CrossRef]
46. Tian, J.; Su, S.; Tian, Q.; Zhan, W.; Xi, Y.; Wang, N. A novel spectral index for estimating fractional cover of non-photosynthetic vegetation using near-infrared bands of Sentinel satellite. *Int. J. Appl. Earth Obs. Geoinf.* **2021**, *101*, 102361. [CrossRef]
47. Guerschman, J.P.; Hill, M.J.; Renzullo, L.J.; Barrett, D.J.; Marks, A.S.; Botha, E.J. Estimating fractional cover of photosynthetic vegetation, non-photosynthetic vegetation and bare soil in the Australian tropical savanna region upscaling the EO-1 Hyperion and MODIS sensors. *Remote Sens. Environ.* **2009**, *113*, 928–945. [CrossRef]
48. Muir, J.; Schmidt, M.; Tindall, D.; Trevithick, R.; Scarth, P.; Stewart, J.B. *Field Measurement of Fractional Ground Cover*; Australian Bureau of Agricultural and Resource Economics and Sciences: Canberra, Australian, 2011.
49. Zheng, G.; Bao, A.; Li, X.; Jiang, L.; Chang, C.; Chen, T.; Gao, Z. The potential of multispectral vegetation indices feature space for quantitatively estimating the photosynthetic, non-photosynthetic vegetation and bare soil fractions in northern china. *Photogramm. Eng. Remote Sens.* **2019**, *85*, 65–76. [CrossRef]
50. World Bank. Djibouti. 2021. Available online: https://data.worldbank.org/country/djibouti (accessed on 22 September 2022).
51. CIA. *The World Factbook Country Summary*; CIA: Langley, VA, USA, 2022.
52. United Nations Development Programme. *Djibouti: Strengthening Resilience to Build Food Security*; United Nations Development Programme: New York, NY, USA, 2018.
53. United Nations Development Programme. *Djibouti*; United Nations Development Programme: New York, NY, USA, 2021.
54. United Nations Statistics. Djibouti Environment Statistics Country Snapshot. 2016. Available online: https://unstats.un.org/unsd/ENVIRONMENT/envpdf/Country_Snapshots_Dec_2016/Djibouti.pdf (accessed on 15 November 2022).
55. US Department of State. Djibouti. 2023. Available online: https://travel.state.gov/content/travel/en/international-travel/International-Travel-Country-Information-Pages/Djibouti.html (accessed on 14 October 2023).
56. World Bank. Djibouti. 2023. Available online: https://data.worldbank.org/country/djibouti (accessed on 14 September 2023).
57. United Nations Development Programme—Djibouti. 2023. Available online: https://www.dj.undp.org/ (accessed on 14 November 2023).
58. African Union. Djibouti. 2022. Available online: https://au.int/en/countryprofiles-3 (accessed on 30 March 2024).
59. World Health Organization. Djibouti. 2023. Available online: https://www.who.int/countries/dji/en/ (accessed on 15 November 2023).
60. United Nations Educational, Scientific and Cultural Organization. Djibouti. 2023. Available online: https://en.unesco.org/countries/djibouti (accessed on 14 November 2023).

61. United Nations Conference on Trade and Development. *Djibouti: Harnessing Trade for Human Development in the Digital Age*; United Nations Conference on Trade and Development: Geneva, Switzerland, 2019.
62. African Development Bank Group. Djibouti Economic Outlook. 2021. Available online: https://www.afdb.org/en/countries/east-africa/djibouti/djibouti-economic-outlook (accessed on 14 November 2022).
63. Wardle, J.A.; Sagan, V.; Mohammed, F. Using Open Data Cube on the Cloud to Investigate Food Security by Means of Cropland Changes in Djibouti. *Int. Arch. Photogramm. Remote Sens. Spat. Inf. Sci.* **2022**, *XLIII-B3-2022*, 1039–1044. [CrossRef]
64. Schmidt, M.; Denham, R.; Scarth, P. Fractional ground cover monitoring of pastures and agricultural areas in Queensland. In Proceedings of the 15th Australian Remote Sensing and Photogrammetry Conference (ARSPC), Alice Springs, Australia, 13–17 September 2010; pp. 13–17.
65. Digital Earth Africa. Water Observations from Space. 2022. Available online: https://docs.digitalearthafrica.org/en/latest/data_specs/Landsat_WOfS_specs.html (accessed on 22 September 2022).
66. National Aeronautics and Space Administration. Landsat 7 2021. Available online: https://landsat.gsfc.nasa.gov/satellites/landsat-7/ (accessed on 14 October 2022).
67. Dask. Available online: https://docs.dask.org/en/stable/index.html (accessed on 22 February 2023).
68. Jiang, Z.; Huete, A.R.; Chen, J.; Chen, Y.; Li, J.; Yan, G.; Zhang, X. Analysis of NDVI and scaled difference vegetation index retrievals of vegetation fraction. *Remote Sens. Environ.* **2006**, *101*, 366–378. [CrossRef]
69. Gao, L.; Wang, X.; Johnson, B.A.; Tian, Q.; Wang, Y.; Verrelst, J.; Mu, X.; Gu, X. Remote sensing algorithms for estimation of fractional vegetation cover using pure vegetation index values: A review. *ISPRS J. Photogramm. Remote Sens.* **2020**, *159*, 364–377. [CrossRef] [PubMed]
70. Elmore, A.J.; Mustard, J.F.; Manning, S.J.; Lobell, D.B. Quantifying vegetation change in semiarid environments: Precision and accuracy of spectral mixture analysis and the normalized difference vegetation index. *Remote Sens. Environ.* **2000**, *73*, 87–102. [CrossRef]
71. Kendall, M.G. *Rank Correlation Methods*; Charles Griffin: London, UK, 1955.
72. Kendall Tau Metric. 2023. Available online: https://encyclopediaofmath.org/index.php?title=Kendall_tau_metric (accessed on 22 September 2022).
73. Sen, P.K. Estimates of the regression coefficient based on Kendall's tau. *J. Am. Stat. Assoc.* **1968**, *63*, 1379–1389. [CrossRef]
74. Mann, H.B. Nonparametric tests against trend. *Econ. J. Econ. Soc.* **1945**, *13*, 245–259. [CrossRef]
75. Daniel, W.W.; Cross, C.L. *Biostatistics: A Foundation for Analysis in the Health Sciences*; Wiley: Hoboken, NJ, USA, 2018.
76. De Smith, M.J.; Goodchild, M.F.; Longley, P. *Geospatial Analysis: A Comprehensive Guide to Principles, Techniques and Software Tools*; Troubador Publishing Ltd.: Leicestershire, UK, 2007.
77. Teegavarapu, R. *Trends and Changes in Hydroclimatic Variables: Links to Climate Variability and Change*; Elsevier: Amsterdam, The Netherlands, 2018.
78. Hussain, e.a. pyMannKendall. Available online: https://pypi.org/project/pymannkendall/ (accessed on 18 May 2023).
79. Hussain, e.a. pyMannKendall: A python package for non parametric Mann Kendall family of trend tests. *J. Open Source Softw.* **2019**, *4*, 1556. [CrossRef]
80. Sutton, A.; Fisher, A.; Metternicht, G. Assessing the Accuracy of Landsat Vegetation Fractional Cover for Monitoring Australian Drylands. *Remote Sens.* **2022**, *14*, 6322. [CrossRef]
81. Liu, W.; Chen, Y.; He, X.; Mao, P.; Tian, H. Is Current Research on How Climate Change Impacts Global Food Security Really Objective? *Foods* **2021**, *10*, 2342. [CrossRef]
82. Muluneh, M.G. Impact of climate change on biodiversity and food security: A global perspective—A review article. *Agric. Food Secur.* **2021**, *10*, 1–25. [CrossRef]
83. Gao, B.C. NDWI—A normalized difference water index for remote sensing of vegetation liquid water from space. *Remote Sens. Environ.* **1996**, *58*, 257–266. [CrossRef]
84. Scanlon, T.M.; Albertson, J.D.; Caylor, K.K.; Williams, C.A. Determining land surface fractional cover from NDVI and rainfall time series for a savanna ecosystem. *Remote Sens. Environ.* **2002**, *82*, 376–388. [CrossRef]
85. Jiménez-Muñoz, J.C.; Sobrino, J.A.; Plaza, A.; Guanter, L.; Moreno, J.; Martínez, P. Comparison between fractional vegetation cover retrievals from vegetation indices and spectral mixture analysis: Case study of PROBA/CHRIS data over an agricultural area. *Sensors* **2009**, *9*, 768–793. [CrossRef] [PubMed]
86. Huang, S.; Tang, L.N.; Hupy, J.P.; Wang, Y.; Shao, G.F. A commentary review on the use of normalized difference vegetation index (NDVI) in the era of popular remote sensing. *J. For. Res.* **2021**, *32*, 1–6. [CrossRef]
87. Carlson, T.N.; Ripley, D.A. On the relation between NDVI, fractional vegetation cover, and leaf area index. *Remote Sens. Environ.* **1997**, *62*, 241–252. [CrossRef]
88. Carlson, T.N.; Gillies, R.R.; Perry, E.M. A method to make use of thermal infrared temperature and NDVI measurements to infer surface soil water content and fractional vegetation cover. *Remote Sens. Rev.* **1994**, *9*, 161–173. [CrossRef]
89. Xiao, J.; Moody, A. A comparison of methods for estimating fractional green vegetation cover within a desert-to-upland transition zone in central New Mexico, USA. *Remote Sens. Environ.* **2005**, *98*, 237–250. [CrossRef]
90. Guan, K.; Wood, E.F.; Caylor, K. Multi-sensor derivation of regional vegetation fractional cover in Africa. *Remote Sens. Environ.* **2012**, *124*, 653–665. [CrossRef]

91. Appel, M.; Pebesma, E. On-Demand Processing of Data Cubes from Satellite Image Collections with the gdalcubes Library. *Data* **2019**, *4*, 92. [CrossRef]
92. Killough, B.; Siqueira, A.; Dyke, G. Advancements in the Open Data Cube and Analysis Ready Data—Past, Present and Future. In Proceedings of the IGARSS 2020—2020 IEEE International Geoscience and Remote Sensing Symposium, 26 September–2 October 2020; pp. 3373–3375.
93. Anaconda. Dask Distributed Documentation. Available online: https://distributed.dask.org/en/latest/ (accessed on 13 September 2023).
94. Zhou, J.; Chen, J.; Chen, X.; Zhu, X.; Qiu, Y.; Song, H.; Rao, Y.; Zhang, C.; Cao, X.; Cui, X. Sensitivity of six typical spatiotemporal fusion methods to different influential factors: A comparative study for a normalized difference vegetation index time series reconstruction. *Remote Sens. Environ.* **2021**, *252*, 112130. [CrossRef]
95. African Union, Semi-Arid Food Grain Research and Development. Challenges and Opportunities for Strategic Agricultural Commodity Value Chains Development in the IGAD Region. 2014. Available online: https://archives.au.int/bitstream/handle/123456789/1973/Challenges%20and%20Opportunities%20for%20Strategic%20Agricultural%20Commodity%20Value%20Chains_E.pdf?sequence=1&isAllowed=y (accessed on 22 September 2022).
96. Chandrasekharam, D.; Lashin, A.; Nassir Al, A.; Al-Bassam, A.M.; Varun, C. Geothermal energy for desalination to secure food security: Case study in Djibouti. *Energy Sustain. Soc.* **2019**, *9*, 24. [CrossRef]

remote sensing

MDPI

Article

Multiple Greenness Indexes Revealed the Vegetation Greening during the Growing Season and Winter on the Tibetan Plateau despite Regional Variations

Jinxia Lv [1], Wenwu Zhao [1,*], Ting Hua [1], Lihao Zhang [1] and Paulo Pereira [2]

[1] State Key Laboratory of Earth Surface Processes and Rescurce Ecology, Faculty of Geographical Science, Beijing Normal University, Beijing 100875, China; lvjinxia@mail.bnu.edu.cn (J.L.); 201831051056@mail.bnu.edu.cn (T.H.); lihaozhang@mail.bnu.edu.cn (L.Z.)

[2] Environmental Management Center, Mykolas Romeris University, Ateities g. 20, 08303 Vilnius, Lithuania; pereiraub@gmail.com

* Correspondence: zhaoww@bnu.edu.cn

Abstract: Vegetation is an essential component of terrestrial ecosystems and supplies multiple ecosystem benefits and services. Several indices have been used to monitor changes in vegetation communities using remotely-sensed data. However, only a few studies have conducted a comparative analysis of different indices concerning vegetation greenness variation. Additionally, there have been oversights in assessing the change in greenness of evergreen woody species. In this study, we used the normalized difference vegetation index (NDVI), the enhanced vegetation index (EVI), the near-infrared reflectance of terrestrial vegetation (NIRv), and the leaf area index (LAI) data derived from MODIS data to examine spatial and temporal change in vegetation greenness in the growing season (May–September) and then evaluated the evergreen vegetation greenness change using winter (December–February) greenness using trend analysis and consistency assessment methods between 2000 and 2022 on the Tibetan Plateau, China. The results found that vegetation greenness increased in 80% of pixels during the growing season (northeastern, central-eastern, and northwestern regions). Nevertheless, a decline in the southwestern and central-southern areas was identified. Similar trends in greenness were also observed in winter in about 80% of pixels. Consistency analyses based on the four indexes showed that vegetation growth was enhanced by 29% and 30% of pixels in the growing season and winter, respectively. Further, there was relatively strong consistency among the different vegetation indexes, particularly between the NIRv and EVI. The LAI was less consistent with the other indexes. These findings emphasize the importance of selecting an appropriate index when monitoring long-term temporal trends over large spatial scales.

Keywords: alpine vegetation; greenness; consistency; multiple indexes; evergreen vegetation

1. Introduction

Vegetation is critical in terrestrial ecosystems and provides multiple ecosystem services, such as carbon storage [1,2]. During the past few decades, vegetation greenness showed an increasing trend in high latitude and altitude regions, such as Arctic tundra, boreal forest, and northeastern Tibetan Plateau [3–5]. Vegetation dynamics affect water cycles and the exchange of carbon and energy, impacting the climate [6–9]. The Tibetan Plateau is known as the "Roof of the World" and the "Third Pole", averaging 4000 m above sea level [10]. This region is also the water tower of Asia, with abundant glaciers, lakes, and rivers [11]. The diverse terrestrial ecosystems distributed the forests, shrubs, alpine grasslands, and alpine deserts along the northwestward direction [12]. Furthermore, vegetation variations in the Tibetan Plateau affected the land surface process, ecosystem structure, and functions [13–17]. For example, grassland growth could attenuate warming by increasing evapotranspiration [13]. Also, the enhanced vegetation greenness on the eastern Tibetan Plateau increases the latent and sensible heat fluxes and affects the regional rainfall [18]. On the other hand, alpine meadow degradation reduces soil carbon storage and nutrient cycling capacity [14]. Furthermore, winter evergreen vegetation is also important in the Tibetan plateau ecosystem dynamic. For instance, deforestation by evergreen coniferous trees reduced the dry season (September–February) rainfall in the east of the Tibetan Plateau (Zagunao watershed) [19]. Evergreen vegetation exhibited the highest gross primary productivity per unit area across the plateau [20]. Therefore, vegetation dynamics are important to the Tibetan Plateau's ecological status.

During the past few decades, vegetation greenness has been assessed by using various remote sensing vegetation indexes, such as the leaf area index (LAI), the normalized difference vegetation index (NDVI), and the enhanced vegetation index (EVI) [21–23]. NDVI and EVI are widely adopted due to their ease of calculation from surface reflectance in optical spectral bands and the vegetation cover indices, which are usually represented by the canopy structure parameter LAI [24–26]. From 1982 to 2015, a substantial greening trend was observed in the northeastern and central, southwestern, eastern, and south-central areas of the Tibetan Plateau, while browning was identified in some areas located in the southeastern part [25,27,28]. Other works showed that from 2000 to the late 2010s, vegetation greenness increased in northeastern, northwestern, east-central, and southeastern areas, and browning was observed in southwestern areas [5,29,30]. These results showed some spatial differences in the areas affected by greening or browning. For example, the areas of vegetation using NDVI in the southwest of the Tibetan Plateau were larger than those indicated by LAI and EVI [21]. EVI analysis showed high vegetation browning in central areas and some areas in the southeast of the Tibetan Plateau compared to NDVI [5,22]. Recently, a new vegetation index proxy, the near-infrared reflectance of terrestrial vegetation (NIRv), was developed and applied to assess the trend of vegetation greenness [21,31]. NIRv assessment showed that greening areas were lower than those indicated by NDVI, EVI, and LAI on the Tibetan Plateau [21]. Therefore, vegetation changes could be inconsistent using different indexes.

Vegetation greenness changes on the Tibetan Plateau were mainly focused on the annual or growing season, dominated by deciduous vegetation (grasslands) [22,29]. In the southeast of the Tibetan Plateau, evergreen vegetation is widely distributed and plays a vital role in carbon storage [32]. Previous studies focused more on evergreen trees' radial growth variation, shrubs, treelines, and shrub lines [33,34]. For instance, evergreen trees' radial growth showed an increasing trend in the southeast and south Tibetan Plateau and the south-central Tibetan Plateau from the 1960s to the 2010s [34–39]. Nevertheless, a reduction in evergreen tree and shrub growth was also identified in the northeastern edge of the Tibetan Plateau (during 1950–2013), south-central Tibetan Plateau (during 1957–2010), and central Himalayas (during the 1970s–2010s) [40–43]. Also, treelines shifted upward in the northeastern and southeastern Tibetan Plateau during the 1910s–2000s due to increasing temperatures [33]. Nevertheless, they remained stable on the southern edge of the Tibetan Plateau due to increasing tree density and intraspecific competition (i.e., spatial segregation between seedlings, juveniles, and adults) [44,45]. All these studies were observed at the local scale. Few assessments were conducted on larger scales. Therefore, studies focused on evergreen vegetation greenness changes at larger scales are unclear. The aims of this study are to:

(1) Evaluate the spatial and temporal change characteristics of vegetation greenness on the Tibetan Plateau during the growing season between 2000 and 2022 using four vegetation greenness indexes.
(2) On the basis of an analysis of growing season vegetation, assess the temporal change in winter evergreen vegetation on the Tibetan Plateau.
(3) Analyze the consistency and difference in vegetation greenness variations using four vegetation greenness indexes.

2. Materials and Methods

2.1. Study Area

The Tibetan Plateau is in southwestern China and has an average elevation of over 4000 m above sea level [11] (Figure 1a). The plateau hosts various types of vegetation, including evergreen forests and several alpine ecosystem types (grasslands, shrublands, and cushion vegetation) (Figure 1b). The region is generally characterized by a cold and dry climate [46]. The mean annual temperature increased from −16 °C to 23 °C (Figure 1c). The mean annual precipitation was highest on the southern edge (i.e., the south face of the Himalayas), relatively lower in the southeast, and decreasing northwestward (Figure 1d). The plateau is also the source of several major rivers in Asia, with implications for the water cycle and water supply of millions of people downstream [11]. Furthermore, vegetation on the Tibetan Plateau is a crucial driver of ecosystem and atmospheric dynamics at the regional and continental scales [5,47–50].

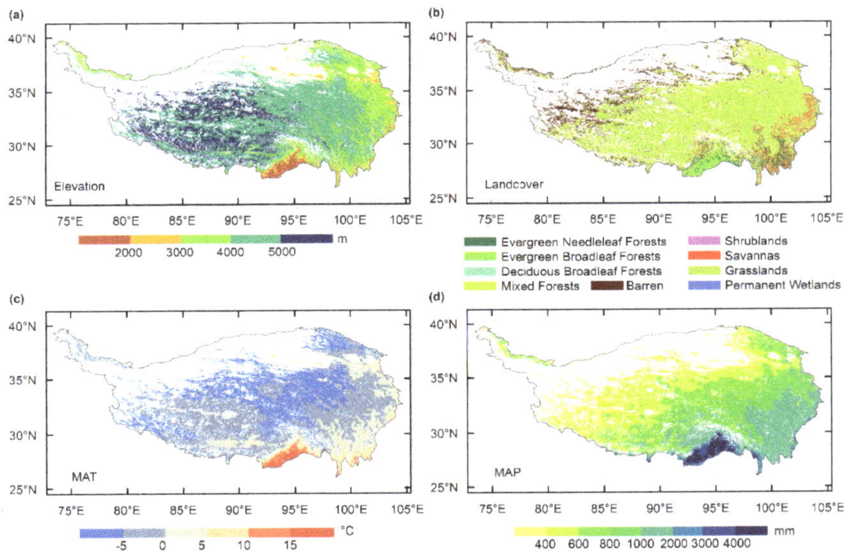

Figure 1. The spatial pattern of elevation, derived from Shuttle Radar Topography Mission version 3 (**a**), the landcover from MCD12Q1 version 6.1 based on the International Geosphere-Biosphere Program classification scheme of 2010 (**b**), the mean annual temperature (**c**), and the mean annual precipitation from the ERA5-Land dataset during 2000–2021 (**d**) on the Tibetan Plateau.

2.2. Framework

Firstly, we evaluated the trend of vegetation greenness during the growing season (May–September) based on four vegetation indices, including NDVI, EVI, NIRv, and LAI. Due to the coexistence of evergreen and winter-deciduous vegetation in numerous regions of the Tibetan Plateau, coupled with uncertainties in existing vegetation type maps [51,52], it is imperative to mitigate the impact of deciduous vegetation on assessing evergreen vegetation greenness. To address this, we conducted an analysis of vegetation greenness trends during the winter, a period when deciduous vegetation lacks greenness. In addition, to exclude the effect of snow cover on winter vegetation greenness, the normalized difference snow index (NDSI) snow cover was used to validate the variation of snow cover in winter. Furthermore, we used the MODIS land cover and GlobeLand30 to exclude water bodies and artificial surfaces in this study area. In summary, we first exclude the low-quality data on winter vegetation greenness and average the greenness in the growing season to evaluate the spatial variation of vegetation greenness in the growing season and winter. Then, we analyze the consistency of the trend of vegetation greenness in the growing season and winter based on different vegetation indices (Figure 2).

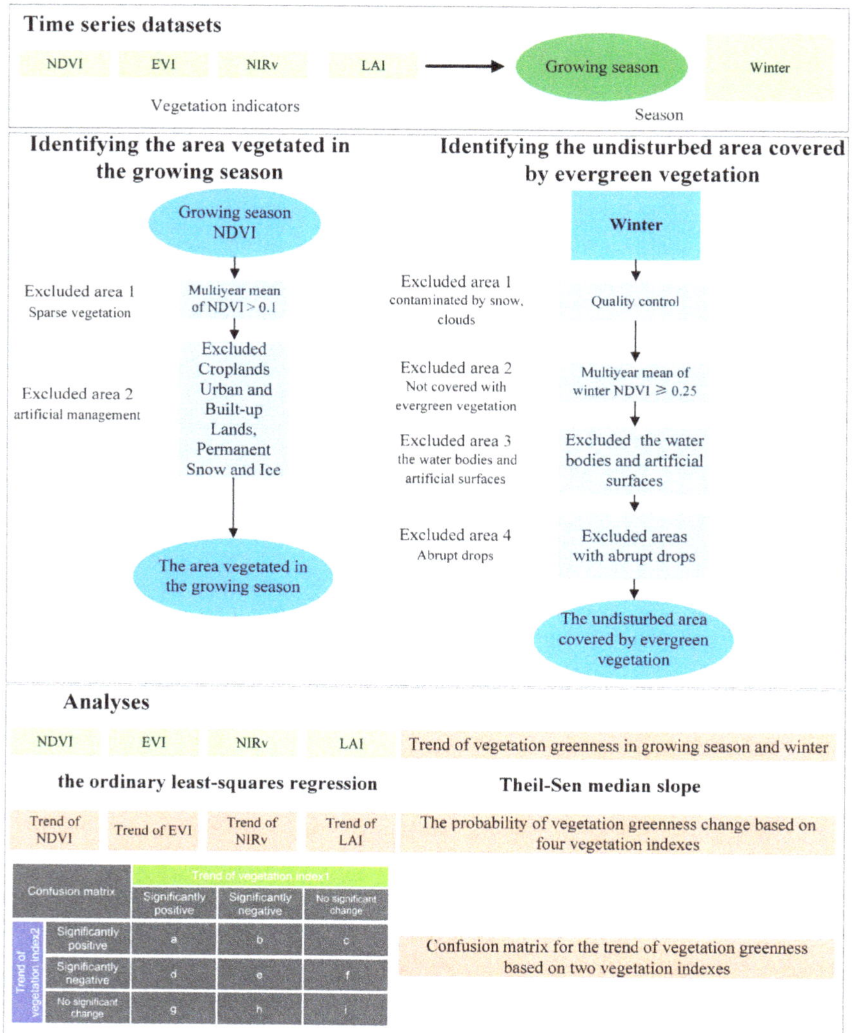

Figure 2. Study framework.

2.3. *Dataset*

2.3.1. Vegetation Greenness

Vegetation greenness was quantified with NDVI, EVI, NIRv, and LAI between 2000 and 2022 [53–55]. The formulas for NDVI, EVI, NIRv, and LAI are as follows:

$$NDVI = \frac{NIR - R}{NIR + R} \tag{1}$$

In the Formula (1), *NIR* and *R* are the spectral reflectances of the near-infrared and red bands, respectively.

$$EVI = G \times \frac{NIR - R}{NIR + C_1 \times R - C_2 \times Blue + L} \quad (2)$$

In Formula (2), *NIR*, *R*, and *Blue* are the spectral reflectances of the near-infrared, red, and blue bands; *G* is the gain factor; *L* is a canopy background adjustment term; and C_1 and C_2 are the coefficients of the aerosol resistance term. The coefficients used in the EVI algorithm are $L = 1$, $C_1 = 6$, $C_2 = 7.5$, and $G = 2.5$ [54].

LAI is defined as the one-sided green leaf area per unit ground area in broadleaf canopies and as the projected needle leaf area in coniferous canopies [55].

$$NIRv = (NDVI - 0.08) \times NIR \quad (3)$$

In the Formula (3), *NIR* is the spectral reflectance of the near-infrared band [53].

NDVI and EVI were extracted from the MOD13A1 version 6.1 product (accessed at https://earthengine.google.com on 10 December 2022). The NIRv was calculated by NDVI and reflectance from near-infrared band extracting from MOD13A1 version 6.1 product, which was better for capturing canopy physiological conditions than traditional vegetation index (i.e., NDVI and EVI) [53]. The MOD13A1 dataset has a spatial resolution of 500 m and a composite period of 16 days [56]. This dataset also provides quality information, including Pixel Reliability and VI (i.e., NDVI) Quality Assessment layers reflecting the status of the overall data quality rating and the conditions under which the data were obtained or processed, including clouds, snow/ice, shadows, and aerosols [54]. LAI was extracted from the MOD15A2H version 6.1 product (accessed at https://earthengine.google.com on 10 December 2022). This dataset has a spatial resolution of 500 m and a composite period of 8 days [57].

2.3.2. Solar-Induced Fluorescence Dataset

The solar-induced fluorescence (SIF) dataset can track the variation in the photosynthetic activity of terrestrial ecosystems [58]. Therefore, we explore the trend of SIF in the growing season to validate the greenness change [59]. The continuous SIF (CSIF) dataset during 2000–2019 was extracted from Orbiting Carbon Observatory-2 observations and MODIS surface reflectance using a neural network algorithm with a spatial resolution of 0.05° and 4 days of temporal resolution, and this dataset has been shown with low uncertainty [60].

2.3.3. Landcover Datasets

The landcover datasets used in this study are the GlobeLand30 dataset, accessed from https://www.nature.com/articles/514434c on 21 July 2021, and the MCD12Q1 version 6.1 dataset, accessed from https://search.earthdata.nasa.gov/ on 10 December 2022. The GlobeLand30 dataset provides landcover type at a spatial resolution of 30 m for 2000, 2010, and 2020, respectively [61]. The landcover types in the MCD12Q1 dataset are based on the International Geosphere-Biosphere Program (IGBP) scheme and have a spatial resolution of 500 m [62].

2.3.4. Snow Cover

The Normalized Difference Snow Index (NDSI) Snow Cover during 2000–2022 was extracted from the MOD10A1 version 6.1 product, with a spatial resolution of 500 m and a temporal resolution of 1 day [63]. The formula for NDSI is as follows:

$$NDSI = \frac{\rho_{VIS} - \rho_{NIR}}{\rho_{VIS} + \rho_{NIR}} \quad (4)$$

In Formula (4), ρ_{VIS} and ρ_{NIR} are the top-of-atmosphere reflectance of the visible band is centered at 0.55 μm and the near-infrared band is centered at 1.66 μm in the MODIS dataset [64].

The MOD10A1 dataset also provides the corresponding quality assessment (NDSI_Snow_Cover_Basic_QA) (accessed at https://search.earthdata.nasa.gov on 15 March 2023).

2.4. Data Preprocessing

The NDVI, EVI, NIRv, and LAI during 2000–2022 were extracted from monthly data based on the maximum value composite (MVC) method, which could eliminate the effects of clouds and atmospheric noise [6,65]. Previous studies revealed some outliers in the growing season due to cloudy contamination [66]. Therefore, we average the monthly data of the growing season (May–September) to derive growing season greenness. The multiyear mean of monthly NDVI < 0.1 was excluded in this study because it was considered covered with sparse vegetation [25]. We also exclude the pixels that are classified as Croplands, Cropland/Natural Vegetation Mosaics, Urban and Built-up Land, Permanent Snow, Ice, and Water based on the MCD12Q1 dataset, which could affect the trend of greenness evaluation due to different artificial management. After this step, we obtained the area of vegetation distribution in the growing season.

In winter, evergreen vegetation is easily contaminated by snow, clouds, and aerosols, which could lead to bias in the assessment of vegetation greenness trends. Therefore, we excluded the low-quality NDVI pixels, which are contaminated by snow, clouds, and aerosols, based on the 'Pixel Reliability' and 'VI Quality Assessment' layers. The EVI, NIRv, and LAI were processed in the same way. Then, to distinguish the area covered by evergreen vegetation, the pixels of the multiyear ($\geq$15) mean of winter NDVI < 0.25 are excluded, considered sparse vegetation or no evergreen vegetation. Further, we excluded the water bodies and artificial surfaces if the proportion of water bodies and artificial surfaces with 30 m pixels in Globland30 in 500 m pixels $\geq$ 5%. Based on the MODIS landcover dataset, we also excluded the 500 m MODIS pixel of Croplands, Cropland/Natural Vegetation Mosaics and Urban and Built-up Land. The pixels with abrupt drops were identified by applying the LandTrendr algorithm [67] to the NDVI time series and excluded. After removing low-quality NDVI, water bodies, artificial surfaces, Croplands, Cropland/Natural Vegetation Mosaics and Urban and Built-up Land, we obtained the area covered by evergreen vegetation on the Tibetan Plateau.

To further validate the trend of vegetation greenness, we calculated the trend of CSIF, as it is considered a reliable proxy for photosynthetic activity. We initially performed quality control on the CSIF time series data from 2000 to 2019 [68]. However, even after preprocessing, we observed outliers and biases in the data. To mitigate this, we divided the annual CSIF time series data into two seasons: photosynthetically active and photosynthetically non-active. The photosynthetically active season was defined as when CSIF values increased by 30% from the beginning to the day when CSIF values decreased by 70% in the latter half of the year. On the other hand, the photosynthetically non-active season was defined as the period outside the photosynthetically active season [69]. We observed that CSIF values were more variable during the photosynthetically active season and less variable during the photosynthetically non-active season. Therefore, we applied different criteria to eliminate outliers and biases in these two seasons [69]. Subsequently, we applied the MVC method and calculated the average monthly CSIF values during the growing season. Due to the difference in spatial resolution between CSIF vegetation greenness data (i.e., NDVI), we spatially aggregated the greenness values by the average method if all the pixels in a corresponding CSIF pixel were valued. All data processing in this study were conducted using MATLAB R2020b and IDL 8.5 software.

2.5. Data Analysis

2.5.1. Temporal Change of Vegetation Greenness

Temporal changes in vegetation greenness over 2000–2022 were assessed using temporal trends in greenness. This study analyzes the temporal changes in vegetation greenness in the growing season and winter. The greenness trend was calculated as the slope in linear regression between greenness and year using the ordinary least-squares regression when the length of vegetation greenness time series ≥ 15, and the significance level was determined by the t-test [25]. In addition, the Theil–Sen median slope for robustness was also used to evaluate the trend of vegetation greenness in the growing season owing to a contiguous time series of 22 years [70,71]. The Theil–Sen median slope did not evaluate the trend of winter vegetation greenness; as a result, the time series were not contiguous for 22 years.

2.5.2. Probability of Vegetation Greenness Change

To evaluate the consistency of different vegetation greenness indices, we first divided the trends of vegetation greenness during 2000–2022 into 3 categories at the 0.05 significance level: significantly positive (↑), significantly negative (↓) and no significant change (–). Then, according to the probabilities or likelihoods of vegetation growth trends [6,72], we summarized the probability categories (Table 1). The criteria were as follows:

(1) If four vegetation indexes (NDVI, EVI, NIRv, and LAI) showed significant change in the same direction, the probability of vegetation greenness change was defined as 'very likely' (i.e., four '↑' is defined as 'Enhanced very likely', and four '↓' is defined as 'Degraded very likely').
(2) If three vegetation indexes showed significant change, the probability of vegetation greenness change was defined as 'likely'.
(3) If two vegetation indexes showed significant change, the probability of vegetation greenness change was defined as 'probably'.
(4) If there was none or only one vegetation index that showed a significant change, the probability of vegetation greenness change was defined as 'Uncertainty'.

Table 1. The probability of vegetation's greenness changing.

Trend of NDVI	Trend of EVI	Trend of NIRv	Trend of LAI	Probability of Vegetation Greenness Change
↑	↑	↑	↑	Enhanced, very likely
↑	↑	↑	–	Enhanced likely
↑	↑	–	↑	
↑	–	↑	↑	
–	↑	↑	↑	
↑	↑	–	–	Enhanced probably
↑	–	–	↑	
–	–	↑	↑	
↓	↓	↓	↓	Degraded very likely
↓	↓	↓	–	Degraded likely
↓	↓	–	↓	
↓	–	↓	↓	
–	↓	↓	↓	
↓	↓	–	–	Degraded probably
↓	–	–	↓	
–	–	↓	↓	
None or only one vegetation index showed a significant change				Uncertainty

Note: significantly positive (↑), significantly negative (↓) and no significant change (–).

Furthermore, we used the confusion matrix to evaluate the difference in vegetation greenness between two vegetation indexes [73]. In this matrix, we expected the proportion to be large in the diagonal positions (a, e, and i in Figure 3), representing large consistency between the changes in two vegetation indexes. The accuracy was used as a proxy to measure the consistency trend of two vegetation indexes, and the expression can be written as:

$$Accuracy = \frac{a+e+i}{a+b+c+d+e+f+g+h+i} \tag{5}$$

The '*Accuracy*' represents the pixel's proportion of consistent change by using two vegetation indexes in total pixels. The '*a*' represents the pixels proportion of the trend of vegetation index1 belonging to the class (significantly positive) and the trend of vegetation index2 belonging to the class (significantly positive) in total pixels. The '*b*'–'*i*' is calculated in the same way. All data analysis in this study were conducted using the MATLAB R2020b software.

Confusion matrix		Trend of vegetation index1		
		Significantly positive	Significantly negative	No significant change
Trend of vegetation index2	Significantly positive	a	b	c
	Significantly negative	d	e	f
	No significant change	g	h	i

Figure 3. A confusion matrix for the trend of vegetation greenness is based on two vegetation indexes. The 'a' represents the pixels proportion of the trend of vegetation index1 belonging to the class (significantly positive) and the trend of vegetation index2 belonging to the class (significantly positive) in total pixels. The 'b–i' is calculated in the same way.

3. Results

3.1. The Trend of Vegetation Greenness during the Growing Season and Winter

Spatially, the temporal trends in growing season NDVI, EVI, NIRv, and LAI ($NDVI_{GS}$, EVI_{GS}, $NIRv_{GS}$, and LAI_{GS}) during 2000–2021 were positive in 75.2%, 76.1%, 75.1%, and 76.0% of pixels covered by vegetation in the growing season on the Tibetan Plateau (Figure 4a–d). The growing season greenness indexes showed a positive trend in the northeastern, central-eastern, and northwestern areas. In contrast, negative trends in growing season greenness were observed in southwestern and central-southern regions. Temporally, the growing season greenness indexes averaged over the region significantly increased ($p < 0.01$) from 2000 to 2021 (Figure 4e,f). Furthermore, for robustness, similar trend patterns of $NDVI_{GS}$, EVI_{GS}, $NIRv_{GS}$, and LAI_{GS} were also observed using the Theil–Sen median slope method (Figure S1).

In addition, the temporal trends in winter NDVI, EVI, NIRv, and LAI ($NDVI_{Winter}$, EVI_{Winter}, $NIRv_{Winter}$, and LAI_{Winter}) from 2000–2001 to 2021–2022 were positive in 85.5%, 80.4%, 81.6%, and 73.4% of pixels covered by evergreen vegetation on the Tibetan Plateau (Figure 5a–d). In most pixels, the evergreen vegetation showed widespread greening trends. The winter greenness indexes averaged over the region significantly increased ($p < 0.01$) from 2000–2001 to 2021–2022 (Figure 5e,f). In this study, the multiyear mean of winter NDSI snow cover from 2000–2001 to 2021–2022 in 96.1% pixels was equal to 0, which indicated that snow cover showed little impact on the change of winter vegetation greenness (Figure S2). In summary, despite the regional variation, the NDVI, EVI, NIRv, and LAI showed similar greening change patterns in the growing season and winter in more than half of the total pixels.

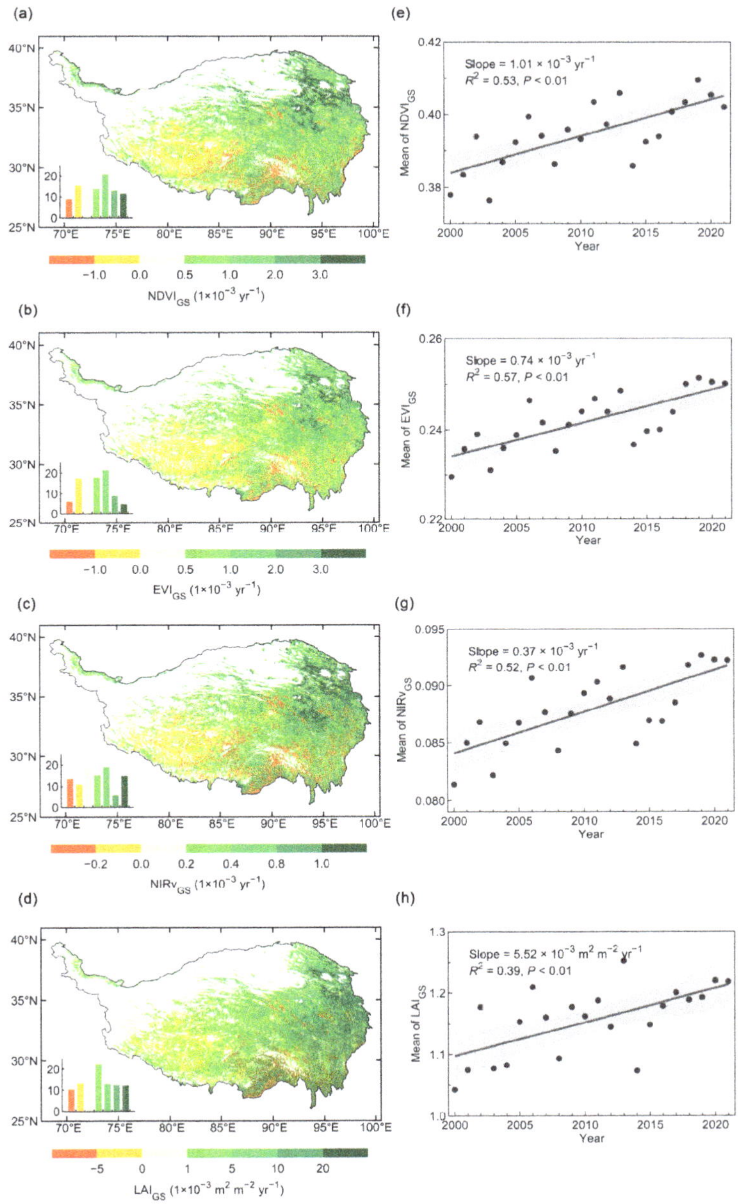

Figure 4. Spatial variation of the trend of growing season (May–September) NDVI, EVI, NIRv, and LAI by using the ordinary least-squares regression, respectively (**a–d**). The bottom-left inset in each map displays the percent of pixels in each trend of vegetation greenness interval with the corresponding interval values depicted by the color in the legend on the bottom. Temporal trend of growing seasons NDVI, EVI, NIRv, and LAI, respectively (**e–h**). $NDVI_{GS}$, EVI_{GS}, $NIRv_{GS}$, and LAI_{GS} are the abbreviations of NDVI, EVI, NIRv, and LAI in the growing season.

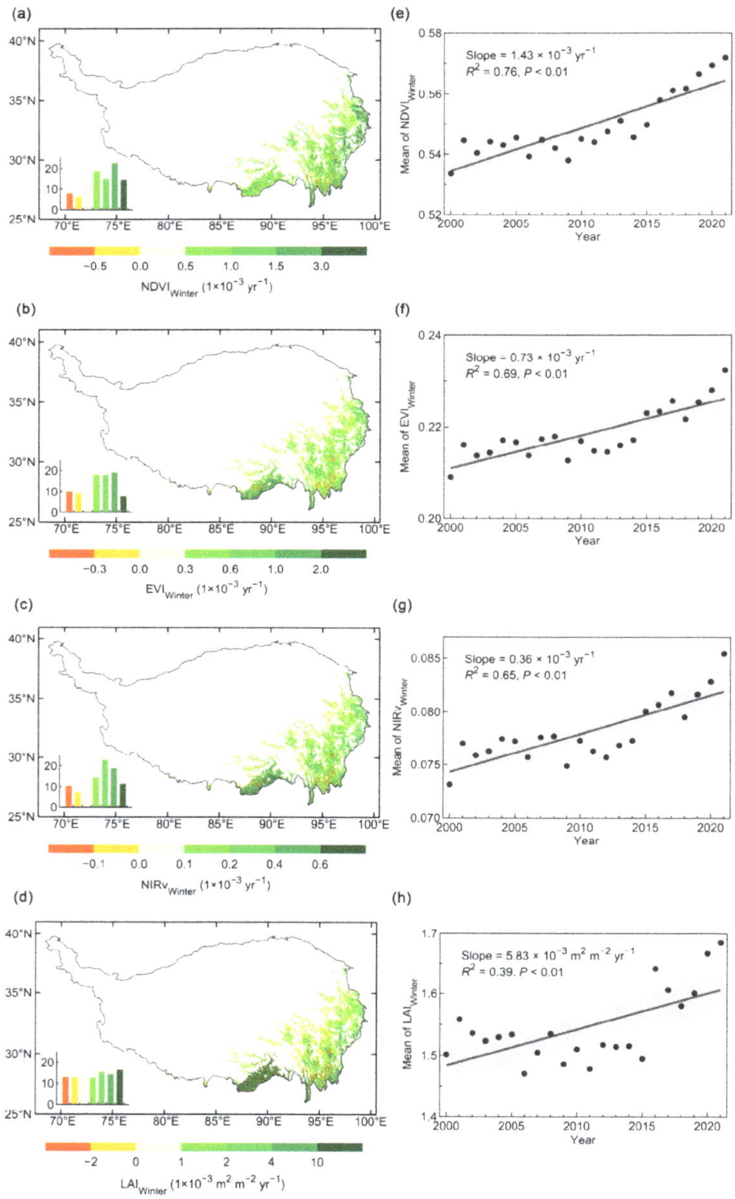

Figure 5. Spatial variation of the trend of winter (December–February) NDVI, EVI, NIRv, and LAI by using the ordinary least-squares regression, respectively (**a**–**d**). The bottom-left inset in each map displays the percent of pixels in each trend of vegetation greenness interval with the corresponding interval values depicted by the color in the legend on the bottom. Temporal trend of growing seasons NDVI, EVI, NIRv, and LAI, respectively (**e**–**h**). $NDVI_{Winter}$, EVI_{Winter}, $NIRv_{Winter}$, and LAI_{Winter} are the abbreviations of NDVI, EVI, NIRv, and LAI in winter.

3.2. Analysis of the Probability of Vegetation Greenness Change

The consistency assessment of the trend of four vegetation greenness indicators in the growing season showed that in 28.5% of pixels, vegetation growth enhanced in pixels covered by vegetation (Figure 6a). The 14.0% and 8.4% pixels were identified as enhanced very likely and enhanced likely, which were distributed in northeastern and northern areas on the Tibetan Plateau. The 6.0% pixels were identified as enhanced and probably distributed in central and central-eastern areas on the Tibetan Plateau. Only 1.5% of pixels showed degraded distribution in southwestern and central-eastern areas. The inconsistent change in four vegetation greenness indexes was observed in only 0.04% of pixels in the growing season. The evergreen vegetation was also enhanced by 29.8% of the pixels covered by evergreen vegetation on the Tibetan Plateau (Figure 6b). The percent of pixels identified with enhanced very likely, enhanced likely, and enhanced probably of pixels covered by evergreen vegetation on the Tibetan Plateau were 10.6%, 11.4%, and 7.8%, which were distributed in eastern edges and southeastern areas. Only 1.3% of pixels showed degraded distribution in the valley of southeastern areas. The 0.64% pixels showed inconsistent change by four greenness indexes. Among multiple greenness indicators, significant greening was observed in 30.1%, 32.7%, 30.8%, and 22.1% of this study area when estimated from $NDVI_{GS}$, EVI_{GS}, $NIRv_{GS}$, and LAI_{GS} (Figure 6c). The $NDVI_{GS}$, EVI_{GS}, and $NIRv_{GS}$ were closest to the probability of vegetation greenness change. The LAI showed the lowest percentage of significantly positive trends in the growing season. The assessment of multiple greenness indicators could balance the percent of significantly positive vegetation trends, especially improving the percent of the significantly positive trend of LAI_{GS}. Whereas, the percent of evergreen vegetation greening indicated by $NDVI_{Winter}$ (42.8%) was higher than that indicated by EVI_{Winter}, $NIRv_{Winter}$, and LAI_{Winter} (28.2%, 30.8%, and 21.5%). Thus, asynchronous changes in different vegetation greenness should be the focus of further study.

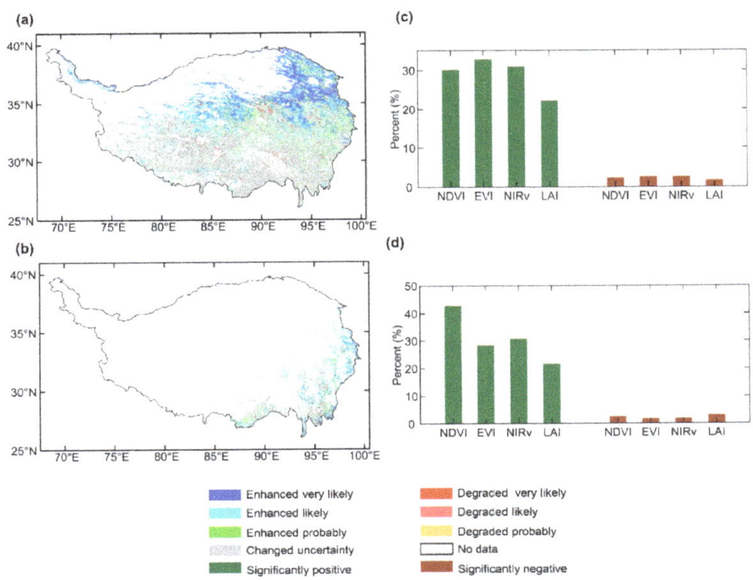

Figure 6. Consistency of trend of vegetation greenness in the growing season (**a**) and winter (**b**) on the Tibetan Plateau; (**c**) the percent of pixels of the trend of growing season vegetation greenness with significantly positive and negative in pixels covered by vegetation in the growing season on the Tibetan Plateau; (**d**) the percent of the trend of winter vegetation greenness pixels with significantly positive and negative in pixels covered by evergreen vegetation on the Tibetan Plateau.

3.3. Comparison of Vegetation Greenness Trends between Two Different Vegetation Indexes

The largest consistency of significantly positive was between $NIRv_{GS}$ and EVI_{GS} in 29.6% pixels, and the following consistency of significantly positive was between EVI_{GS} and $NDVI_{GS}$, $NIRv_{GS}$ and $NDVI_{GS}$ (Figure 7). The lowest consistency of significantly positive results was between EVI_{GS} and LAI_{GS}. The overall accuracy between the two vegetation indexes was about 0.80, and the accuracy between the trend of EVI_{GS} and $NIRv_{GS}$ showed the biggest value. The accuracy between the LAI_{GS} trend and other vegetation indexes was relatively low (Table 2).

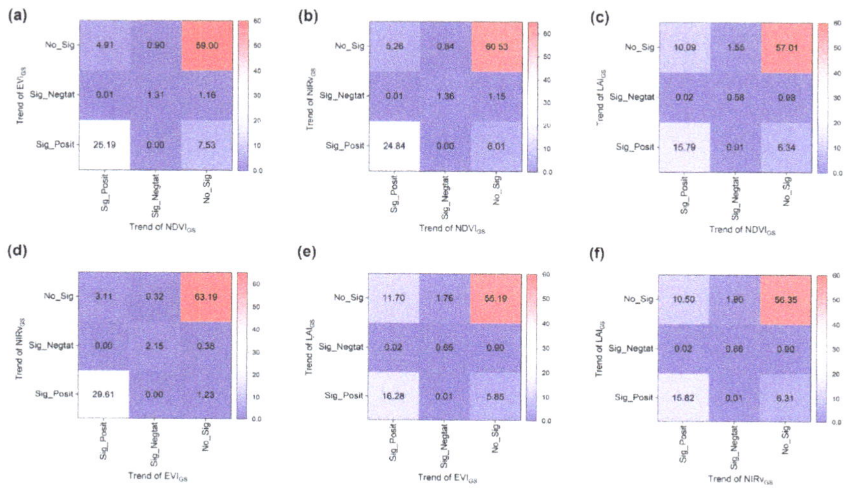

Figure 7. (**a**) The confusion matrix between the trend of growing season EVI (EVI_{GS}) and NDVI ($NDVI_{GS}$); (**b**) the confusion matrix between the trend of growing season NIRv ($NIRv_{GS}$) and $NDVI_{GS}$; (**c**) the confusion matrix between the trend of growing season LAI (LAI_{GS}) and $NDVI_{GS}$; (**d**) the confusion matrix between the trend of growing season LAI (LAI_{GS}) and EVI_{GS}; (**e**) the confusion matrix between the trend of LAI_{GS} and EVI_{GS}; (**f**) the confusion matrix between the trend of LAI_{GS} and $NIRv_{GS}$.

Table 2. The accuracy between the trends of different vegetation greenness indexes in the growing season.

Accuracy	$NDVI_{GS}$	EVI_{GS}	$NIRv_{GS}$	LAI_{GS}
$NDVI_{GS}$	-			
EVI_{GS}	0.86	-		
$NIRv_{GS}$	0.87	0.95	-	
LAI_{GS}	0.79	0.78	0.79	-

In the comparison of the trend of evergreen vegetation, the largest consistency of significantly positive was between $NIRv_{Winter}$ and EVI_{Winter} in 27.0% pixels, and the following consistency of significantly positive was between $NIRv_{Winter}$ and $NDVI_{Winter}$, EVI_{Winter} and $NDVI_{Winter}$ (Figure 8). The lowest consistency of significantly positive results was between EVI_{Winter} and LAI_{Winter}, which showed a similar result with the trend of greenness in the growing season. The overall accuracy between the two vegetation indexes was about 0.65 lower than in the growing season, and the accuracy between the trends of LAI_{Winter} and $NDVI_{Winter}$ was relatively low (Table 3). The overall accuracy between $NDVI_{Winter}$ and EVI_{Winter} was lower than between $NDVI_{GS}$ and EVI_{GS}. We found that the more pixels, the more $NDVI_{Winter}$ trends were significantly positive. However, EVI_{Winter} trends showed no significant change.

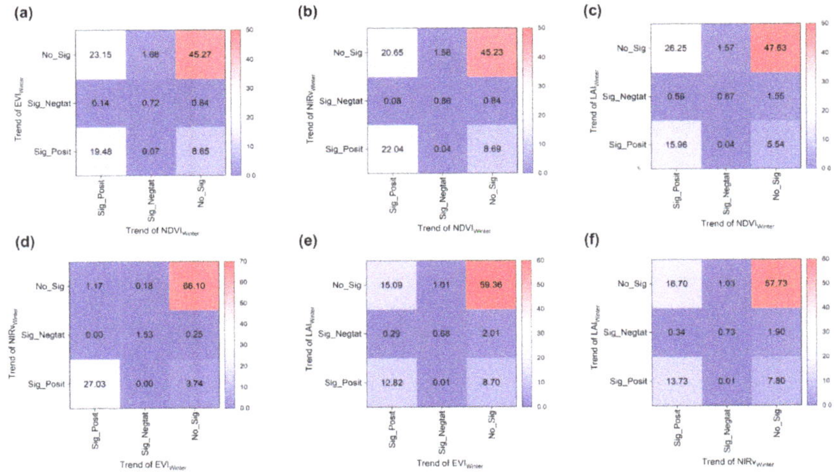

Figure 8. (**a**) The confusion matrix between the trend of winter EVI (EVI_{Winter}) and NDVI ($NDVI_{Winter}$); (**b**) the confusion matrix between the trend of winter NIRv ($NIRv_{Winter}$) and $NDVI_{Winter}$; (**c**) the confusion matrix between the trend of winter LAI (LAI_{Winter}) and $NDVI_{Winter}$; (**d**) the confusion matrix between the trend of winter LAI (LAI_{Winter}) and EVI_{Winter}; (**e**) the confusion matrix between the trend of LAI_{Winter} and EVI_{Winter}; (**f**) the confusion matrix between the trend of LAI_{Winter} and $NIRv_{Winter}$.

Table 3. The accuracy between the trends of different vegetation greennesses in winter.

Accuracy	$NDVI_{Winter}$	EVI_{Winter}	$NIRv_{Winter}$	LAI_{Winter}
$NDVI_{Winter}$	-			
EVI_{Winter}	0.65	-		
$NIRv_{Winter}$	0.68	0.95	-	
LAI_{Winter}	0.64	0.73	0.72	-

4. Discussion

4.1. Significances of the Consistency of Vegetation Greenness Trend

The consistency analysis of four vegetation indexes for vegetation change on the Tibetan Plateau revealed a green trend during the growing and winter seasons. The EVI_{GS} and $NIRv_{GS}$, as well as EVI_{Winter} and $NIRv_{Winter}$, demonstrated the greatest consistency (0.95) in vegetation change trends, consistent with previous studies [31]. The consistency between $NDVI_{GS}$ and other vegetation greenness indexes (EVI_{GS}, $NIRv_{GS}$, and LAI_{GS}) was higher in the growing season, with a value of about 0.80, but lower in $NDVI_{Winter}$ with EVI_{Winter}, $NIRv_{Winter}$, and LAI_{Winter}, with a value of about 0.65. This could be explained by dataset accuracy due to the contamination by clouds and aerosols in winter, despite removing low-quality datasets. The LAI exhibited lower consistency with the other three vegetation indices, particularly in LAI_{Winter} and $NDVI_{Winter}$, as NDVI may not be suitable due to the lack of sensitivity in high LAI conditions [74]. To further explore the physiological activity of vegetation, we used the CSIF to assess the change in vegetation growth. The CSIF also showed a greening trend in the eastern and northeastern areas, and browning was observed in the southwestern and some southeastern areas (Figure S3). In most areas, the CSIF showed a consistent change with greenness (Figures S3 and S4). However, there were some differences in trends in SIF and greenness. The CSIF showed more browning areas (accounting for 34.8%) in the northwestern and southeastern parts of the Tibetan Plateau than $NDVI_{GS}$, EVI_{GS}, $NIRv_{GS}$, and LAI_{GS} (accounting for 24.4%, 25.7%, 27.0%, and 23.0%, respectively). When the leaf area remains stable, the change in greenness (i.e., NDVI) may be small, while the change in photosynthesis could provide more information about influencing factors such as solar radiation and drought conditions [75,76]. Therefore, in most areas, vegetation greenness is consistent with SIF in monitoring vegetation dynamics, but CSIF is more sensitive to greenness in the physiological activity of vegetation.

4.2. The Driving Factors of Vegetation Greenness Change

Compared with previous studies, the change in vegetation greenness in the growing season based on satellite observations showed increasing trends in northeastern, northwestern, central-eastern, and northwestern areas since 2000 as a result of increasing precipitation, increasing temperature, and solar radiation, respectively [5,77,78]. Browning was observed in the southwestern Tibetan Plateau due to decreasing precipitation from 2000 to 2018 [5]. On the other hand, since 2000, the Chinese government has implemented some ecological restoration projects, and vegetation growth has benefited from these projects [79–81]. Some studies revealed that establishing the protected areas increased vegetation greenness [79,80]. In addition, grazing reduction has been identified as an important and effective strategy in these projects on the Tibetan Plateau, such as fencing, degrading grassland, and ecological compensation [80]. Implementing these projects has mitigated grassland degradation in the central Tibetan Plateau [81].

For winter evergreen vegetation, similar to our findings, vegetation greenness in the southeastern Tibetan Plateau showed an increasing trend over the past 20 years due to increasing temperatures [78,82,83]. On the other hand, upward treelines and tree recruitment could lead to an increase in evergreen vegetation greenness [33,84]. For example, in an analysis of treeline dynamics on the Tibetan Plateau by collecting 59 treeline sites, 67% showed an upward trend from 1901 to 2017, and 73% of the tree recruitments showed an increasing trend [84]. In addition, ecological restoration projects such as afforestation also caused an increase in the greenness of evergreen vegetation [85].

4.3. Uncertainty and Limitations

In our analysis, we carefully considered the robustness of vegetation indices and chose to analyze four widely utilized ones. Nonetheless, there remains potential for future expansion by incorporating a broader range of indices. It is worth noting that our current focus is on identifying regions characterized by evergreen vegetation cover. However, the distinction between deciduous and evergreen vegetation can sometimes be challenging, warranting dedicated efforts to enhance accuracy in identifying areas with deciduous vegetation. The current study provides a concise overview of the factors contributing to variations in vegetation greenness. Recognizing the intricate nature of these factors, we should strive for a more refined and nuanced understanding of the underlying causes driving these changes.

4.4. Implications and Future Research Directions

A widespread greening of vegetation influences the structure and function of the ecosystem. For example, greening increases vegetation's photosynthetic carbon absorption capacity, which in turn increases vegetation productivity, and enhanced productivity can increase terrestrial carbon storage [3]. Conversely, greening also causes some negative impacts on the ecosystem. For example, shrub encroachment decreased herbaceous coverage, density, and species richness and led to the loss of endemic species in the meadow community [86]. Vegetation greening also impacts the land surface process. For example, vegetation greening could reduce the temperature [78]. However, vegetation greening can only indirectly indicate vegetation expansion, and further exploration is necessary, combined with ground observations and other investigation data. Furthermore, when we assessed the change in vegetation greenness, more indexes could enhance the robustness of the results, especially in winter evergreen vegetation. In addition, vegetation growth in deciduous and winter evergreen vegetation is different. Evergreen vegetation should be considered to improve process-based land surface models, as most of these models use scenarios specified for vegetation types [87]. Further research is needed to elucidate the underlying drivers of these trends and their potential impacts on ecosystem services and land surface processes on the Tibetan Plateau.

5. Conclusions

In this study, we first assessed the change in growing-season vegetation greenness. Then, we identified the area covered by evergreen vegetation and evaluated the change in evergreen vegetation greenness from 2000 to 2022 based on satellite observations. Four vegetation greenness indexes were selected to show vegetation change and consistency assessment. This study found positive temporal trends in growing season vegetation greenness using NDVI, EVI, NIRv, and LAI on the Tibetan Plateau between 2000 and 2021. These positive trends were mainly observed in the northeastern, central-eastern, and northwestern areas, while negative trends were observed in the southwestern and central-southern regions. Widespread green trends in winter greenness indexes were observed during 2000–2022. The consistency assessment of different vegetation greenness indexes showed relatively high consistency in positive trends, particularly between NIRv and EVI. However, the consistency was relatively low between LAI and other vegetation indexes. These findings suggest the application of multiple indexes to study the dynamics of vegetation on the Tibetan Plateau synthetically.

Supplementary Materials: The following supporting information can be downloaded at: https://www.mdpi.com/article/10.3390/rs15245697/s1. Figure S1: Spatial variation of trend of growing season (May–September) NDVI, EVI, NIRv, and LAI by using the Theil–Sen median slope method, respectively (a, b, c, and d); Figure S2: Spatial distribution of multiyear mean of winter normalized difference snow index (NDSI) snow cover; Figure S3: Spatial variation of trend of growing season CSIF during 2000–2019 by using the ordinary least-squares regression; Figure S4: Spatial variation of trend of growing season NDVI, EVI, NIRv, and LAI during 2000–2019 by using the ordinary least-squares regression (a, b, c, and d).

Author Contributions: Conceptualization, J.L. and W.Z.; methodology, J.L.; writing—original draft preparation, J.L.; writing—review and editing, J.L., W.Z., T.H., L.Z. and P.P.; funding acquisition, W.Z. All authors have read and agreed to the published version of the manuscript.

Funding: This work was supported by the Second Tibetan Plateau Scientific Expedition and Research Program (No. 2019QZKK0405).

Data Availability Statement: The data supporting this study are available from the following resources in the public domain. The MOD13A1 V061 NDVI and EVI dataset and MOD15A2H v061 LAI dataset were downloaded from Google Earth Engine at https://developers.google.com/earth-engine/datasets/ (accessed on 10 December 2022). The NIRv dataset was calculated by NDVI and reflectance from near-infrared band extracting from MOD13A1 v061 product at https://earthengine.google.com (accessed on 10 December 2022). The Contiguous solar-induced fluorescence (CSIF) dataset was downloaded from https://figshare.com/articles/dataset/CSIF/6387494 (accessed on 10 February 2023). The GlobeLand30 and MCD12Q1 v061 landcover dataset were downloaded from https://www.nature.com/articles/514434c (accessed on 21 July 2021) and https://developers.google.com/earth-engine/datasets/ (accessed on 10 December 2022). The MOD10A1 v061 snow cover dataset was downloaded from https://developers.google.com/earth-engine/datasets/ (accessed on 15 March 2023).

Acknowledgments: We would like to thank David Eldridge at the University of NSW for his assistance with the English language and editing of this manuscript.

Conflicts of Interest: The authors declare that they have no conflict of interest.

References

1. García-Pardo, K.A.; Moreno-Rangel, D.; Domínguez-Amarillo, S.; García-Chávez, J.R. Remote Sensing for the Assessment of Ecosystem Services Provided by Urban Vegetation: A Review of the Methods Applied. *Urban For. Urban Green.* **2022**, *74*, 127636. [CrossRef]
2. Zhao, Y.; Liu, Z.; Wu, J. Grassland Ecosystem Services: A Systematic Review of Research Advances and Future Directions. *Landsc. Ecol.* **2020**, *35*, 793–814. [CrossRef]
3. Piao, S.; Wang, X.; Park, T.; Chen, C.; Lian, X.; He, Y.; Bjerke, J.W.; Chen, A.; Ciais, P.; Tømmervik, H.; et al. Characteristics, Drivers and Feedbacks of Global Greening. *Nat. Rev. Earth Environ.* **2019**, *1*, 14–27. [CrossRef]
4. Berner, L.T.; Massey, R.; Jantz, P.; Forbes, B.C.; Macias-Fauria, M.; Myers-Smith, I.; Kumpula, T.; Gauthier, G.; Andreu-Hayles, L.; Gaglioti, B.V.; et al. Summer Warming Explains Widespread but Not Uniform Greening in the Arctic Tundra Biome. *Nat. Commun.* **2020**, *11*, 4621. [CrossRef]
5. Li, P.; Hu, Z.; Liu, Y. Shift in the Trend of Browning in Southwestern Tibetan Plateau in the Past Two Decades. *Agric. For. Meteorol.* **2020**, *287*, 107950. [CrossRef]
6. Ding, Z.; Peng, J.; Qiu, S.; Zhao, Y. Nearly Half of Global Vegetated Area Experienced Inconsistent Vegetation Growth in Terms of Greenness, Cover, and Productivity. *Earth's Future* **2020**, *8*, e2020EF001618. [CrossRef]
7. Bonan, G.B. Forests and Climate Change: Forcings, Feedbacks, and the Climate Benefits of Forests. *Science* **2008**, *320*, 1444–1449. [CrossRef]
8. Zheng, L.; Zhao, G.; Dong, J.; Ge, Q.; Tao, J.; Zhang, X.; Qi, Y.; Doughty, R.B.; Xiao, X. Spatial, Temporal, and Spectral Variations in Albedo Due to Vegetation Changes in China's Grasslands. *ISPRS J. Photogramm. Remote Sens.* **2019**, *152*, 1–12. [CrossRef]
9. Anwar, S.A. Understanding the Contribution of the Vegetation-Runoff System for Simulating the African Climate Using the Regcm4 Model. *Theor. Appl. Climatol.* **2019**, *138*, 1219–1230. [CrossRef]
10. Liu, Z.; Duan, Q.; Fan, X.; Li, W.; Yin, J. Bayesian Retro-and Prospective Assessment of Cmip6 Climatology in Pan Third Pole Region. *Clim. Dyn.* **2022**, *60*, 767–784. [CrossRef]
11. Yao, T.; Bolch, T.; Chen, D.; Gao, J.; Immerzeel, W.; Piao, S.; Su, F.; Thompson, L.; Wada, Y.; Wang, L.; et al. The Imbalance of the Asian Water Tower. *Nat. Rev. Earth Environ.* **2022**, *3*, 618–632. [CrossRef]
12. Zhong, L.; Ma, Y.; Xue, Y.; Piao, S. Climate Change Trends and Impacts on Vegetation Greening over the Tibetan Plateau. *J. Geophys. Res. Atmos.* **2019**, *124*, 7540–7552. [CrossRef]
13. Shen, M.G.; Jeong, S.J.; Zhou, L.M.; Zeng, Z.Z.; Ciais, P.; Chen, D.L.; Huang, M.T.; Jin, C.S.; Li, L.Z.X.; Li, Y.; et al. Evaporative Cooling over the Tibetan Plateau Induced by Vegetation Growth. *Proc. Natl. Acad. Sci. USA* **2015**, *112*, 9299–9304. [CrossRef]
14. Wu, J.; Wang, H.; Li, G.; Ma, W.; Wu, J.; Gong, Y.; Xu, G. Vegetation Degradation Impacts Soil Nutrients and Enzyme Activities in Wet Meadow on the Qinghai-Tibet Plateau. *Sci. Rep.* **2020**, *10*, 21271. [CrossRef] [PubMed]
15. Zuo, Z.; Zhang, R.; Zhao, P. The Relation of Vegetation over the Tibetan Plateau to Rainfall in China During the Boreal Summer. *Clim. Dyn.* **2011**, *36*, 1207–1219. [CrossRef]
16. Wen, L.; Dong, S.; Li, Y.; Wang, X.; Li, X.; Shi, J.; Dong, Q. The Impact of Land Degradation on the C Pools in Alpine Grasslands of the Qinghai-Tibet Plateau. *Plant Soil* **2012**, *368*, 329–340. [CrossRef]
17. She, Y.; Zhang, Z.; Ma, L.; Xu, W.; Huang, X.; Zhou, H. Vegetation Attributes and Soil Properties of Alpine Grassland in Different Degradation Stages on the Qinghai-Tibet Plateau, China: A Meta-Analysis. *Arab. J. Geosci.* **2022**, *15*, 193. [CrossRef]
18. Zhang, J.; Wu, L.; Huang, G.; Zhu, W.; Zhang, Y. The Role of May Vegetation Greenness on the Southeastern Tibetan Plateau for East Asian Summer Monsoon Prediction. *J. Geophys. Res. Atmos.* **2011**, 116. [CrossRef]
19. Hou, Y.; Zhang, M.; Meng, Z.; Liu, S.; Sun, P.; Yang, T. Assessing the Impact of Forest Change and Climate Variability on Dry Season Runoff by an Improved Single Watershed Approach: A Comparative Study in Two Large Watersheds, China. *Forests* **2018**, *9*, 46. [CrossRef]
20. Ma, M.; Yuan, W.; Dong, J.; Zhang, F.; Cai, W.; Li, H. Large-Scale Estimates of Gross Primary Production on the Qinghai-Tibet Plateau Based on Remote Sensing Data. *Int. J. Digit. Earth* **2017**, *11*, 1166–1183. [CrossRef]
21. Wang, Z.; Cui, G.; Liu, X.; Zheng, K.; Lu, Z.; Li, H.; Wang, G.; An, Z. Greening of the Qinghai–Tibet Plateau and Its Response to Climate Variations Along Elevation Gradients. *Remote Sens.* **2021**, *13*, 3712. [CrossRef]
22. Diao, C.; Liu, Y.; Zhao, L.; Zhuo, G.; Zhang, Y. Regional-Scale Vegetation-Climate Interactions on the Qinghai-Tibet Plateau. *Ecol. Inform.* **2021**, *65*, 101413. [CrossRef]
23. Samra, A.; Rasha, M.; El-Barbary, S.M.A. The Use of Remote Sensing Indices for Detecting Environmental Changes: A Case Study of North Sinai, Egypt. *Spat. Inf. Res.* **2018**, *26*, 679–689. [CrossRef]
24. Zeng, Y.; Hao, D.; Huete, A.; Dechant, B.; Berry, J.; Chen, J.M.; Joiner, J.; Frankenberg, C.; Bond-Lamberty, B.; Ryu, Y. Optical Vegetation Indices for Monitoring Terrestrial Ecosystems Globally. *Nat. Rev. Earth Environ.* **2022**, *3*, 477–493. [CrossRef]
25. Zhang, L.; Shen, M.; Shi, C.; Shi, F.; Jiang, N.; Yang, Z.; Ji, Z. Local Climatic Factors Mediated Impacts of Large-Scale Climate Oscillations on the Growth of Vegetation across the Tibetan Plateau. *Front. Environ. Sci.* **2021**, *9*, 597971. [CrossRef]
26. Zhou, D.; Zhang, L.; Hao, L.; Sun, G.; Xiao, J.; Li, X. Large Discrepancies among Remote Sensing Indices for Characterizing Vegetation Growth Dynamics in Nepal. *Agric. For. Meteorol.* **2023**, *339*, 109546. [CrossRef]
27. Jiao, K.; Gao, J.; Liu, Z. Precipitation Drives the Ndvi Distribution on the Tibetan Plateau While High Warming Rates May Intensify Its Ecological Droughts. *Remote Sens.* **2021**, *13*, 1305. [CrossRef]

28. Wang, C.; Huang, M.; Zhai, P. Change in Drought Conditions and Its Impacts on Vegetation Growth over the Tibetan Plateau. *Adv. Clim. Change Res.* **2021**, *12*, 333–341. [CrossRef]
29. Chen, J.; Yan, F.; Lu, Q. Spatiotemporal Variation of Vegetation on the Qinghai–Tibet Plateau and the Influence of Climatic Factors and Human Activities on Vegetation Trend (2000–2019). *Remote Sens.* **2020**, *12*, 3150. [CrossRef]
30. Li, L.; Zhang, Y.; Liu, L.; Wu, J.; Wang, Z.; Li, S.; Zhang, H.; Zu, J.; Ding, M.; Paudel, B. Spatiotemporal Patterns of Vegetation Greenness Change and Associated Climatic and Anthropogenic Drivers on the Tibetan Plateau During 2000–2015. *Remote Sens.* **2018**, *10*, 1525. [CrossRef]
31. Qiu, B.; Yan, X.; Chen, C.; Tang, Z.; Wu, W.; Xu, W.; Zhao, Z.; Yan, C.; Berry, J.; Huang, W.; et al. The Impact of Indicator Selection on Assessment of Global Greening. *GIScience Remote Sens.* **2021**, *58*, 372–385. [CrossRef]
32. Sun, X.; Wang, G.; Huang, M.; Chang, R.; Ran, F. Forest Biomass Carbon Stocks and Variation in Tibet's Carbon-Dense Forests from 2001 to 2050. *Sci. Rep.* **2016**, *6*, 34687. [CrossRef] [PubMed]
33. Liang, E.; Wang, Y.; Piao, S.; Lu, X.; Camarero, J.J.; Zhu, H.; Zhu, L.; Ellison, A.M.; Ciais, P.; Peñuelas, J. Species Interactions Slow Warming-Induced Upward Shifts of Treelines on the Tibetan Plateau. *Proc. Natl. Acad. Sci. USA* **2016**, *113*, 4380–4385. [CrossRef]
34. Shi, C.; Schneider, L.; Hu, Y.; Shen, M.; Sun, C.; Xia, J.; Forbes, B.C.; Shi, P.; Zhang, Y.; Ciais, P. Warming-Induced Unprecedented High-Elevation Forest Growth over the Monsoonal Tibetan Plateau. *Environ. Res. Lett.* **2020**, *15*, 054011. [CrossRef]
35. Huang, R.; Zhu, H.; Liu, X.; Liang, E.; Grießinger, J.; Wu, G.; Li, X.; Bräuning, A. Does Increasing Intrinsic Water Use Efficiency (Iwue) Stimulate Tree Growth at Natural Alpine Timberline on the Southeastern Tibetan Plateau? *Glob. Planet. Chang.* **2017**, *148*, 217–226. [CrossRef]
36. Keyimu, M.; Li, Z.; Zhang, G.; Fan, Z.; Wang, X.; Fu, B. Tree Ring–Based Minimum Temperature Reconstruction in the Central Hengduan Mountains, China. *Theor. Appl. Climatol.* **2020**, *141*, 359–370. [CrossRef]
37. Liu, B.; Wang, Y.; Zhu, H.; Liang, E.; Camarero, J.J. Topography and Age Mediate the Growth Responses of Smith Fir to Climate Warming in the Southeastern Tibetan Plateau. *Int. J. Biometeorol.* **2016**, *60*, 1577–1587. [CrossRef]
38. Liu, J.; Yang, B.; Qin, C. Tree-Ring Based Annual Precipitation Reconstruction since Ad 1480 in South Central Tibet. *Quat. Int.* **2011**, *236*, 75–81. [CrossRef]
39. Shi, C.; Shen, M.; Wu, X.; Cheng, X.; Li, X.; Fan, T.; Li, Z.; Zhang, Y.; Fan, Z.; Shi, F.; et al. Growth Response of Alpine Treeline Forests to a Warmer and Drier Climate on the Southeastern Tibetan Plateau. *Agric. For. Meteorol.* **2019**, *264*, 73–79. [CrossRef]
40. Liang, E.; Leuschner, C.; Dulamsuren, C.; Wagner, B.; Hauck, M. Global Warming-Related Tree Growth Decline and Mortality on the North-Eastern Tibetan Plateau. *Clim. Chang.* **2016**, *134*, 163–176. [CrossRef]
41. Liang, E.; Lu, X.; Ren, P.; Li, X.; Zhu, L.; Eckstein, D. Annual Increments of Juniper Dwarf Shrubs above the Tree Line on the Central Tibetan Plateau: A Useful Climatic Proxy. *Ann. Bot.* **2012**, *109*, 721–728. [CrossRef] [PubMed]
42. Schwab, N.; Kaczka, R.; Janecka, K.; Böhner, J.; Chaudhary, R.; Scholten, T.; Schickhoff, U. Climate Change-Induced Shift of Tree Growth Sensitivity at a Central Himalayan Treeline Ecotone. *Forests* **2018**, *9*, 267. [CrossRef]
43. Sigdel, S.R.; Dawadi, B.; Camarero, J.J.; Liang, E.; Leavitt, S.W. Moisture-Limited Tree Growth for a Subtropical Himalayan Conifer Forest in Western Nepal. *Forests* **2018**, *9*, 340. [CrossRef]
44. Liang, E.; Wang, Y.; Eckstein, D.; Luo, T. Little Change in the Fir Tree-Line Position on the Southeastern Tibetan Plateau after 200 Years of Warming. *New Phytol.* **2011**, *190*, 760–769. [CrossRef] [PubMed]
45. Wang, Y.; Pederson, N.; Ellison, A.M.; Buckley, H.L.; Case, B.S.; Liang, E.; Camarero, J.J. Increased Stem Density and Competition May Diminish the Positive Effects of Warming at Alpine Treeline. *Ecology* **2016**, *97*, 1668–1679. [CrossRef] [PubMed]
46. Liu, D.; Wang, T.; Yang, T.; Yan, Z.; Liu, Y.; Zhao, Y.; Piao, S. Deciphering Impacts of Climate Extremes on Tibetan Grasslands in the Last Fifteen Years. *Sci. Bull.* **2019**, *64*, 446–454. [CrossRef] [PubMed]
47. Duan, A.M.; Wu, G.X. Role of the Tibetan Plateau Thermal Forcing in the Summer Climate Patterns over Subtropical Asia. *Clim. Dyn.* **2005**, *24*, 793–807. [CrossRef]
48. Wu, G.; Duan, A.; Liu, Y.; Mao, J.; Ren, R.; Bao, Q.; He, B.; Liu, B.; Hu, W. Tibetan Plateau Climate Dynamics: Recent Research Progress and Outlook. *Natl. Sci. Rev.* **2015**, *2*, 100–116. [CrossRef]
49. Wu, G.; Liu, Y.; Zhang, Q.; Duan, A.; Wang, T.; Wan, R.; Liu, X.; Li, W.; Wang, Z.; Liang, X. The Influence of Mechanical and Thermal Forcing by the Tibetan Plateau on Asian Climate. *J. Hydrometeorol.* **2007**, *8*, 770–789. [CrossRef]
50. Hua, W.; Lin, Z.; Guo, D.; Fan, G.; Zhang, Y.; Yang, K.; Hu, Q.; Zhu, L. Simulated Long-Term Vegetation–Climate Feedbacks in the Tibetan Plateau. *Asia-Pac. J. Atmos. Sci.* **2018**, *55*, 41–52. [CrossRef]
51. Su, Y.; Guo, Q.; Hu, T.; Guan, H.; Jin, S.; An, S.; Chen, X.; Guo, K.; Hao, Z.; Hu, Y.; et al. An Updated Vegetation Map of China (1:1000000). *Sci. Bull.* **2020**, *65*, 1125–1136. [CrossRef]
52. Sulla-Menashe, D.; Gray, J.M.; Abercrombie, S.P.; Friedl, M.A. Hierarchical Mapping of Annual Global Land Cover 2001 to Present: The Modis Collection 6 Land Cover Product. *Remote Sens. Environ.* **2019**, *222*, 183–194. [CrossRef]
53. Badgley, G.; Field, C.B.; Berry, J.A. Canopy near-Infrared Reflectance and Terrestrial Photosynthesis. *Sci. Adv.* **2017**, *3*, e1602244. [CrossRef] [PubMed]
54. Huete, A.; Didan, K.; Miura, T.; Rodriguez, E.P.; Gao, X.; Ferreira, L.G. Overview of the Radiometric and Biophysical Performance of the Modis Vegetation Indices. *Remote Sens. Environ.* **2002**, *83*, 195–213. [CrossRef]
55. Myneni, R.B.; Hoffman, S.; Knyazikhin, Y.; Privette, J.L.; Glassy, J.; Tian, Y.; Wang, Y.; Song, X.; Zhang, Y.; Smith, G.R. Global Products of Vegetation Leaf Area and Fraction Absorbed Par from Year One of Modis Data. *Remote Sens. Environ.* **2002**, *83*, 214–231. [CrossRef]

56. Didan, K. MODIS/Terra Vegetation Indices 16-Day L3 Global 500 m SIN Grid V061 [Data set]. NASA EOSDIS Land Processes Distributed Active Archive Center. 2021. Available online: https://lpdaac.usgs.gov/products/mod13a1v061/ (accessed on 12 October 2022).
57. Myneni, R.; Knyazikhin, Y.; Park, T. MODIS/Terra Leaf Area Index/FPAR 8-Day L4 Global 500 m SIN Grid V061 [Data Set]. NASA EOSDIS Land Processes Distributed Active Archive Center. 2021. Available online: https://lpdaac.usgs.gov/products/mod15a2hv061/ (accessed on 12 October 2022).
58. Magney, T.S.; Bowling, D.R.; Logan, B.A.; Grossmann, K.; Stutz, J.; Blanken, P.D.; Burns, S.P.; Cheng, R.; Garcia, M.A.; Köhler, P.; et al. Mechanistic Evidence for Tracking the Seasonality of Photosynthesis with Solar-Induced Fluorescence. *Proc. Natl. Acad. Sci. USA* **2019**, *116*, 11640–11645. [CrossRef]
59. Turner, A.J.; Köhler, P.; Magney, T.S.; Frankenberg, C.; Fung, I.; Cohen, R.C. A Double Peak in the Seasonality of California's Photosynthesis as Observed from Space. *Biogeosciences* **2020**, *17*, 405–422. [CrossRef]
60. Zhang, Y.; Joiner, J.; Alemohammad, S.H.; Zhou, S.; Gentine, P. A Global Spatially Contiguous Solar-Induced Fluorescence (Csif) Dataset Using Neural Networks. *Biogeosciences* **2018**, *15*, 5779–5800. [CrossRef]
61. Chen, J.; Ban, Y.; Li, S. Open Access to Earth Land-Cover Map. *Nature* **2014**, *514*, 434.
62. Friedl, M.; Sulla-Menashe, D. MODIS/Terra+Aqua Land Cover Type Yearly L3 Global 500 m SIN Grid V061 [Data Set]. NASA EOSDIS Land Processes Distributed Active Archive Center. 2022. Available online: https://lpdaac.usgs.gov/products/mcd12q1v061/ (accessed on 12 October 2022).
63. Hall, D.K.; Riggs, G.A. MODIS/Terra Snow Cover Daily L3 Global 500 m SIN Grid, Version 61 [Data Set]. Boulder, Colorado USA. NASA National Snow and Ice Data Center Distributed Active Archive Center. 2021. Available online: https://nsidc.org/data/mod10a1/versions/61 (accessed on 15 March 2023).
64. Riggs, G.; Hall, D. Continuity of Modis and Viirs Snow Cover Extent Data Products for Development of an Earth Science Data Record. *Remote Sens.* **2020**, *12*, 3781. [CrossRef]
65. Peng, W.; Kuang, T.; Tao, S. Quantifying Influences of Natural Factors on Vegetation Ndvi Changes Based on Geographical Detector in Sichuan, Western China. *J. Clean. Prod.* **2019**, *233*, 353–367. [CrossRef]
66. Jiang, N.; Shen, M.; Ciais, P.; Campioli, M.; Peñuelas, J.; Körner, C.; Cao, R.; Piao, S.; Liu, L.; Wang, S.; et al. Warming Does Not Delay the Start of Autumnal Leaf Coloration but Slows Its Progress Rate. *Glob. Ecol. Biogeogr.* **2022**, *31*, 2297–2313. [CrossRef]
67. Kennedy, R.E.; Yang, Z.; Cohen, W.B. Detecting Trends in Forest Disturbance and Recovery Using Yearly Landsat Time Series: 1. Landtrendr—Temporal Segmentation Algorithms. *Remote Sens. Environ.* **2010**, *114*, 2897–2910. [CrossRef]
68. Zhang, Y.; Commane, R.; Zhou, S.; Williams, A.P.; Gentine, P. Light Limitation Regulates the Response of Autumn Terrestrial Carbon Uptake to Warming. *Nat. Clim. Chang.* **2020**, *10*, 739–743. [CrossRef]
69. Zhang, L.; Shen, M.; Jiang, N.; Lv, J.; Liu, L.; Zhang, L. Spatial Variations in the Response of Spring Onset of Photosynthesis of Evergreen Vegetation to Climate Factors across the Tibetan Plateau: The Roles of Interactions between Temperature, Precipitation, and Solar Radiation. *Agric. For. Meteorol.* **2023**, *335*, 109440. [CrossRef]
70. Sen, P.K. Robustness of Some Nonparametric Procedures in Linear Models. *Ann. Math. Stat.* **1968**, *39*, 1913–1922. [CrossRef]
71. Theil, H. A Rank-Invariant Method of Linear and Polynomial Regression Analysis. *Indag. Math.* **1950**, *12*, 173.
72. Hua, T.; Zhao, W.; Cherubini, F.; Hu, X.; Pereira, P. Effectiveness of Protected Areas Edges on Vegetation Greenness, Cover and Productivity on the Tibetan Plateau, China. *Landsc. Urban Plan.* **2022**, *224*, 104421. [CrossRef]
73. Luque, A.; Carrasco, A.; Martín, A.; de las Heras, A. The Impact of Class Imbalance in Classification Performance Metrics Based on the Binary Confusion Matrix. *Pattern Recognit.* **2019**, *91*, 216–231. [CrossRef]
74. Towers, P.C.; Strever, A.; Poblete-Echeverría, C. Comparison of Vegetation Indices for Leaf Area Index Estimation in Vertical Shoot Positioned Vine Canopies with and without Grenbiule Hail-Protection Netting. *Remote Sens.* **2019**, *11*, 1073. [CrossRef]
75. Chen, S.; Huang, Y.; Wang, G. Detecting Drought-Induced Gpp Spatiotemporal Variabilities with Sun-Induced Chlorophyll Fluorescence During the 2009/2010 Droughts in China. *Ecol. Indic.* **2021**, *121*, 107092. [CrossRef]
76. Zhou, Z.; Ding, Y.; Liu, S.; Wang, Y.; Fu, Q.; Shi, H. Estimating the Applicability of Ndvi and Sif to Gross Primary Productivity and Grain-Yield Monitoring in China. *Remote Sens.* **2022**, *14*, 3237. [CrossRef]
77. Hao, A.; Duan, H.; Wang, X.; Zhao, G.; You, Q.; Peng, F.; Du, H.; Liu, F.; Li, C.; Lai, C.; et al. Different Response of Alpine Meadow and Alpine Steppe to Climatic and Anthropogenic Disturbance on the Qinghai-Tibetan Plateau. *Glob. Ecol. Conserv.* **2021**, *27*, e01512. [CrossRef]
78. Liu, Y.; Li, Z.; Chen, Y.; Kayumba, P.M.; Wang, X.; Liu, C.; Long, Y.; Sun, F. Biophysical Impacts of Vegetation Dynamics Largely Contribute to Climate Mitigation in High Mountain Asia. *Agric. For. Meteorol.* **2022**, *327*, 109233. [CrossRef]
79. Hua, T.; Zhao, W.; Cherubini, F.; Hu, X.; Pereira, P. Upgrading Protected Areas Can Improve or Reverse the Decline in Conservation Effectiveness: Evidence from the Tibetan Plateau, China. *Sci Total Env.* **2023**, *873*, 162345. [CrossRef] [PubMed]
80. Zhang, B.; Zhang, Y.; Wang, Z.; Ding, M.; Liu, L.; Li, L.; Li, S.; Liu, Q.; Paudel, B.; Zhang, H. Factors Driving Changes in Vegetation in Mt. Qomolangma (Everest): Implications for the Management of Protected Areas. *Remote Sens.* **2021**, *13*, 4725. [CrossRef]
81. Cai, H.; Yang, X.; Xu, X. Human-Induced Grassland Degradation/Restoration in the Central Tibetan Plateau: The Effects of Ecological Protection and Restoration Projects. *Ecol. Eng.* **2015**, *83*, 112–119. [CrossRef]
82. Anderson, K.; Fawcett, D.; Cugulliere, A.; Benford, S.; Jones, D.; Leng, R. Vegetation Expansion in the Subnival Hindu Kush Himalaya. *Glob. Chang. Biol.* **2020**, *26*, 1608–1625. [CrossRef]

83. Maina, F.Z.; Kumar, S.V.; Albergel, C.; Mahanama, S.P. Warming, Increase in Precipitation, and Irrigation Enhance Greening in High Mountain Asia. *Commun. Earth Environ.* **2022**, *3*, 43. [CrossRef]
84. Wang, Y.; Li, D.; Ren, P.; Sigdel, S.R.; Camarero, J.J. Heterogeneous Responses of Alpine Treelines to Climate Warming across the Tibetan Plateau. *Forests* **2022**, *13*, 788. [CrossRef]
85. Wenqi, S.; Yuhao, F.; Zhiheng, W. Ecological Restoration Programs Dominate Vegetation Greening in China. *Sci. Total Environ.* **2022**, *848*, 157729.
86. Zhang, Z.; Liu, Y.-F.; Cui, Z.; Huang, Z.; Liu, Y.; Leite, P.A.M.; Zhao, J.; Wu, G.-L. Shrub Encroachment Impaired the Structure and Functioning of Alpine Meadow Communities on the Qinghai–Tibetan Plateau. *Land Degrad. Dev.* **2022**, *33*, 2454–2463. [CrossRef]
87. Sitch, S.; Friedlingstein, P.; Gruber, N.; Jones, S.D.; Murray-Tortarolo, G.; Ahlström, A.; Doney, S.C.; Graven, H.; Heinze, C.; Huntingford, C.; et al. Recent Trends and Drivers of Regional Sources and Sinks of Carbon Dioxide. *Biogeosciences* **2015**, *12*, 653–679. [CrossRef]

remote sensing MDPI

Article

Analysis of Spatial and Temporal Changes in Vegetation Cover and Driving Forces in the Yan River Basin, Loess Plateau

Zhilin He [1,2,3,4], Tianming Yue [1,2,3,4], Yanglong Chen [1,2,3,4], Weichen Mu [1,2,3,4], Mengfei Xi [1,2,3,4] and Fen Qin [1,2,3,4,*]

1 College of Geography and Environmental Science, Henan University, Kaifeng 475004, China; hzlgeo@henu.edu.cn (Z.H.); tianming@henu.edu.cn (T.Y.); chenyanlglong@henu.edu.cn (Y.C.); mwc104754210193@henu.edu.cn (W.M.); xmf@henu.edu.cn (M.X.)
2 Key Laboratory of Geospatial Technology for the Middle and Lower Yellow River Regions, Ministry of Education, Henan University, Kaifeng 475004, China
3 Henan Industrial Technology Academy of Spatio-Temporal Big Data, Henan University, Kaifeng 475004, China
4 Henan Technology Innovation Center of Spatial-Temporal Big Data, Henan University, Kaifeng 475004, China
* Correspondence: qinfen@henu.edu.cn; Tel.: +86-135-0378-5302

Abstract: The Yan River Basin of the Loess Plateau is a key region for ensuring the environmental protection and sustainable development of the Yellow River Basin. Therefore, it is essential to identify how vegetation cover has changed and determine the factors that have driven these changes. In this study, we applied a three-dimensional vegetation cover model to examine the spatiotemporal variation characteristics of vegetation cover at the watershed scale in the Yan River Basin from 2001 to 2020 and forecast future trends. Subsequently, the driving forces of fractional vegetation cover (FVC) change were quantified based on meteorological, surface, and anthropogenic factors to explore the common driving relationships among these factors. (1) The accuracy of 3DFVC is better than that of FVC in the Yanhe River Basin, where the terrain is complex. (2) The temporal change trends indicated that the vegetation cover in the Yan River Basin significantly recovered and the basin FVC increased rapidly from 2001 to 2013 (S = 0.0152/a, $p < 0.01$) and increased gradually from 2013 to 2020 (S = 0.0015/a). The main reason for the increase in vegetation cover was the enhanced growth of medium FVC. (3) The vegetation spatial distribution showed that the FVC values varied substantially from north to south, indicating spatial heterogeneity, and 83.9% of the area presented a trend of increasing vegetation. Furthermore, vegetation cover was predicted to improve in the future. (4) The spatial heterogeneity of FVC was mainly influenced by relative humidity and rainfall, and the spatial variations in FVC were mainly determined by climate factors. Land use and cover change variations, which are influenced by human activities, represent major factors underlying the observed spatial heterogeneity. Most interactions between driving factors showed two-way enhancement or non-linear enhancement, with relative humidity and land use patterns presenting the strongest explanatory power. This study provides a scientific basis for vegetation conservation in the Yan River Basin and contributes theoretical support for decision-making regarding ecological environmental protection in the Loess Plateau and sustainable development in the Yellow River Basin.

Keywords: 3DFVC; spatiotemporal analysis; human activities; Geodetector; GEE; Yan River Basin

1. Introduction

Ecosystems represent the foundation of civilization and promote socioeconomic development. However, over the past 50 years, climate change and socioeconomic development have considerably threatened natural ecosystems worldwide [1]. Fractional vegetation cover (FVC) is a measure of plant community structure that expresses the proportion of an observation area occupied by the vertical projection of vegetation, and it indicates the degree of ground cover by plants in a given area [2]. Vegetation is a crucial component of ecosystems that mediates the interactions among the atmospheric [3,4], hydrological [5,6], and pedological processes [7,8]; moreover, it regulates the health of ecosystems and represents an indicator of the health and characteristics of the natural environment. Vegetation reflects the direct impact of socioeconomic activities on environmental change; thus, determining the interaction mechanism between vegetation and its drivers offers a scientific foundation for directing regional ecological development [9].

For the acquisition of vegetation cover, surface measurements and remote sensing measurements are usually used. Due to the spatial heterogeneity of vegetation cover, the surface measurement method has obvious scale limitations and it is difficult to carry out measurements on a large scale, while the remote sensing method can efficiently obtain continuous data with high accuracy on a large scale [10]. Remote sensing methods include the estimation method based on the vegetation index and the estimation method based on data mining technology. Among them, the estimation method based on the vegetation index is simpler and more efficient than the other methods, and it is more generalized because it is not restricted by any conditions [11,12]. The image dichotomous model posits that sensor-obtained information includes both vegetation and soil data, with the effects of atmospheric and soil background as well as vegetation type attenuated by the linear stretching of modifiers [13,14]. Although vegetation occupies a three-dimensional surface in reality, the image meta-dichotomous model considers it a two-dimensional plane within an image, which results in an inaccurate representation of true vegetation cover. To estimate vegetation cover more accurately, this study adopted a three-dimensional vegetation cover model (3DFVC) that accounts for the effects of topography and terrain complexity. Three-dimensional FVC has more physical meaning than FVC and can be applied to various landscapes [15].

The Loess Plateau, a region of high ecological vulnerability and soil erosion in central North China, responds strongly to climatic variations [16]. China has implemented various conservation practices on the Loess Plateau over the past five decades to reduce soil erosion and improve ecological conditions [17]. Among the most prominent ecological restoration initiatives is the "Returning Cultivated Land to Forests and Grasses Project", which was initiated in 1999. This policy has resulted in a significant enhancement of vegetation cover and quality on the Loess Plateau. Zhang et al. showed that the proportion of the Loess Plateau region presenting an increasing trend in vegetation cover from 1981 to 2016 was 90% and revealed a high overall rate of increase and a significant ecological restoration effect, with Yulin and Yan'an in northern Shaanxi exhibiting rapid increases in vegetation coverage [18]. Zhang et al. showed that from 2001 to 2018, anthropogenic activity had a positive effect on Loess Plateau vegetation cover and corresponded to a trend of 0.36×10^{-2} a^{-1} [19]. Ye et al. showed that the vegetation cover of the Loess Plateau in northern Shaanxi increased from 31.7% to 47.1% on average, with some fluctuations, between 2000 and 2020; moreover, they observed significant improvements in vegetation cover and significant spatial dependence between vegetation cover and ecosystem services [20]. Previous studies on vegetation cover changes on the Loess Plateau presented limitations in their design. For instance, some studies only examined the correlation between vegetation and climate change while neglecting the effects of multiple factors on vegetation growth [21,22]. This may lead to biased conclusions because climate is not the only factor influencing vegetation growth. Furthermore, most studies have concentrated on large-scale spatial and temporal changes in vegetation and have not considered the finer-scale variations [23,24]; thus, changes in vegetation at the watershed scale have not been sufficiently studied to provide detailed recommendations on soil and water conservation measures and other aspects.

A basin is a unique geographic unit in nature with a clear physical boundary and a high degree of comprehensiveness. The world is composed of different basins of different sizes [25], and regional ecological and environmental problems are all related to the destruction of basin resources and irrational management, so solving environmental problems and realizing sustainable socio-economic development from the perspective of a basin is a more effective and comprehensive systematic method. Currently, China's basin-scale ecological and environmental monitoring network is not sound, and the results are scattered, so it is urgent to carry out systematic and comprehensive research [26]. In view of the above existing research deficiencies, this study used the modified 3D vegetation cover model, 3DFVC, to calculate the interannual FVC of the Yan River Basin from 2001 to 2020 and carefully explore the spatial and temporal processes of vegetation cover. The Hurst index was then used to predict the future trend of vegetation development. Our analysis, which was facilitated by a geographic probe, centers on the driving relationships between FVC and an array of factors that include meteorological, surface, and anthropogenic elements. The evaluation considers both climatic and anthropogenic factors that affect plant growth. Our findings may provide a conceptual basis for achieving environmental protection and sustainable economic growth in the Yellow River basin. They also have important implications for enhancing and maintaining regional ecological stability and promoting sustainable socio-economic development.

2. Study Area and Data Sources

2.1. Study Area

The Yan River Basin, located on the Loess Plateau in northern Shaanxi between 108°45′–110°32′E and 36°23′–37°25′N, has a total area of 7687 km^2 (Figure 1). The watershed has a warm temperate continental semi-arid monsoon climate, with spring and winter being cold and dry and summer being hot and rainy. Over 75% of the total annual rainfall occurs in summer, while low temperatures and mild rainfall occur in autumn. The average annual rainfall is approximately 500 mm, and it exhibits minor spatial variations and decreases gradually from southeast to northwest. The average annual temperature is approximately 9 °C, and the average daily temperature difference is 13 °C. The soil type is mainly cultivated loessial soils. Vegetation in the region is influenced by topography and climate and varies significantly from north to south, thus showing a vertical geographical variation pattern.

2.2. Data Sources

Data from 2000 to 2020 were used in this study and include satellite images, meteorological data, surface data, and human activity data (Table 1).

Table 1. Data sources.

Data Category	Data Name	Data Products	Resolution	Data Source
Image	MODIS	MOD13Q1-NDVI	250 m	Google Earth Engine (https://earthengine.google.com/, accessed on 22 March 2023)
	Landsat	Landsat 5/7/8 Surface Reflectance Tier 1	30 m	
Meteorological data	Temperature Precipitation	GPRChinaTemp1 km [27]	1 km	Zenodo (https://zenodo.org/, accessed on 22 March 2023)
	Relative humidity	China Relative Humidity Dataset		National Earth System Science Data Center (www.geodata.cn, accessed on 22 March 2023)
	Sunshine hours	China Sunshine Hours Dataset		
Surface data	Soil	Soil Type	1 km	Resource and Environmental Science Data Center (www.resdc.cn, accessed on 22 March 2023)
	DEM	NASADEM	30 m	EARTHDATA (www.earthdata.nasa.gov, accessed on 22 March 2023)
Human activity data	Population density	LandScan Global	1 km	LandScan (landscan.ornl.gov, accessed on 22 March 2023)
	GDP	ChinaGDP [28]		Zenodo (https://zenodo.org/, accessed on 22 March 2023)
	LUCC	CLCD [29]	30 m	

Note: DEM, digital elevation model; LUCC, land use and cover change; CLCD, China Land Cover Dataset; GDP, gross domestic product.

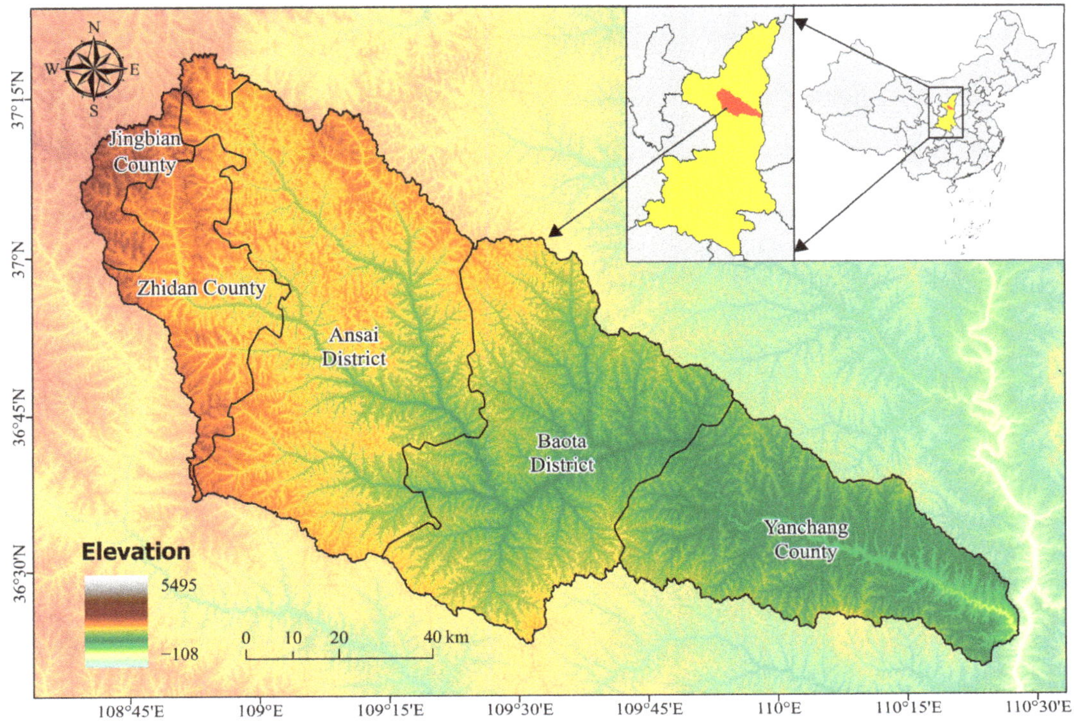

Figure 1. Location of the study area.

3. Methodology

In this study, interannual NDVI data were inferred based on the maximum value compositing (MVC) method, which compensates for the shortcomings of the NDVI, for which vegetation becomes easily saturated in high-cover areas and is difficult to distinguish in low-cover areas [30], and the modified three-dimensional vegetation cover model 3DFVC was used to calculate the FVC. We compare the applicability of FVC and 3DFVC in the Yan River Basin. The spatiotemporal dynamics of FVC were analyzed using the Theil–Sen median method for trend estimation, the Mann–Kendall test for trend significance, and the coefficient of variation for FVC fluctuation. The Hurst index was used to predict the future trend of FVC development. Additionally, we probed the non-linear spatial response of climatic (average annual temperature, average annual precipitation, relative humidity, and sunshine hours), surface (soil type, elevation, and slope), and human activity factors (land use and cover change (LUCC), population density, and gross domestic product (GDP)) to FVC in the Yan River Basin using a geographic probe. This allowed us to identify the dominant factors underlying spatial variations in FVC, quantify their influence, and examine multifactor interactions. The specific methods are described below.

3.1. Three-Dimensional Vegetation Cover Model

This study applied calculus to calibrate the image element dichotomous model. FVC is calculated using a dichotomous model based on the NDVI [11,15,31]. The formula is shown in Equations (1) and (2):

$$FVC = \frac{NDVI - NDVI_{soil}}{NDVI_{veg} - NDVI_{soil}} \quad (1)$$

$$3DFVC = FVC \times cos(\pi \times slope/180) \tag{2}$$

where FVC is the vegetation cover, NDVI is the normalized difference vegetation index, $NDVI_{veg}$ is the vegetation index value of pure vegetation soil image elements, $NDVI_{soil}$ is the vegetation index value of pure bare soil image elements, and 3DFVC is the three-dimensional vegetation cover model.

The 3DFVC values were categorized into 5 classes based on the equally spaced classification method [32,33] (Table 2). For the sake of science and simplicity in the Section 4, FVC was used uniformly instead of 3DFVC in the Section 4.

Table 2. Vegetation cover class classification standard.

Class	Classification Criteria
Low	$0 \leq$ 3DFVC < 0.2
Relatively low	$0.2 \leq$ 3DFVC < 0.4
Medium	$0.4 \leq$ 3DFVC < 0.6
Relatively high	$0.6 \leq$ 3DFVC < 0.8
High	$0.8 \leq$ 3DFVC < 1

3.2. Accuracy Assessment Method of Vegetation Cover

In this study, the sub-pixel comparison method was used to compare the obtained vegetation cover with the actual vegetation cover (MODIS13Q1-NDVI) so that the accuracies of 3DFVC and FVC could be evaluated. The closer the regression coefficient between the vegetation cover to be verified and the actual vegetation cover is to 1, the higher the accuracy of the vegetation cover model [15]. In addition, the root mean square error (RMSE) was used to test the accuracy of the vegetation cover model [34]. The equations are as follows:

$$RMSE = \sqrt{\sum_{i=1}^{n} (X_i - Y_i)^2 / n} \tag{3}$$

where X_i is the observed vegetation cover; Y_i is the actual vegetation cover; and n denotes the number of samples. The smaller the value of $RMSE$, the higher the model accuracy.

3.3. Theil–Sen Median and Mann–Kendall Methods

The Theil–Sen median is a non-parametric statistical method of trend calculation. The Mann–Kendall trend test method is a non-parametric statistical technique [35]. The calculation formula is as follows:

$$\beta = median\left(\frac{FVC_p - FVC_i}{p - i}\right), \forall\, i < p \tag{4}$$

where FVC_p and FVC_i are FVC time series data. When $\beta > 0$, the FVC time series has an increasing trend; when $\beta < 0$, the time series has a decreasing trend.

$$S = \sum_{i=1}^{n-1} \sum_{p=i+1}^{n} sign(FVC_p - FVC_i) \tag{5}$$

$$var(S) = \frac{n(n-1)(2n+5) - \sum_{i=1}^{n} t_i(t_i - 1)(2t_i + 5)}{18} \tag{6}$$

$$Z = \begin{cases} \frac{S-1}{\sqrt{var(S)}}\ S > 0 \\ 0\ S = 0 \quad (6) \\ \frac{S+1}{\sqrt{var(S)}}\ S < 0 \end{cases} \tag{7}$$

where S is the statistical quantity, FVC_p and FVC_i are the NDVI values of the studied time series under the corresponding year, n is the number of study years, $var(S)$ is the variance of statistic S, and Z is the test statistic.

In this paper, the significance level α is 0.05.

3.4. Coefficient of Variation

The coefficient of variation (C_v) can reflect the degree of dispersion in a set of data [36]. The calculation equation is as follows:

$$C_v = \frac{\sqrt{\frac{1}{n}\sum_{i=1}^{n}\left(C_i - \overline{C}\right)^2}}{\overline{C}} \tag{8}$$

where C_i is the FVC value in year I, $\overline{C}$ is the multi-year average of FVC in the monitoring period, and n is the number of study years.

They are categorized into five categories based on the actual situation [37], as shown in Table 3.

Table 3. The coefficient of variation classification standard.

Class	Classification Criteria
Low variation	$0 < C_v \le 0.05$
Relatively low variation	$0.05 < C_v \le 0.10$
Medium variation	$0.10 < C_v \le 0.15$
Relatively high variation	$0.15 < C_v \le 0.20$
High variation	$C_v > 0.20$

3.5. Hurst Index

The Hurst index is widely used to determine the strength of the persistence or inverse persistence of time series trends [38].

$$\overline{FVC}_{(\tau)} = \frac{1}{\tau}\sum_{t=1}^{\tau} FVC_{(\tau)}\ \tau = 1, 2, 3, \ldots, n \tag{9}$$

where $FVC_{(\tau)}$ is the FVC time series.

The cumulative deviation series $X_{(t,\tau)}$ is defined as follows:

$$X_{(t,\tau)} = \sum_{j=1}^{t}\left(FVC_{(t)} - \overline{FVC}_{(\tau)}\right)\ 1 \le t \le \tau\ 1 \tag{10}$$

The extreme difference is defined as follows:

$$R_{(\tau)} = \max_{1\le t\le \tau} X_{(t,\tau)} - \min_{1\le t\le \tau} X_{(t,\tau)} \tag{11}$$

The standard deviations are defined as follows:

$$S_{(\tau)} = \sqrt{\frac{1}{\tau}\sum_{t=1}^{\tau}\left(FVC_{(t)} - \overline{FVC}_{(\tau)}\right)^2} \tag{12}$$

The Hurst index is as follows:

$$\frac{R_{(\tau)}}{S_{(\tau)}} = (c\tau)^H \tag{13}$$

where H is the Hurst index.

The meaning of the Hurst index is shown in Table 4.

Table 4. Hurst index classification criteria.

Type of Trend Change	Judgment Basis
Anti-Continuity	$0 < H < 0.5$
Independence	$H = 0.5$
Sustainability	$0.5 < H < 1$

3.6. Geodetector

The Geodetector method can directly and reliably measure the interactions and effects of driving forces and is not constrained by the assumptions of traditional statistical methods (Table 5). Moreover, it does not require any assumptions of linearity and is not affected by linearity [39,40]. The influencing factors selected in this paper include three categories—climate, land surface, and human activities (Table 6)—in order to explore the response relationship between vegetation cover changes and factors in the Yanhe River Basin.

Table 5. Role of the Geodetector modules.

Detector Type	Function
Factor detector	Investigating the spatial heterogeneity of Y and its dependence on factor X
Interaction detectors	Interaction between different influences on X and independent variables was identified
Risk detector	Determining the significance of the difference in the mean values of the attributes between the two subregions
Ecological detector	Comparing the significance of the difference between the effects of factors X1 and X2 on the spatial distribution of attribute Y

Table 6. Drivers of vegetation cover.

Factor Category	Driving Factor	Unit
Meteorological factor	Temperature	°C
	Precipitation	mm
	Relative humidity	%
	Sunshine hours	h
Surface factor	Soil	-
	DEM	m
	Slope	°
Human activity factor	Population density	persons/km^2
	GDP	million CNY
	LUCC	-

4. Results and Analysis

4.1. Quantitative Comparative Analysis of 3DFVC and FVC

In this study, the annual average vegetation cover of FVC and 3DFVC in the Yanhe River Basin from 2001 to 2020 was calculated separately, and the difference between FVC and 3DFVC was obtained by spatially stacking and subtracting (Figure 2). In general, the more complex the terrain, the stronger the differences. The Yanhe River Basin watershed is characterized by complex topography, numerous rivers, and undulating beams and mounts. By comparison, we found that the spatial distribution of high variability showed high similarity with that of a high slope, indicating that the difference between FVC and 3DFVC was not significant in certain areas with relatively simple topography. However, with the complexity of the terrain, the difference between FVC and 3DFVC becomes more obvious.

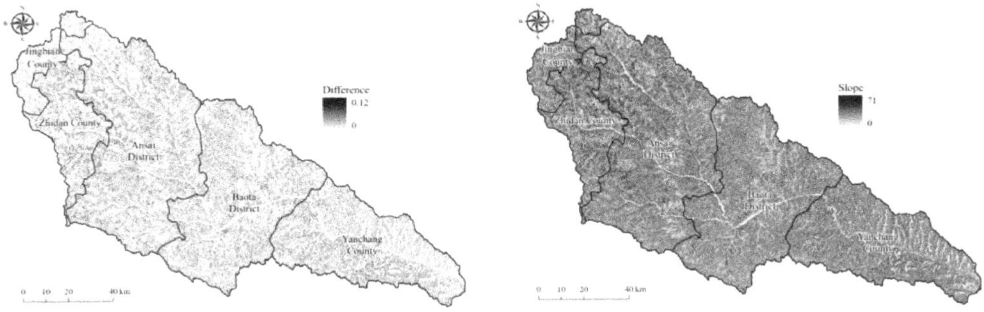

Figure 2. Differential distribution of FVC and 3DFVC and slope distribution in the Yanhe River Basin.

To further verify the extraction accuracy of the model, we used the subpixel comparison method to verify the extraction accuracy of FVC and 3DFVC (Figure 3). The results show that the regression coefficient of 3DFVC is 0.036 higher than that of FVC, which is closer to 1. In addition, the root mean square error (RMSE) of 3DFVC is reduced by 0.0175, which is lower than that of FVC. Obviously, 3DFVC reduces the effect of complex topography on vegetation extraction compared to FVC, which makes it have higher extraction accuracy and stronger applicability.

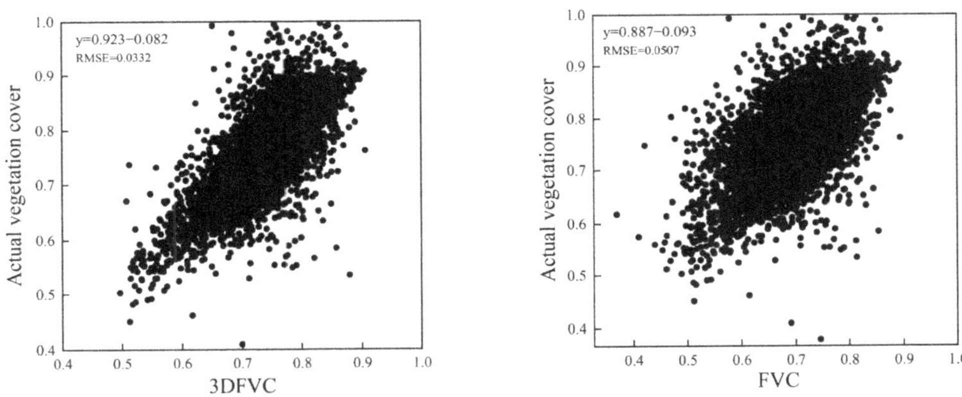

Figure 3. Accuracy comparison between 3DFVC and FVC.

4.2. Characteristics of Temporal Changes in Vegetation Cover in the Yanhe River Basin and Its Counties, 2001–2020

4.2.1. Temporal Trends in Vegetation Cover in the Yanhe River Basin

We used linear regression and *t*-test methodologies to analyze FVC trends in the Yan River Basin from 2001 to 2020 (Figure 4). The analysis showed a highly significant increase in overall FVC over the 20-year period, with an average annual growth rate of 0.0112/a ($p < 0.01$) and an average FVC of 0.48. During this time frame, the annual average FVC increased from 0.30 to 0.57, the minimum value occurred in 2001 (0.30), and the maximum value occurred in 2018 (0.59).

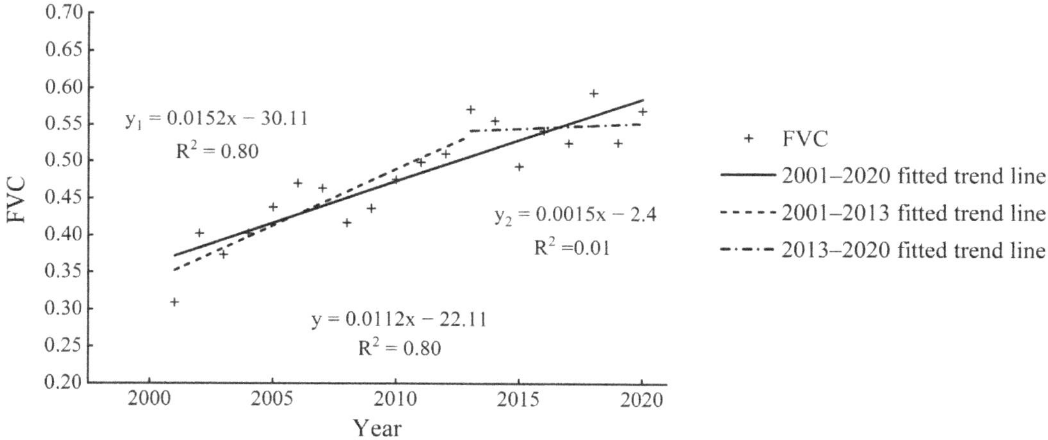

Figure 4. Temporal trends in vegetation cover in the Yanhe River Basin, 2001–2020.

Based on the partial linear regression (PLR) of the lowest error variance ($p < 0.01$), FVC can be divided into two phases, with 2013 as the time node. From 2001 to 2013, the average annual FVC exhibited a highly significant and rapid growth trend, as evidenced by an average annual growth rate S of 0.0152/a ($p < 0.01$), an average FVC of 0.44, and an increase in FVC of 84.9% relative to its value in 2001. From 2013 to 2020, the FVC in the basin showed a highly significant slow growth trend, with an average annual growth rate of 0.0015/a. Moreover, the average value of FVC gradually stabilized above and below 0.55, with a large interannual FVC value change and a significant fluctuating change trend from year to year.

Overall, the FVC in the Yan River Basin experienced two distinct periods in its time series. After reaching its peak growth rate in 2013, the FVC underwent a "rapid growth period" from 2000 to 2013, followed by a "fluctuating change period" from 2013 to 2020, with 2013 serving as the inflection point.

4.2.2. Temporal Trends in Vegetation Classes in the Yanhe River Basin

In order to further analyze the vegetation cover in the Yanhe River Basin, the vegetation cover was classified into five categories based on Table 2. Figure 5 shows the temporal distribution of different FVC classes in the Yan River Basin from 2001 to 2020 as a percentage of the total area. In the study area, the percentages of low, relatively low, and medium FVC classes declined over time. Conversely, the percentages of relatively high and high FVC classes increased over time. These findings indicate an overall enhancement in vegetation cover, suggesting a progressive improvement in the vegetation conditions of the Yan River Basin. In addition, the area shares of low and relatively low FVC classes exhibited dynamic changes with a 6-year cycle of rising and then falling, although the overall trend was decreasing. The area shares of the medium FVC class showed a fluctuating decreasing trend from 69.5% in 2001 to 38.3% in 2020. The area shares of relatively high and significantly high classes increased, with the proportion of relatively high areas increasing from 21.3% in 2001 to 35.2% in 2020. Overall, the vegetation cover of the Yan River Basin improved, as evidenced by the increasing proportion of relatively high and high FVC classes. This improvement was primarily due to the enhancement of low- and relatively low-grade vegetation.

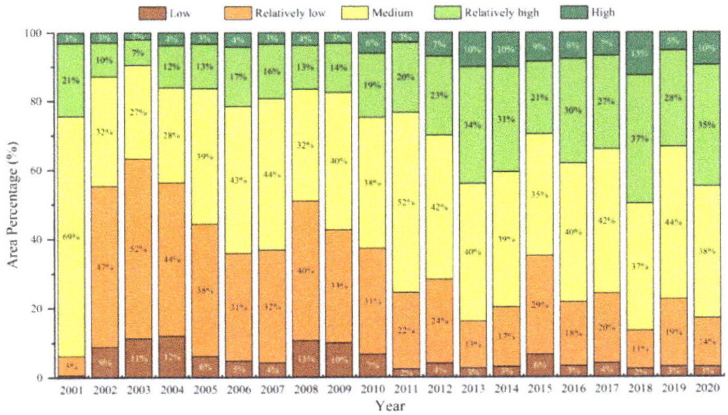

Figure 5. Area percentage of different classes of FVC in the Yan River Basin for 2001–2020.

In order to understand the details of the changes in each vegetation class, Table 7 presents more statistical data on the transition of different FVC classes in the Yan River Basin from 2001 to 2020. In 2020, it was observed that the areas of low, relatively low, relatively high, and high FVC classes within the Yan River Basin increased by 178.21 km^2, 643.76 km^2, 1067.29 km^2, and 493.91 km^2, respectively. Conversely, the area of the medium FVC class decreased by 2383.18 km^2. The areas with low, relatively low, relatively high, and high FVC classes in the study area were primarily shifted from areas with medium, relatively low, relatively high, and high FVC classes, with transfers of 118.72 km^2, 775.34 km^2, 1788.05 km^2, and 424.64 km^2, respectively. The area with medium FVC was mostly transferred to relatively high FVC, indicating that the enhancement of vegetation cover was mainly driven by the improvement in the medium class.

Table 7. Transfer matrix for different classes of FVC in the Yan River Basin, 2001–2020 (km^2).

2001	2020						
	Low	**Relatively Low**	**Medium**	**Relatively High**	**High**	**Total**	**Transfer out**
Low	36.80	4.19	5.17	2.83	0.56	49.55	12.75
Relatively low	17.67	123.39	153.56	102.62	22.15	419.40	296.00
Medium	118.72	775.34	2204.43	1788.05	424.64	5311.18	3106.76
Relatively high	52.45	150.26	520.22	694.47	208.22	1625.62	931.15
High	2.11	9.97	44.63	104.94	79.20	240.85	161.65
Total	227.76	1063.16	2928.00	2692.92	734.76		
Transfer to	190.96	939.77	723.58	1998.45	655.57		
Amount of change	178.21	643.76	−2383.18	1067.29	493.91		

4.2.3. Temporal Trends in Vegetation Cover of Counties in the Yanhe River Basin

We used linear regression and *t*-tests to analyze the temporal trends of vegetation cover in the counties (districts) in the Yanhe River Basin from 2001 to 2020 (Figure 6). The FVC in all counties (districts) showed a highly significant growth trend ($p < 0.01$), with Ansai District having the highest growth rate of 0.0132/a ($p < 0.01$). In addition, Jingbian County had the slowest growth rate of FVC at 0.0064/a ($p < 0.01$). On the other hand, Zhidan County, Baota District, and Yanchang County had higher growth rates of 0.0128/a ($p < 0.01$), 0.0109/a ($p < 0.01$), and 0.081/a ($p < 0.01$), respectively.

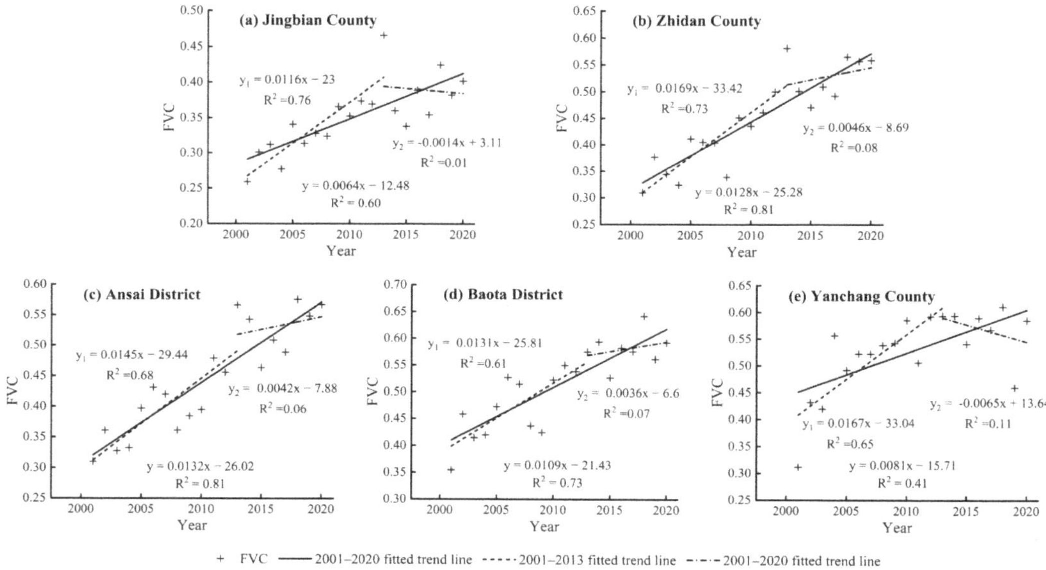

Figure 6. Temporal trends of vegetation cover in counties (districts) of the Yanhe River Basin, 2001–2020.

Based on partial linear regression (PLR) of the lowest error variance ($p < 0.01$), FVC can be divided into two phases, with 2013 as the time node. From 2001 to 2013, The FVC in all counties (districts) showed a highly significant growth trend ($p < 0.01$), with Zhidan County having the highest growth rate of 0.0169/a ($p < 0.01$). In addition, Jingbian County had the slowest growth rate of FVC at 0.0116/a ($p < 0.01$). On the other hand, Ansai District, Baota District, and Yanchang County had higher growth rates of 0.0145/a ($p < 0.01$), 0.0131/a ($p < 0.01$), and 0.0167/a ($p < 0.01$), respectively. From 2013 to 2020, all counties (districts) except Jingbian and Yanchang showed a highly significant growth trend, with the highest growth rate in Zhidan County at 0.0046/a. Moreover, the growth rate S in the Ansai and Baota districts was 0.0042/a and 0.0036/a, respectively. However, Jingbian and Yanchang counties had a negative growth rate S of −0.0014/a and −0.0065/a, respectively.

Overall, among the counties (districts), the FVC in Ansai District had the fastest growth rate in 20 years and presented a highly significant growth trend, while Jingbian County had the slowest growth rate.

4.3. Spatial Distribution and Change Characteristics of Vegetation Cover in the Yanhe River Basin, 2001–2020

4.3.1. Spatial Distribution Patterns of Vegetation Cover in the Yanhe River Basin, 2001–2020

The Yan River Basin was divided into five different regions based on administrative divisions for this study: Jingbian County, Zhidan County, Ansai District, Baota District, and Yanchang County. Figure 7 shows that the spatial distribution pattern of FVC within the Yan River Basin from 2001 to 2020 exhibited considerable variations from north to south, indicating significant spatial heterogeneity. The central region of Baota District had a high level of urbanization, suggesting that urban development had a substantial impact on vegetation cover. The areas with high FVC values were primarily located in Zhidan County, the southern regions of Ansai District and Baota District, and the eastern and western regions of southern Yanchang County. These areas were characterized by tall tree vegetation types, such as white birch, mountain poplar, and oil pine.

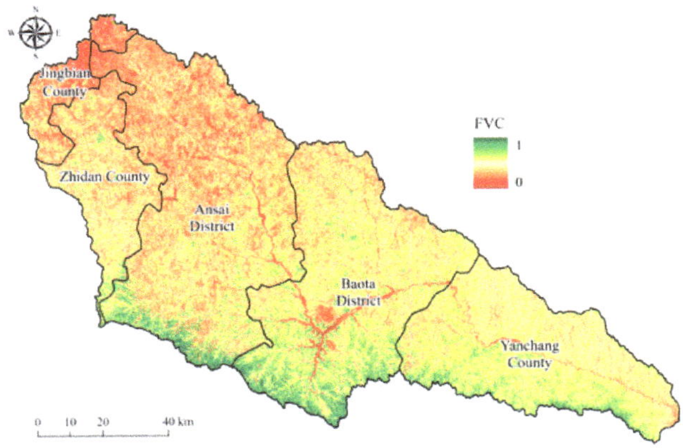

Figure 7. Spatial distribution pattern of mean vegetation cover in the Yanhe River Basin, 2001–2020.

To examine the distribution of vegetation in the Yan River Basin in depth, the 20-year average FVC was classified according to the classification criteria (Table 2). The statistical results of different FVC types are shown in Figure 8, which shows that medium FVC was predominant (54.9%), followed by relatively low FVC (24.4%), while low, relatively high, and high FVC covered 1.2%, 17.1%, and 2.4% of the Yanhe River Basin, respectively. Among the counties (districts), Zhidan County, Ansai District, Baota District, and Yanchang County were all mainly characterized by medium vegetation cover, with area shares of 58.2%, 48.8%, 56.1%, and 67.1%, respectively. Yanchang County is situated in the lower reaches of the Yan River Basin and has a relatively low elevation and gentle slopes that are favorable for vegetation. Tall trees such as birch, mountain poplar, and oleander are found in southern Yanchang County. Jingbian County is dominated by medium to low vegetation cover (62.99%) followed by medium vegetation cover (29%). This county is situated in the transition zone from desert to loess hills, where the surface is largely covered by grass, shrubs, and other low vegetation.

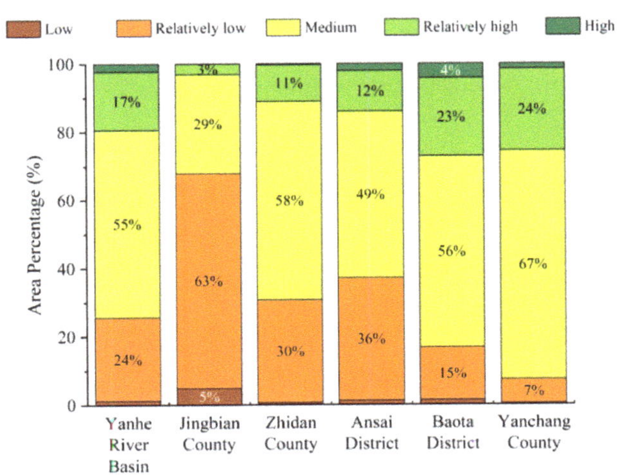

Figure 8. Average area shares of vegetation cover classes in the Yanhe River Basin and its counties (districts), 2001–2020.

4.3.2. Spatial and Temporal Pattern of Vegetation Change and Its Future Trend in the Yanhe River Basin, 2001–2020

The Theil–Sen median and Mann–Kendall trend analysis methods were used to analyze the spatial and temporal variation characteristics of FVC in the Yan River Basin from 2001 to 2020, as shown in Figure 9. The analysis results show that FVC increased in 83.9% of the Yan River Basin, with a significant increase observed in 55.7% of the area, mainly in the upper and middle reaches. This suggests that the project of returning farmland to forest and grassland was effective in enhancing vegetation growth. However, in 15.9% of the Yan River Basin, FVC decreased, and in 3.9% of the area, FVC showed a significant decreasing trend, mainly in some regions of Jingbian County, Ansai District, Baota District, and Yanchang County. This may indicate that rapid urbanization has adversely affected regional vegetation conditions [41].

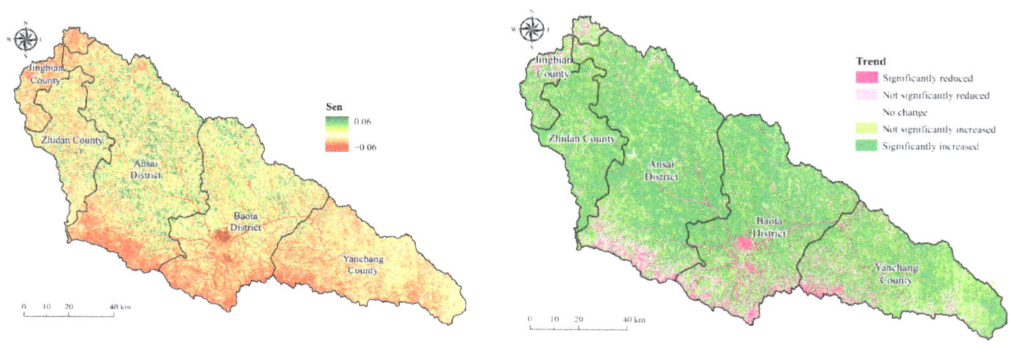

Figure 9. Spatial distribution of average growth rates and overall trends in vegetation cover in the Yanhe River Basin, 2001–2020.

After categorizing the coefficients of variation according to Table 3, Figure 10 shows the fluctuations in FVC in the Yan River Basin from 2001 to 2020. Areas with high FVC fluctuations covered 30.26% of the total area, and they were mainly observed in some regions of the Ansai and Baota Districts. This may be due to the combined influence of human activities, such as ecological management and resource development. Additionally, areas with low and medium fluctuations accounted for 16.28% of the total and were primarily located in the southern and southeastern border regions of the Yan River Basin.

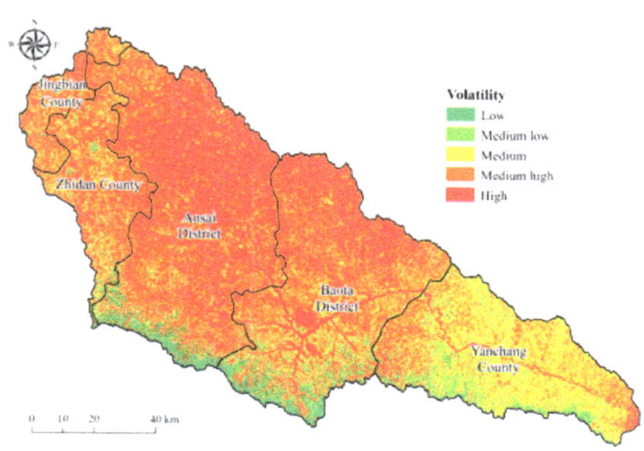

Figure 10. Spatial distribution pattern of the degree of volatility of vegetation cover change in the Yanhe River Basin, 2001–2020.

Despite the usefulness of the Theil–Sen median and Mann–Kendall trend analysis methods for analyzing the spatial and temporal evolution of vegetation, it should be noted that these methods cannot forecast future trends. To address this, the Hurst and Sen indexes were overlaid to explore future vegetation change. Judging the results of the Hurst index according to Table 4, Figure 11 presents the Hurst index for FVC within the Yan River Basin from 2001 to 2020. The analysis revealed a maximum value of 0.99, a minimum value of 0.08, and a mean value of 0.5. The results indicated that most regions (50.77%) may show an anti-sustainability trend in the future, particularly in the Ansai, Baota, and Yanchang districts. Furthermore, 49.23% of the vegetation cover within the basin exhibited persistence, particularly in regions such as Zhidan County, northwest and northeast Ansai District, central Baota District, and east and southeast western Yanhe County. Further analysis of the future trend of FVC combined with Sen's rate of change was performed, and the results are shown in Figure 11. The FVC showed a future growth trend in 51.9% of the region, with 42.5% showing continuous growth (improvement—improvement), mainly in Zhidan County and northwest of Ansai District. Sustained growth (degradation—improvement) was observed in 9.4% of the area, mainly in Jingbian County, southern Ansai District, southeast and southwest Baota District, and southwest Yanhe County. A decline in anti-sustained conditions (improvement—degradation) was observed in 41.5% of the area under study. This phenomenon was primarily evident in northern and south-central Ansai District, northwestern Baota District, and central and eastern Yanchang County. In contrast, persistent decline (degradation—degradation) was observed in 6.6% of the area, mainly in the townships of Baota District and their surrounding areas and the southeast and southwest Yanchang County.

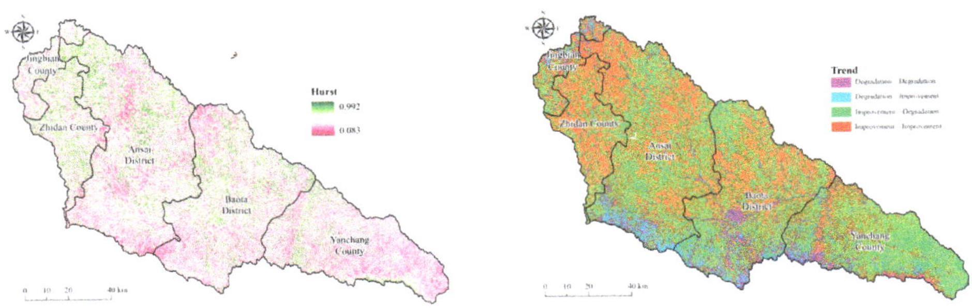

Figure 11. Spatial distribution patterns of the Hurst index of vegetation cover and its future trends in the Yanhe River Basin, 2001–2020.

4.4. Analysis of Factors Influencing Vegetation Cover

4.4.1. Factor Detector

The spatiotemporal analysis revealed that vegetation cover in the Yan River Basin is spatially heterogeneous. However, the presence of severe multicollinearity among the explanatory variables may hinder the model's capacity to effectively account for this heterogeneity. On the other hand, the geographic detector used in this study is robust to the effects of multicollinearity among independent variables, which allows it to accurately assess the individual impact of each factor on the spatial distribution of vegetation cover. An analysis using the factor detector to examine the influence of various factors on the spatial variation in FVC at different stages showed that although all factors had a significant impact on the spatial variation in FVC ($p < 0.01$), the extent to which each factor could explain this variation varied (Table 8). In descending order of explanatory power, the factors are relative humidity, precipitation, sunshine hours, elevation, temperature, LUCC, population density, soil type, GDP, and slope. The spatial distribution of vegetation cover in the Yan River Basin was mainly influenced by climatic factors related to moisture, as evidenced by the high q values (>20%) of relative humidity (29%), precipitation (26.2%), and sunshine hours (23.8%). Elevation, temperature, and LUCC had moderate effects on the spatial variation in FVC, with q values of 11.2%, 9.6%, and 9.4%, respectively. Thus, they were classified as secondary factors. Conversely, population density, soil type, GDP, and slope had relatively low explanatory power for vegetation cover, as evidenced by their q values, which were all below 5%.

Table 8. Decision value q of each driving factor.

Factor	Tem	Pre	RH	SD	Soil	Elevation	Slope	Pop	LUCC	GDP
q	0.096	0.262	0.290	0.238	0.026	0.112	0.006	0.026	0.094	0.014
p	0.000	0.000	0.000	0.000	0.000	0.000	0.000	0.000	0.000	0.000

Note: Tem, temperature; Pre, precipitation, RH, relative humidity; SD, sunshine duration; Pop, population density; LUCC, land use and cover change; GDP, gross domestic product.

4.4.2. Interaction Detector and Ecological Detector

The interaction detector method revealed a significant interrelationship between the driving factors underlying changes in vegetation FVC, as shown in Figure 12. After two interactions, the factors had an enhanced effect on FVC, indicating that their effects were synergistic rather than independent. The effects of interaction were classified into two categories, two-factor enhancement and nonlinear enhancement, which refer to the different ways that two factors can amplify each other's impact. Nonlinear enhancement was found to be more prevalent and influential than two-factor enhancement. We applied the ecological detector method to examine the spatial variation in FVC across different combinations of factors (Table 9).

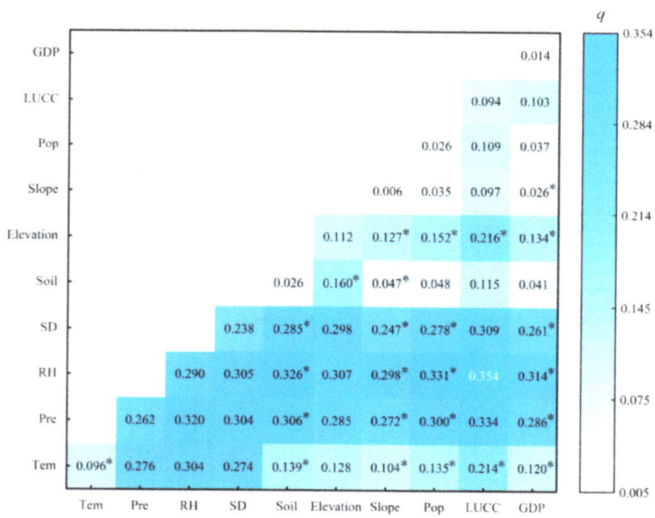

Figure 12. Interaction detector results. Values marked with * are non-linear enhancements; the remaining values are two-factor enhancements.

The strongest explanatory power occurred for the interaction between relative humidity and land use, which reached a two-factor interaction q value of 0.354 and there is a significant difference in the effect of their combination on the spatial distribution of FVC. The worst explanatory power was observed for the interaction of slope and GDP, which may have been due to the poor explanatory power of slope and GDP for FVC, and there is no significant difference in the effect of their combination on the spatial distribution of FVC; thus, insufficient explanatory power was observed for this two-factor interaction. Meanwhile, the combination of relative humidity with most of the factors produced higher q values, and there are significant differences in the effects of its combination with each of the other factors on the spatial distribution of FVC, with the lowest observed for the combination of relative humidity and slope, which reached 0.298. Slope and GDP interacted with most of the factors with low explanatory power.

Table 9. Statistical significance of the detection factors.

Factor	Tem	Pre	RH	SD	Soil	Elevation	Slope	Pop	LUCC	GDP
Tem										
Pre	Y									
RH	Y	Y								
SD	Y	N	Y							
Soil	Y	Y	Y	Y						
Elevation	N	Y	Y	Y	Y					
Slope	Y	Y	Y	Y	Y	Y				
Pop	Y	Y	Y	Y	N	Y	Y			
LUCC	Y	Y	Y	Y	Y	Y	Y	Y		
GDP	Y	Y	Y	Y	N	Y	N	Y	Y	

Note: Tem, temperature; Pre, precipitation, RH, relative humidity; SD, sunshine duration; Pop, population density; LUCC, land use and cover change; GDP, gross domestic product. Y stands for significant; N stands for not significant.

The effects of each factor on the FVC spatial distribution were significantly different (Table 9). The spatial distribution of vegetation cover significantly differed between each pair of factors except for temperature with elevation, precipitation, and sunshine hours; soil type with GDP; elevation with temperature; slope with GDP; and population density with soil type, which did not lead to significant differences.

4.4.3. Risk Detector

The risk detector methodology can determine the average FVC value that corresponds to a specific range or category of factors known to exert an influence on vegetation growth (Figure 13). Among meteorological factors, increases in temperature led to a gradual increase in FVC and stable values, and the maximum FVC was observed at 9.9–10.5 °C. As precipitation levels increased, an initial corresponding rise in FVC values was observed, followed by a decline. The maximum FVC value was recorded within a precipitation range of 620–634 mm. Similarly, increases in relative humidity led to an increase in FVC, with the maximum reached at 56–57% relative humidity. Increases in sunshine hours led to a decrease in FVC, and the maximum was reached at 18.6–19.3 h of sunshine. Thus, the suitable growth conditions for vegetation in the Yan River Basin include low insolation, high rainfall, and high humidity.

Among surface factors, different soil types corresponded to the FVC in the Yanhe basin. Vegetation cover was greatest for black clay soil, indicating that black clay soil is most suitable for vegetation growth.

As elevation increased, the FVC fluctuated and then decreased, and the maximum was reached at 1088–1205 m. As the slope increased, FVC fluctuated, and the maximum was reached at 22–27°. The fact that elevation possesses a larger q value in comparison to slope suggests that among surface factors, elevation is the primary driver of vegetation change within the Yan River Basin.

Among human activity factors, as population density increased, FVC showed a decreasing trend followed by a small increase, and the maximum was reached at a population density of 0–1122 persons/km^2. With different land use types, the FVC in the Yan River Basin differed, and the largest FVC value corresponded to the shrub land use type. With increases in GDP, FVC initially decreased and then gradually fluctuated, and the maximum was observed when the GDP value was 0–18.7 million CNY. Therefore, the suitable growth conditions for vegetation in the Yan River Basin include a low population density, low GDP, and the shrub land use type.

Figure 13. Statistical results of FVC at each level for different impact factors.

5. Discussion

5.1. Impacts of Climate Change on Vegetation Cover Change

Prior research has established that vegetation cover has significantly improved within the Yan River Basin over time [42,43]. Water scarcity limits the vegetation development in the arid region of the Loess Plateau [44]. As the fundamental source of water for vegetation growth, precipitation exerts a delayed influence on vegetation and plays a crucial role in determining vegetation cover [45]. The semi-arid climate of the Yan River Basin results in vegetation that is sensitive to changes in water supply. The factor detector methodology revealed that relative humidity and precipitation in the study region presented q values of 29% and 26.2%, respectively, with both factors exhibiting a gradual increase in influence from northwest to southeast. Precipitation enhances plant growth by activating plant physiology and biochemistry, increasing soil moisture, and improving plant health [46]. As a result, the vegetation in the basin exhibited a gradual geographical differentiation pattern and transitioned from forest to forest-steppe, to typical grassland, and finally to desert grassland based on precipitation levels. Moreover, the results obtained through the application of the interaction detector revealed that the combined influence of relative humidity and temperature, when considered in conjunction with other factors, served to amplify the overall impact of these factors on vegetation. In areas with relatively high precipitation levels, favorable moisture conditions were provided for vegetation growth. Sunshine hours had a q value of 23.8%, which is because sunshine is essential for photosynthesis. Therefore, the duration of light exposure can be considered to exert a direct influence on the growth and development of vegetation. The most important factor affecting FVC was the interaction of precipitation, humidity, and sunshine. Sunlight could also affect precipitation, which could change the relative humidity and the spatial variation and development of vegetation, with implications for the ecosystem dynamics [47].

Temperature is well known to have a positive effect on vegetation cover. This influence is primarily manifested through the extension of the growing period and the enhancement of photosynthetic processes and water use efficiency in plants. These factors collectively contribute to promoting plant growth and overall ecosystem productivity. For instance, the average temperature in Yanchang County was higher than that of Jingbian County, which could account for the observed differences in vegetation cover between the two counties. Moreover, areas with excessively high temperatures may experience negative impacts on vegetation growth because higher temperatures can accelerate plant transpiration, reduce soil moisture levels, increase evaporation rates, and inhibit photosynthesis, all of which can be detrimental to vegetation growth [48]. Previous research has indicated that rainfall is the main driver of vegetation growth in arid regions, while temperature is more influential in semi-humid regions. In semi-arid regions, both temperature and rainfall play significant roles in influencing vegetation growth [49].

Elevation, as a key topographic factor, has been shown to influence plant growth and physiological metabolism through its effects on environmental variables, such as light, temperature, and precipitation. These changes can subsequently impact vegetation cover [50]. In the Yan River Basin, for example, vegetation cover exhibited an initial gentle increase followed by a significant downward trend with increasing elevation. This indicates that elevation influences vegetation development.

5.2. Impact of Human Activities on Changes in Vegetation Cover

Besides meteorological factors, vegetation cover change is also driven by human activities, such as land use change due to arable land expansion, urbanization, and afforestation. Land use analysis in the Yan River Basin from 2001 to 2020 showed significant changes in forests, grasslands, and croplands, with a forest area increase of 776.29 km^2 (Table 10). The majority of the increase in forested areas was transferred from grasslands and croplands. Vegetation cover increased significantly in most parts of the basin, especially in the central and northern areas of the middle reaches, after the implementation of the farmland-to-forest and grassland conversion project in 1999. This indicates that human activities have enhanced vegetation cover, as supported by previous studies. For example, He et al. found that urbanization and economic development in the Yellow River Basin of Shaanxi improved the ecological environment and vegetation growth. The most notable increase in vegetation cover was observed in the northern Shaanxi areas, where forest and grassland restoration efforts were undertaken [43]. The Yellow River Basin of Shaanxi has experienced a major trend of synergistic development between economic and ecological quality. This indicates that human interventions, such as farmland-to-forest and grassland conversion and soil and water conservation projects, have been essential for improving vegetation growth and ecosystem health in northern Shaanxi [34]. However, it is important to note that human activities can also negatively impact vegetation cover. For example, Yan'an has implemented a project to expand and develop the city by creating a flat mountain [51]. This project is the world's largest geotechnical project, and it is being conducted in a wet sink loess gully area. The speed and intensity of action far exceed the geological force; thus, it has had a large impact on the geological environment and greatly impacted vegetation cover. Land use change due to reforestation projects and urban expansion will disturb and alter the regional ecological environment to varying extents in the long term.

Table 10. LUCC transfer matrix for the Yan River Basin from 2001 to 2020.

2001	2020								
	Cropland	Forest	Shrub	Grassland	Water	Barren	Impervious	Total	Transfer out
Cropland	499.24	43.50	0.20	383.36	0.75	0.15	23.23	950.43	451.18
Forest	2.74	502.26	0.20	1.25	0.00	0.00	0.03	506.47	4.22
Shrub	0.07	6.87	0.97	5.45	0.00	0.00	0.00	13.36	12.39
Grassland	315.55	730.08	5.30	5060.30	1.03	0.50	10.06	6122.83	1062.53
Water	0.48	0.05	0.00	0.09	2.62	0.00	0.31	3.56	0.94
Barren	0.01	0.00	0.00	0.04	0.00	0.00	0.01	0.07	0.06
Impervious	0.01	0.00	0.00	0.00	0.02	0.00	26.58	26.61	0.03
Total	818.10	1282.76	6.66	5450.50	4.43	0.66	60.21		
Transfer to	318.85	780.51	5.70	390.19	1.81	0.66	33.64		
Amount of change	−132.33	776.29	−6.69	−672.34	0.87	0.59	33.61		

5.3. Suitable Intervals for Vegetation Growth

Using the risk detector, we determined the optimal range or type of environmental factors for vegetation growth based on the highest mean FVC (Table 11). The factor interval with the highest mean FVC indicated the best condition for enhancing vegetation growth [52]. The suitable growth conditions for vegetation in the Yan River Basin included low sunlight, high rainfall, and high humidity. The FVC was greatest at elevations of 1088–1205 m and slopes of 22–27° when the soil type was black clay, which is probably due to the reduced human activity in areas with high altitude and steep terrain. In addition, the three anthropogenic factors also showed significant differences in their effects on FVC, with the vegetation being suitable for low population density and low GDP environments and the shrub land use type being most suitable for vegetation growth.

Table 11. Factor suitability interval (at 95% confidence level).

Factor	Suitable Type or Range of FVC	FVC
Tem	9.9–10.5 °C	0.797
Pre	620–634 mm	0.811
RH	56–57%	0.834
SD	18.6–19.3 h	0.821
Soil	Black clay	0.824
Elevation	1088–1205 m	0.794
Slope	22–27°	0.766
Pop	0–1122 persons/km^2	0.762
LUCC	Shrub	0.834
GDP	0–18.7 million CNY	0.760

Note: FVC, fractional vegetation cover; Tem, temperature; Pre, precipitation, RH, relative humidity; SD, sunshine duration; Pop, population density; LUCC, land use and cover change; GDP, gross domestic product.

5.4. Limitations and Prospects

We analyzed the spatial and temporal variation in FVC in the Yan River Basin at the basin scale, using the GEE platform, Landsat series Earth observation data, and Geodetectors. We also quantitatively explored the driving relationships among meteorological, surface, and human activity factors. Despite utilizing high-quality remote sensing data and employing well-established analytical models, however, the results of this study may have some errors due to certain factors, such as the displacement of remote sensing sensors, complex meteorological conditions, topographical variations, and model selection biases [53]. Although this study included difficult-to-quantify human activity variables (LUCC, population density, and GDP) in the factor analysis, the diversity of human activities and the difficulty of quantification mean that the driving relationship of these factors on FVC and their interaction with other factors have not been fully revealed in the research results. This study concludes that vegetation restoration in the Yan River Basin must consider local natural conditions and socioeconomic development under future climate change scenarios. Moreover, protection and restoration plans should be formulated in a scientifically rigorous and objective manner and incorporate robust legal constraints to ensure a balance among water, food, and ecological security and realize the national "carbon neutral" strategy for promoting sustainable regional development.

6. Conclusions

We analyzed the spatial and temporal patterns and changes in vegetation cover in the Yan River Basin over the past two decades using an enhanced three-dimensional vegetation cover model. We also used a geographic probe to quantify the spatial heterogeneity of FVC in the Yan River Basin. Our study incorporated human activity variables, including land use patterns, population density, and GDP, which are often challenging to quantify, into the factor analysis process and considered the interactions among these factors. We summarize the results as follows.

1. Three-dimensional FVC has better accuracy than FVC. Three-dimensional FVC has higher regression coefficients and lower RMSE, which indicates that 3DFVC is better than FVC in vegetation cover extraction. In addition, 3DFVC is better than FVC for the Yanhe River Basin, which has complex topography.
2. The FVC in the Yan River Basin varied significantly over time from 2001 to 2020, with an overall significant but gradually increasing trend (S = 0.01/a, $p < 0.01$). The year 2013 was identified as the time change point. The overall FVC increased significantly and rapidly before 2013 (S = 0.0152/a, $p < 0.01$) and showed a significant slow-increasing trend after 2013 (S = 0.0015/a). We found that the vegetation cover in the Yan River Basin improved significantly, as shown by the increasing proportion of areas with high and relatively high FVC. The main reason for the increase in vegetation cover was the enhanced growth of medium FVC. Among the counties (districts), the FVC in Ansai District had the fastest growth rate in 20 years and presented a highly significant growth trend, while Jingbian County had the slowest growth rate.
3. Our analysis of the spatial distribution of FVC within the Yan River Basin from 2001 to 2020 revealed a pattern of gradually increasing spatial distribution from north to south, with significant differences observed between these regions. High-value areas were primarily located in Zhidan County, southern Ansai District, southern Baota District, and the eastern and western parts of southern Yanchang County, with Yanchang County exhibiting the most favorable vegetation cover. We found that the FVC of all counties (districts) in the Yan River Basin mainly increased, as shown in 83.9% of the area, especially in the upper and middle reaches. This indicates the effectiveness of the project of returning farmland to forest and grassland. The FVC in the river basin fluctuated considerably over the past two decades, and our analysis suggests that the FVC in the Yan River Basin will likely increase further in the future.
4. The main factors causing spatial variation in FVC from 2001 to 2020 were relative humidity and rainfall, which had an explanatory power of over 25%. The spatial

variation in the imaged FVC was mainly influenced by climate factors. Anthropogenic LUCC (mainly decreasing arable and grassland areas and increasing forest areas) was also a key driver of spatial and temporal changes in vegetation cover and differences. Most of the interactions between drivers were linear or nonlinear enhancement, with the strongest interaction being relative humidity and land use, which explained 35.4% of the variation. The suitable growth conditions for Yan River Basin vegetation included low sunlight, high rainfall, high humidity, 1088–1205 m elevation, 22–27° slope, black clay soil, low population density, low GDP, and shrub land use type.

This paper can quantify the effects of multiple factors on the spatial heterogeneity of vegetation, thus offering a scientific reference for realizing ecological civilizations, sustainable development, and ecological protection in watersheds.

Author Contributions: Project administration and funding acquisition, F.Q.; writing—review and editing, F.Q.; writing—original draft and methodology, Z.H.; investigation, T.Y., Y.C., W.M. and M.X. All authors have read and agreed to the published version of the manuscript.

Funding: This study was supported by the High-Resolution Satellite Project of the State Administration of Science, Technology, and Industry for National Defense of the PRC, grant number 80-Y50G19-9001-22/23; the National Science and Technology Platform Construction Project, grant number 2005DKA32300; the Key Projects of National Regional Innovation Joint Fund, grant number U21A2014; the Major Research Projects of the Ministry of Education, grant number 16JJD770019; and the Open Program of Collaborative Innovation Center of Geo-Information Technology for Smart Central Plains Henan Province, grant number G202006.

Data Availability Statement: Not applicable.

Conflicts of Interest: The authors declare no conflict of interest.

References

1. Jin, K.; Wang, F.; Han, J.; Shi, S.; Ding, W. Contribution of climatic change and human activities to vegetation NDVI change over China during 1982–2015. *Acta Geogr. Sin.* **2020**, *75*, 961–974.
2. Gitelson, A.A.; Kaufman, Y.J.; Stark, R.; Rundquist, D. Novel algorithms for remote estimation of vegetation fraction. *Remote Sens. Environ.* **2002**, *80*, 76–87. [CrossRef]
3. Liu, M.; Liu, Y.; Chen, M.; Li, Q.; Liang, Q.; Zou, J.; Qiaolipanguli, T. Spatiotemporal Evolution of Vegetation Coverage and Its Response to Climate Change in Upper Reaches of Ganjiang River Basin during 2000–2018. *Bull. Soil Water Conserv.* **2020**, *40*, 284–290.
4. Guo, M.; Zhang, T.; Zhang, J.; Chen, L.; Zhang, X. Response of Vegetation Coverage to Climate Change in the Loess Plateau in 1982–2006. *Res. Soil Water Conserv.* **2014**, *21*, 35–40, 48.
5. Wang, H.; Chen, W.; He, L.; Li, H. Responses of aquatic vegetation coverage to interannual variations of water level in different hydrologically connected sub -lakes of Poyang Lake, China. *Yingyong Shengtai Xuebao* **2022**, *33*, 191–200. [PubMed]
6. Wang, W.; Hu, P.; Wang, J.; Yang, Z.; Liu, H.; Yang, Q. Response of Vegetation Cover and Structure to Meteorological and Hydrologic Factors in Zhalong Wetland. *J. Hydroecol.* **2020**, *41*, 89–97.
7. Deng, B.; Chen, H.; Li, H.; Lei, S. Influence of dump vegetation coverage and topographic changes on soil and water loss in drainage basin. *Coal Sci. Technol.* **2022**, *50*, 299–308.
8. Xu, H.; He, H.; Huang, S. Analysis of fractional vegetation cover change and its impact on thermal environment in the Hetian basinal area of County Changting, Fujian Province, China. *Acta Ecol. Sin.* **2013**, *33*, 2954–2963.
9. Liu, R.; Xu, L.; Feng, F.; Liu, Y.; Zhang, X. Spatiotemporal Variation of Vegetation Coverage in Wuhai City from 2000 to 2018. *Res. Soil Water Conserv.* **2022**, *29*, 265–273.
10. Qin, W.; Zhu, Q.; Zhang, X.; Li, W.; Fang, B. Review of vegetation covering and its measuring and calculating method. *J. Northwest Sci-Tech Univ. Agric. For.* **2006**, *34*, 163–170.
11. Li, M.; Wu, B.; Yan, C.; Zhou, W. Estimation of Vegetation Fraction in the Upper Basin of Miyun Reservoir by Remote Sensing. *Resour. Sci.* **2004**, *26*, 153–159.
12. Cheng, H.; Zhang, W.; Chen, F. Advances In Researches on Application of Remote Sensing Method to Estimating Vegetation Coverage. *Remote Sens. Land Resour.* **2008**, *20*, 13–18.
13. Peng, W.; Wang, G.; Zhou, J.; Xu, X.; Luo, H.; Zhao, J.; Yang, C. Dynamic monitoring of fractional vegetation cover along Minjiang River from Wenchuan County to Dujiangyan City using multi-temporal landsat 5 and 8 images. *Acta Ecol. Sin.* **2016**, *36*, 1975–1988.
14. Li, D.; Fan, J.; Wang, J. Change characteristics and their causes of fractional vegetation coverage(FVC) in Shaanxi Province. *Chin. J. Appl. Ecol.* **2010**, *21*, 2896.

15 Li, M.; Yan, Q.; Li, G.; Yi, M.; Li, J. Spatio-Temporal Changes of Vegetation Cover and Its Influencing Factors in Northeast China from 2000 to 2021. *Remote Sens.* **2022**, *14*, 5720. [CrossRef]
16 Chen, S.; Wen, Z. Zonal Species Distribution Response to Climate Change in Yanhe River Catchment. *J. Soil Water Conserv.* **2011**, *25*, 157–161.
17 Zhao, J.; Van Oost, K.; Chen, L.; Govers, G. Moderate topsoil erosion rates constrain the magnitude of the erosion-induced carbon sink and agricultural productivity losses on the Chinese Loess Plateau. *Biogeosciences* **2016**, *13*, 4735–4750. [CrossRef]
18 Zhang, F.; Yang, L.; Yang, Y. Change of Vegetation NDVI and Its Response to Climatic and Human Activities in the Loess Plateau During 1981–2016. *Res. Soil Water Conserv.* **2023**, *30*, 230–237.
19 Zhang, C.; Bai, Z.; Li, X.; Ran, Q.; Wei, Z.; Lei, T.; Wang, N. Spatio-temporal evolution and attribution analysis of human effects of vegetation cover on the Loess Plateau from 2001 to 2018. *Arid Land Geogr.* **2021**, *44*, 188–196.
20. Xuan, Y.; Shuai-zhi, K.; Yong-hua, Z.; Lei, H.; Xi-ming, X.; Fan, L. Spatio-temporal relationship between vegetation restoration and ecosystem services in the Loess Plateau of Northern Shaanxi, China. *Chin. J. Appl. Ecol.* **2022**, *33*, 2760–2768.
21. Xie, H.; Tong, X.; Li, J.; Zhang, J.; Liu, P.; Yu, P. Changes of NDVI and EVI and their responses to climatic variables in the Yellow River Basin during the growing season of 2000–2018. *Acta Ecol. Sin.* **2022**, *42*, 4536–4549.
22. Chen, C.; Wang, Y.; Li, Y.; Zhou, S. Vegetation Changes and Influencing Factors in Different Climatic Regions of Yellow River Basin from 1982 to 2015. *J. Yangtze River Sci. Res. Inst.* **2022**, *39*, 56–62, 81.
23. Zhang, J.; Li, C.; Wang, T. Dynamic Changes of Vegetation Coverage on the Loess Plateau and Its Factors. *Res. Soil Water Conserv.* **2022**, *29*, 224–230, 241.
24. Zhang, J.; Hu, A.; Zhang, D. Vegetation coverage variation in Loess Plateau Area of Shaanxi-Gansu-Ningxia. *J. Northwestern Teach. Univ. Nat. Sci. Ed.* **2021**, *57*, 71–76.
25. Maes, J.; Egoh, B.; Willemen, L.; Liquete, C.; Vihervaara, P.; Schagner, J.P.; Grizzetti, B.; Drakou, E.G.; La Notte, A.; Zulian, G.; et al. Mapping ecosystem services for policy support and decision making in the European Union. *Ecosyst. Serv.* **2012**, *1*, 31–39. [CrossRef]
26. Qian, C.; Gong, J.; Zhang, J.; Liu, D.; Ma, X. Change and tradeoffs-synergies analysis on watershed ecosystem services: A case study of BailongjiangWatershed, Gansu. *Acta Geogr. Sin.* **2018**, *73*, 868–879.
27. He, Q.; Wang, M.; Liu, K.; Jiang, Z. GPRChinaTemp1km: A high-resolution monthly air temperature dataset for China (1951–2020) based on machine learning. *Copernic. GmbH* **2021**, 1–29. [CrossRef]
28. Zhao, N.; Liu, Y.; Cao, G.; Samson, E.L.; Zhang, J. Forecasting China's GDP at the pixel level using nighttime lights time series and population images. *Mapp. Sci. Remote Sens.* **2017**, *54*, 407–425. [CrossRef]
29. Yang, J.; Huang, X. The 30 m annual land cover dataset and its dynamics in China from 1990 to 2019. *Earth Syst. Sci. Data* **2021**, *13*, 3907–3925. [CrossRef]
30. Yin, D.; Wang, Y. Temporal and spatial changes of vegetation coverage and its topographic differentiation in temperate continental semi-arid monsoon climate region. *Acta Ecol. Sin.* **2021**, *41*, 1158–1167.
31. Qi, J.; Marsett, R.C.; Moran, M.S.; Goodrich, D.C.; Zhang, X.X. Spatial and temporal dynamics of vegetation in the San Pedro River basin area. *Agric. For. Meteorol* **2000**, *105*, 55–68. [CrossRef]
32. Fu, H.; Wang, R.; Wang, X. Analysis of Spatiotemporal Variations and Driving Forces of NDVI in the Yellow River Basin during 1999–2018. *Res. Soil Water Conserv.* **2022**, *29*, 145–153, 162.
33. Nie, T.; Dong, G.; Jiang, X.; Gu, J. Spatiotemporal Variation and Driving Forces of Vegetation Coverage in Yan'an Area. *Res. Soil Water Conserv.* **2021**, *28*, 340–346.
34. Chen, Y.; Wang, W.; Guan, Y.; Liu, F.; Zhang, Y.; Du, J.; Feng, C.; Zhou, Y. An integrated approach for risk assessment of rangeland degradation: A case study in Burqin County, Xinjiang, China. *Ecol. Indic.* **2020**, *113*, 106203. [CrossRef]
35. Sen, P.K. Estimates of the Regression Coefficient Based on Kendall's Tau. *J. Am. Stat. Assoc.* **1968**, *63*, 1379–1389. [CrossRef]
36. He, T.; Shao, Q. Spatial-temporal Variation of Terrestrial Evapotranspiration in China from 2001 to 2010 Using MOD16 Products. *J. Geo-Inf. Sci.* **2014**, *16*, 979–988.
37. Wang, X.; Shi, S.; Chen, J. Change and driving factors of vegetation coverage in the Yellow River Basin. *China Environ. Sci.* **2022**, *42*, 5358–5368.
38. Hurst, H. Methods of Using Long-Term Storage in Reservoirs. *ICE Proc.* **1956**, *5*, 519–543. [CrossRef]
39. Nie, T.; Dong, G.; Jiang, X.; Lei, Y. Spatio-Temporal Changes and Driving Forces of Vegetation Coverage on the Loess Plateau of Northern Shaanxi. *Remote Sens.* **2021**, *13*, 613. [CrossRef]
40. Wang, J.; Xu, C. Geodetector: Principle and prospective. *Acta Geogr. Sin.* **2017**, *72*, 116–134.
41. Du, J.; Fu, Q.; Fang, S.; Wu, J.; He, P.; Quan, Z. Effects of rapid urbanization on vegetation cover in the metropolises of China over the last four decades. *Ecol. Indic.* **2019**, *107*, 105458. [CrossRef]
42. Yue, M.; Geng, G.; Wang, T.; Yang, R.; Gu, Q. Spatiotemporal Variation of Vegetation NDVI and Its Driving Factors in theShaanxi Section of the Yellow River Basin from 2000 to 2019. *Res. Soil Water Conserv.* **2023**, *30*, 238–246, 255.
43. He, H.; Wang, Z.; Dong, J.; Wang, J.; Zou, J. Synergy and trade-off between vegetation change and urbanization development in the Yellow River Basin of Shaanxi Province based on satellite remote sensing data. *Acta Ecol. Sin.* **2022**, *42*, 3536–3545.
44. BAO, C.; FANG, C. Impact of Water Resources Exploitation and Utilization on Eco-environment in Arid Area: Progress and Prospect. *Prog. Geogr.* **2008**, *27*, 38–46.

45. Xie, Y.; Xia, Z.; Wang, T.; Zhang, Z.; Zhu, D. Temporal and spatial variation of vegetation net primary product and its response to hydrothermal conditions and grain for green project in the Yellow River basin. *Bull. Surv. Mapp.* **2023**, *0*, 15–20.
46. Chen, S.; Zhang, Q.; Chen, Y.; Zhou, H.; Xiang, Y.; Liu, Z.; Hou, Y. Vegetation Change and Eco-Environmental Quality Evaluation in the Loess Plateau of China from 2000 to 2020. *Remote Sens.* **2023**, *15*, 424. [CrossRef]
47. Guo, B.; Wang, Y.; Pei, L.; Yu, Y.; Liu, F.; Zhang, D.; Wang, X.; Su, Y.; Zhang, D.; Zhang, B.; et al. Determining the effects of socioeconomic and environmental determinants on chronic obstructive pulmonary disease (COPD) mortality using geographically and temporally weighted regression model across Xi'an during 2014–2016. *Sci. Total Environ.* **2021**, *756*, 143869. [CrossRef]
48. Dong, Y.; Yin, D.; Li, X.; Huang, J.; Su, W.; Li, X.; Wang, H. Spatial–Temporal Evolution of Vegetation NDVI in Association with Climatic, Environmental and Anthropogenic Factors in the Loess Plateau, China during 2000–2015: Quantitative Analysis Based on Geographical Detector Model. *Remote Sens.* **2021**, *13*, 4380. [CrossRef]
49. Xie, B.; Jia, X.; Qin, Z.; Shen, J.; Chang, Q. Vegetation dynamics and climate change on the Loess Plateau, China: 1982–2011. *Reg. Environ. Chang.* **2016**, *16*, 1583–1594. [CrossRef]
50. Bueno, M.L.; Rezende, V.L.; De Paula, L.F.A.; Meira-Neto, J.A.A.; Pinto, J.R.R.; Neri, A.V.; Pontara, V. Understanding how environmental heterogeneity and elevation drives the distribution of woody communities across vegetation types within the campo rupestre in South America. *J. Mt. Sci.-Engl.* **2021**, *18*, 1192–1207. [CrossRef]
51. Pu, C.; Xu, Q.; Zhao, K.; Jiang, Y.; Guo, P.; Du, P.; Yuan, S. Remote sensing analysis of land subsidence and vegetation restoration characteristics in excavation and filling areas of mountain region for urban extension in Yan'an. *J. Eng. Geol.* **2020**, *28*, 597–609.
52. Meng, X.; Gao, X.; Li, S.; Lei, J. Spatial and Temporal Characteristics of Vegetation NDVI Changes and the Driving Forces in Mongolia during 1982–2015. *Remote Sens.* **2020**, *12*, 603. [CrossRef]
53. Dumka, U.C. A Long-Term Spatiotemporal Analysis of Vegetation Greenness over the Himalayan Region Using Google Earth Engine. *Climate* **2021**, *9*, 109.

remote sensing

MDPI

Article

Response of Vegetation to Different Climate Extremes on a Monthly Scale in Guangdong, China

Leidi Wang [1,†], Fei Hu [1,†], Caiyue Zhang [1], Yuchen Miao [2], Huilin Chen [1], Keyou Zhong [1] and Mingzhu Luo [1,*]

1 College of Agriculture, South China Agricultural University, Guangzhou 510642, China
2 Faculty of Veterinary and Agricultural Sciences, University of Melbourne, Melbourne, VIC 3010, Australia
* Correspondence: lmzhd2701@scau.edu.cn
† These authors contributed equally to this work.

Abstract: Climate extremes, particularly drought, often affect the ecosystem. Guangdong Province is one of the most vulnerable areas in China. Using the normalized difference vegetation index (NDVI) to capture vegetation dynamics, this study investigated vegetation responses to drought, temperature, and precipitation extremes on a monthly scale in the vegetation area of Guangdong without vegetation type changes from 1982 to 2015. As extreme temperatures rose, a drought trend occurred in most months, with a higher rate in February and April. The vegetation evenly showed a significant greening trend in all months except June and October. The vegetation activity was significantly positively correlated with the increased extreme temperatures in most months. However, it exerted a negative correlation with drought in February, April, May, June, and September, as well as precipitation extremes in February, April, and June. The response of vegetation to drought was the most sensitive in June. The vegetation tended to be more sensitive to short-term droughts (1–2 months) and had no time lag in response to drought. The results are helpful to provide references for ecological management and ecosystem protection.

Keywords: vegetation dynamics; monthly scale; climate extremes; drought; NDVI; Guangdong

1. Introduction

Vegetation, the plant community covering the ground's surface, is an irreplaceable component of terrestrial ecosystems and is very sensitive to climate change [1,2]. Changes in vegetation can alter the ecology and reflect natural evolution [3–5]. Climate extremes have profound impacts on ecosystems and vegetation patterns, such as productivity reduction and ecosystem degeneration [6–15]. Among all climate extremes, drought is the most common natural hazard that occurs owing to an imbalance between precipitation and high temperatures [16,17]. The variability of climate extremes has great spatial inhomogeneity [18–20]. In the few past decades, high-temperature extremes mostly showed an increasing trend in China, while precipitation extremes showed various trends in different directions [21]. Drought has become increasingly severe since the late 1990s in China [22]. Areas affected by drought have doubled in China during the last three decades [23]. Understanding the vegetation response to climate extremes can help evaluate hydrological and ecological responses, identify the most vulnerable ecosystems, and guide adaptation decisions [8,19,20,24]. Therefore, it is important to study the variations in vegetation dynamics and their responses to climate extremes.

The influence of climate extremes on vegetation dynamics differs in different regions [5,13,25]. Extreme temperature shows a complex effect on vegetation in China [26]. Extreme precipitation generally promotes vegetation growth in most arid areas of China but inhibits it in humid areas [26,27]. Water shortage caused by drought can trigger leaf damage and substantial reductions in ecosystem productivity [7,10]. Vegetation recovery is usually slow after severe damage [8]. The effect of drought on terrestrial ecosystems is becoming increasingly acute and will worsen and intensify with global warming [28]. Thus, the response of vegetation to climate extremes, especially drought, is a crucial scientific issue in regions sensitive to climate warming [29].

The normalized difference vegetation index (NDVI) has been widely used to represent vegetation coverage and investigate vegetation responses to climate extremes [13,26,30–33]. An increase in NDVI generally indicates enhanced vegetation growth, whereas a decrease in NDVI displays reduced vegetation growth. Many studies investigated the spatial heterogeneity of the NDVI trends in different regions, pointing out that the NDVI has been increasing in recent years in China at both the national and regional scales [34]. Nevertheless, due to uneven changes in the NDVI in different growing periods [19,35], the analyses focusing only on the annual or longer scales rather than the monthly scale are insufficient to reflect variations in the NDVI.

In addition, many studies respectively identified the temporal differences in vegetation responses to climate extremes, as well as different time-lag effects in different spaces [8,19,34–36]. The timescales at which different plants respond to drought can also be substantially different [8]. Temperature and water requirements of vegetation vary in different months [29], and the influence of extreme factors on vegetation differs in different growing periods [35,37]. In terms of exploring the relationship between vegetation dynamics and climate extremes, the analysis on a monthly scale is more helpful to understand the main limiting factors of vegetation growth than that on a longer time scale [35,37]. Hence, it is of great significance to study this relationship on a monthly scale.

Guangdong Province, one of China's southernmost provinces, is a typical subtropical monsoon region. Strong monsoon, uneven terrain, dense river net, and negative effects of erosion and deposition have resulted in Guangdong being one of the most vulnerable areas to climate change. Vegetation is a significant ecological barrier in Guangdong. Extreme temperature and precipitation indices in Guangdong were much higher than those in most areas of China [38]. The fragile environment makes it vitally important to understand the impact of climate extremes on vegetation in this region. Some studies have addressed the influences of mean climate variations and anthropogenic activities on vegetation in Guangdong [39,40]. Nevertheless, the impacts of climate extremes, especially drought, on vegetation dynamics remain unclear in Guangdong.

A monthly scale analysis is essential for the vegetation variations and their responses to different climate extremes (temperature and precipitation extremes, and drought) in Guangdong. This kind of study can determine the critical period and the sensitivity of vegetation responses to climate extremes, and provide knowledge for meteorological disaster forecasts, vegetation conservation, and ecosystem restoration in both Guangdong and other similar regions. Considering the different climate extremes and focusing on the effect of drought during 1982–2015, this study tried to identify the leading extreme indicators driving vegetation changes in each month, the most sensitive months, and the drought timescales at which vegetation highly responded to drought in the vegetation area of Guangdong.

2. Materials and Methods

2.1. Study Area and Data

Figure 1 shows the geographic location of Guangdong. As the largest province in the southern coastal area of China, Guangdong has a subtropical monsoon climate with an average temperature of around 21 °C and annual total precipitation of approximately 1800 mm (Figure 1a). Guangdong usually experiences wet conditions from April to September and arid conditions from November to January. It is an important crop production base in China, and its main crops include rice, vegetables, peanuts, tuber crops, sugar cane, corn, and soybeans. In view of a strong monsoon climate, uneven terrain, and dense river net, Guangdong is vulnerable to climate change.

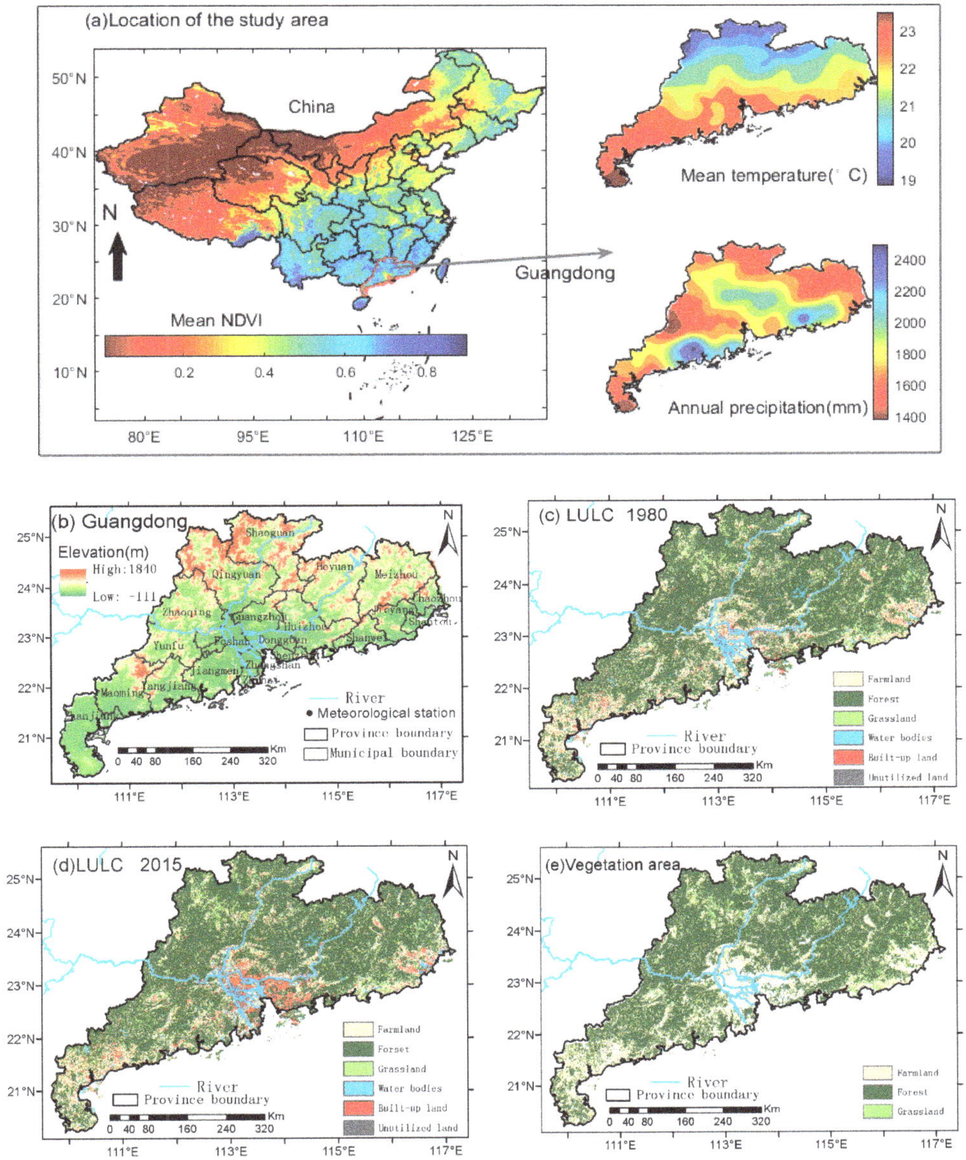

Figure 1. The geographic location of the study area, and spatial distribution of meteorological stations and land use land cover classes (LULC). Note: the normalized difference vegetation index (NDVI) data originated from the advanced very-high-resolution radiometer sensors of the National Oceanic and Atmospheric Administration (https://www.nasa.gov/nex (accessed on 1 January 2021)); the digital elevation data and the LULC data were provided by the Resource and Environment Science and Data Center, Chinese Academy of Sciences (https://www.resdc.cn/ (accessed on 10 October 2021)); only the vegetation areas where the vegetation types did not change during 1982–2018 are colored in (**e**).

Table 1 shows the datasets used in this study. The daily maximum temperature (TX), daily minimum temperature (TN), and daily precipitation data were obtained from the China Meteorological Administration (http://data.cma.cn/ (accessed on 15 October 2019)). This study selected 85 meteorological stations with high quality data records. Based on the climate characteristics of Guangdong (a subtropical monsoon climate with generally warm temperature and sufficient rainfall) (Figure 1a), this study selected six extreme temperature indices and four extreme precipitation indices from the Expert Team on Climate Change Detection and Indices (ETCCDI) [41] to reflect the frequency and intensity of temperature and preconception extremes (Table 2).

Table 1. Datasets used in this study.

Name	Data Source	Spatial Scale
Daily weather data	China Meteorological Administration (http://data.cma.cn/, accessed on 15 October 2019)	-
SPEI data	The Global SPEI database (SPEIbase v2.6) (https://spei.csic.es/database.html, accessed on 15 October 2021)	0.5°
LULC data	Resource and Environment Science and Data Center, Chinese Academy of Sciences (https://www.resdc.cn/, accessed on 10 October 2021)	1 km
Digital Elevation data	Resource and Environment Science and Data Center, Chinese Academy of Sciences (https://www.resdc.cn/, accessed on 10 October 2021)	250 m
GIMMS NDVI3g data	National Oceanic and Atmospheric Administration (https://www.nasa.gov/nex, accessed on 1 January 2021)	1/12° (approximately 8 km)

Table 2. Definitions of climate indices in this study.

Indices	Indicator Name	Definition	Unit
TXm	Maximum temperature	Monthly mean value of daily maximum temperature	°C
TNm	Minimum temperature	Monthly mean value of daily minimum temperature	°C
TXx	Max TX	Monthly maximum value of daily maximum temperature	°C
TNx	Max TN	Monthly maximum value of daily minimum temperature	°C
TXn	Min TX	Monthly minimum value of daily maximum temperature	°C
TNn	Min TN	Monthly minimum value of daily minimum temperature	°C
Pt	Total precipitation	Monthly total values of daily precipitation	mm
Rx1day	Max 1-day precipitation amount	Monthly maximum 1-day precipitation	mm
Rx5day	Max 5-day precipitation amount	Monthly maximum consecutive 5-day precipitation	mm
SDII	Simple daily intensity index	Total precipitation divided by the number of wet days (defined as PRCP $\geq$ 1.0 mm) in the month	mm day^{-1}
SPEI	Standardized precipitation evapotranspiration index	The difference between monthly precipitation and potential evapotranspiration	1

Note: PRCP represents the daily precipitation.

The standardized precipitation evapotranspiration index (SPEI), which is sensitive to temperature and precipitation, integrates the effect of evapotranspiration and precipitation. The SPEI is a standardized metric that allows comparisons across regions [42] and can indicate drought on multiple timescales [43]. Meanwhile, it is advantageous in quantifying drought compared with many other drought indices (i.e., the standardized precipitation index and the Palmer drought severity index) and studying responses of vegetation to drought [8,19,29,44,45]. The Global SPEI database (https://spei.csic.es/database.html (accessed on 15 October 2021)) has been widely used worldwide [8,29,44]. Therefore, the SPEI from 1-month scale to 24-month scale obtained from the Global SPEI database (SPEIbase v.2.6) was chosen to characterize drought in this research. The SPEI values less than or equal to –0.5 indicate drought conditions.

The land use and land cover (LULC) data and the digital elevation data were obtained from the Resource and Environment Science and Data Center, Chinese Academy of Sciences (https://www.resdc.cn/ (accessed on 10 October 2021)). The sub-regions of the three vegetation types (farmlands, forests, and grasslands) were extracted from the LULC data (Figure 1c–e). This study only analyzed the vegetation areas that did not change in vegetation type from 1982 to 2018 to avoid the impact of land use and land cover change, and, therefore, excluded urban, water, and barren areas from the spatial analysis.

At present, although a series of vegetation indices, such as the NDVI, the vegetation condition index (VCI), and the vegetation health index (VHI), have been developed to reflect changes in vegetation activities [46], the NDVI is still a good indicator when dealing with vegetation coverage and greenness in China and many other regions [13,26,29–33]. Thus, the NDVI was employed as an indicator to monitor vegetation dynamics in this work. This study used the NDVI remote sensing data from the Global Inventory Monitoring and Modeling Studies (GIMMS) NDVI3g dataset [47], which originated from the advanced very-high-resolution radiometer sensors of the National Oceanic and Atmospheric Administration (https://www.nasa.gov/nex (accessed on 1 January 2021)). The GIMMS NDVI3g dataset has been corrected to remove non-vegetation effects, such as atmospheric attenuation, cloud cover, volcanic aerosol, radiometric calibration, sensor degradation, view, and illumination geometry [48]. Although the spatial resolution of the GIMMS NDVI3g dataset is relatively low, the GIMMS NDVI3g dataset has the longest time series (1982–2015). The dataset has been verified to be reliable through comparisons with ground-based validations or other NDVI products [49,50] and possesses sufficient quality in identifying vegetation dynamics and its relationship with climate indices [47,51]. Considering the above, this study used the GIMMS NDVI data directly. The monthly NDVI values were obtained using the maximum-value composites (MVCs) method.

2.2. Methods

Since the GIMMS NDVI3g dataset only covers data from 1982 to 2015, this research conducted the analysis based on the data of this period. The flowchart of this study is illustrated in Figure 2. All climate indices were generally calculated on a monthly scale. The climate data of the meteorological stations were resampled to a spatial resolution of 1/12° to spatially match the NDVI dataset via the kriging method. Additionally, the SPEI data were rescaled to the spatial resolution of the NDVI dataset by the bilinear interpolation method. The Theil–Sen (TS) median trend analysis method has low sensitivity to both the influence of missing time series observations and the outliers in the time series [52]. Consequently, in this study, the TS slope estimator was used to explore the variation rate β within the time series and calculated using Equation (1):

$$\beta = \text{Median}[\frac{x_i - x_j}{i - j}] \text{ for all } j < i,\ 1 < j < i < n \quad (1)$$

where β is the variation rate within the time series; n is the number of years (equal to 34 in this study); i and j are the ordinal numbers of years; x_i and x_j are the sequential data at times i and j, respectively. A positive β-value indicates an upward trend, and vice versa. $\beta = 0$ means no change.

Since the Mann–Kendall (MK) test [53] does not require the data to be distributed normally or linearly, in this study the MK trend test was used to detect significant trends. Pearson correlation analysis was applied to investigate the strength of the correlations between extreme indices and vegetation dynamics. The significance level of $p = 0.05$ was used as the threshold to distinguish significance. In addition, this research analyzed the time-lag effects of climate extremes on vegetation. The time lag for a climate index was defined as the number of months after which the vegetation showed the strongest significant correlation with this climate index [17,36,54]. The time lag is generally shorter than six months [36,54]. Thus, this paper only took into account time lags of 0–6 months.

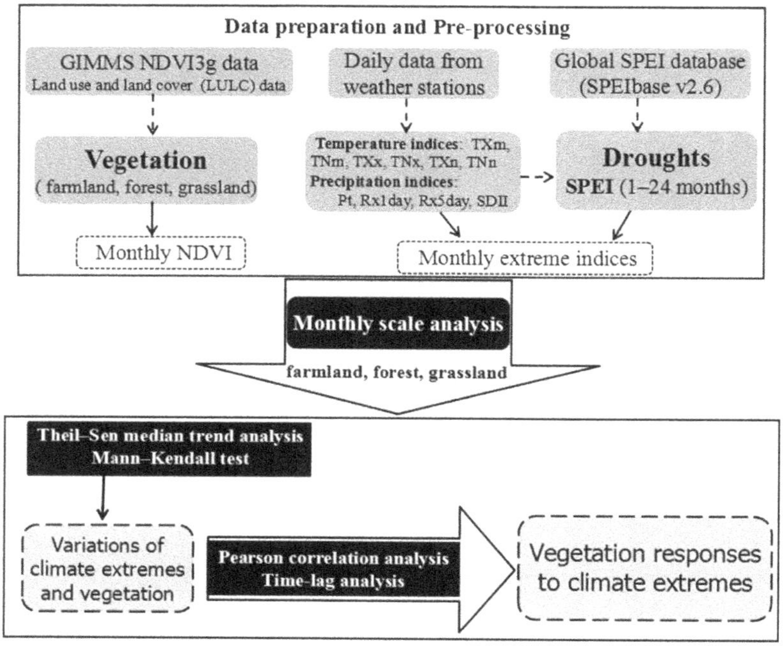

Figure 2. Flowchart of this study.

The SPEI time series for different timescales represent the cumulative water balance over the previous corresponding number of months [8]. For each pixel, this study calculated the maximum value ($R_{\max}$) of the correlation coefficient between the monthly NDVI series (Jan–Dec) and the different timescales SPEI (from 1-month scale to 24-month scale).

$$R_{i,j} = cor\left(NDVI_i, SPEI_{i,j}\right) \text{ for all } 1 \leq i \leq 12,\ 1 \leq j \leq 24 \quad (2)$$

$$R_{max} = \max_{1 \leq i \leq 12\ ,\ 1 \leq j \leq 24}\left(R_{i,j}\right) \quad (3)$$

where i is the ith month, ranging from 1st to 12nd month; j is the SPEI timescale, ranging from 1 to 24 months; $NDVI_i$ is the NDVI series of ith month; $SPEI_{i,j}$ is the ith month SPEI with the timescale of j months; $R_{i,j}$ is the correlation coefficient between $NDVI_i$ and $SPEI_{i,j}$; $R_{\max}$ is the maximum correlation coefficient, indicating the strength of drought impacts on vegetation.

The month corresponding to $R_{\max}$ is the sensitive month, indicating the period when vegetation is the most sensitive to drought. The SPEI timescale corresponding to $R_{\max}$ indicates the dominant drought timescale, representing the speed at which vegetation responds to drought. The longer sensitive timescale implies a slower vegetation response and a stronger vegetation resistance or resilience, partly indicating the low sensitivity of vegetation to drought. On the contrary, the shorter timescale demonstrates the higher sensitivity of vegetation to drought [29].

3. Results

3.1. Variability of Climate Extremes and Vegetation Dynamics

Figure 3 illustrates the regional mean of the monthly drought frequency and the trend rates of each climate index in each month in the whole Guangdong region. The frequencies of drought on the SPEI timescales of 13–24 months are not shown (Figure 3a), as they were similar to the results for the timescales less than 12 months. It is clear that drought has occurred every month and on different timescales over the past 34 years. June, October, and November witnessed a relatively higher frequency of 1-month timescale drought. With regard to the trend rate, the extreme temperatures showed a significant increasing trend in all months except January and December (Figure 3b). The warming rates were higher in February, October, and November. The highest warming rate was 0.124 °C year^{-1} for TXm in February.

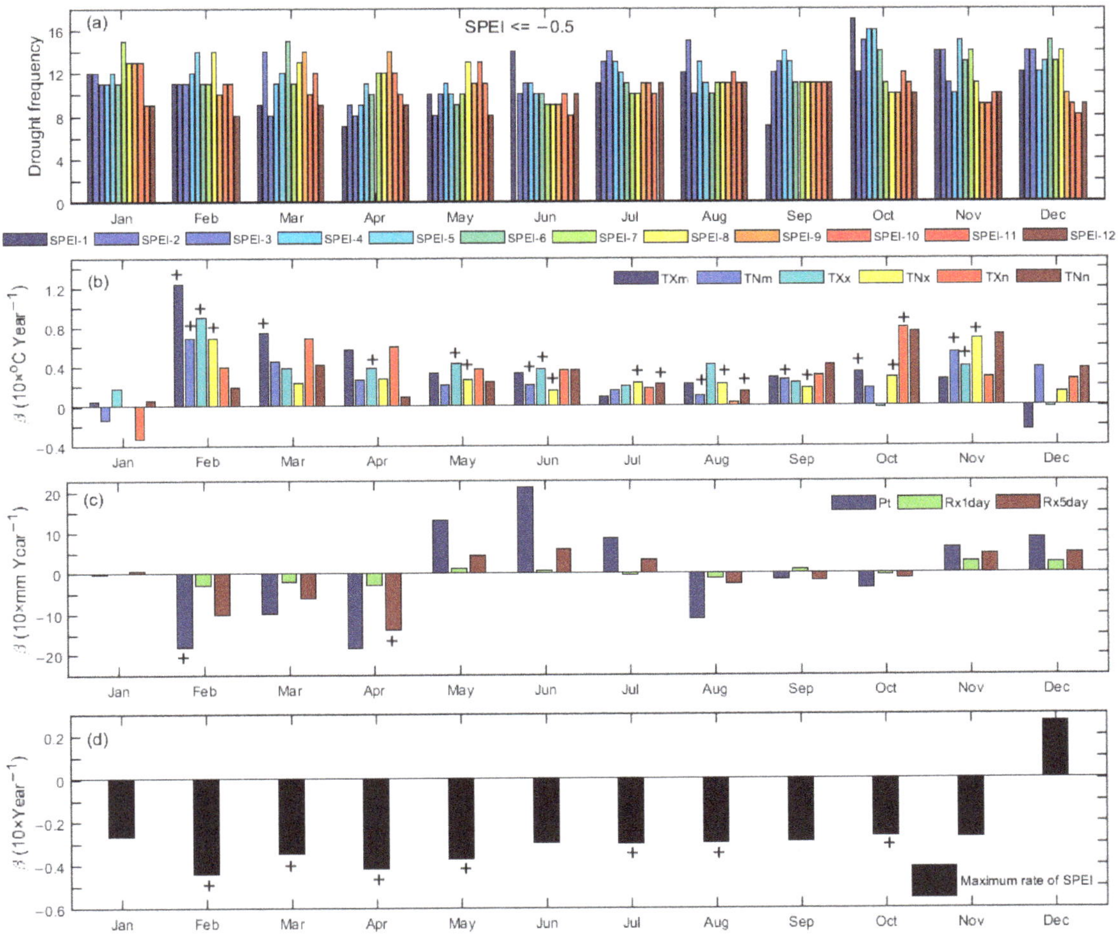

Figure 3. Drought frequency (SPEI ≤ -0.5) under the different timescales SPEI (from 1-month scale to 12-month scale) (**a**); regional trend rates of extreme temperatures during 1982–2015 (**b**); regional trend rates of extreme precipitations during 1982–2015 (**c**); maximum trend rates among the different timescales SPEI during 1982–2015 (**d**). Note: '+' significant at $p < 0.05$.

Nevertheless, as for the precipitation extremes in the whole Guangdong region (Figure 3c), only Rx5day experienced a significant decreasing trend in April, with a rate of −1.389 mm year^{-1}. In February, Pt significantly decreased with a rate of −1.933 mm year^{-1} in the vegetation area. According to the maximum rate of the different timescales SPEI (Figure 3d), the SPEI significantly decreased in February (−0.044 year^{-1}), April (−0.042 year^{-1}), May (−0.037 year^{-1}), March (−0.035 year^{-1}), July (−0.030 year^{-1}), August (−0.030 year^{-1}), and October (−0.026 year^{-1}). This indicates that the frequency of drought has increased in the past few decades. In February and April, the greater warming rate and decreasing precipitation rate may give rise to the relatively stronger drought trend.

With a focus on a monthly scale, Figure 4 investigates the variation of the regional averaged NDVI in different vegetation areas. The NDVI was around 0.6 in all months, and generally higher in the second half of the year than in the first half of the year. Farmlands had the lowest NDVI, whereas forests had the highest in each month. In terms of the variation trends during 1982−2015, the NDVI was relatively stable in June and October, while it showed a significant increase in other months. For all vegetation, the highest greening rate was 0.0028 year^{-1} in February, followed by 0.0023 year^{-1} in January, 0.0018 year^{-1} in August, and 0.0016 year^{-1} in July and December. In May, September, and November, the NDVI increased at a rate of around 0.0013 year^{-1}. For each type of vegetation, the highest greening rate also occurred in February, peaking at 0.0031 year^{-1} in the farmland area. Generally, the greening rate of forests was the lowest, while that of farmlands was the highest in each month.

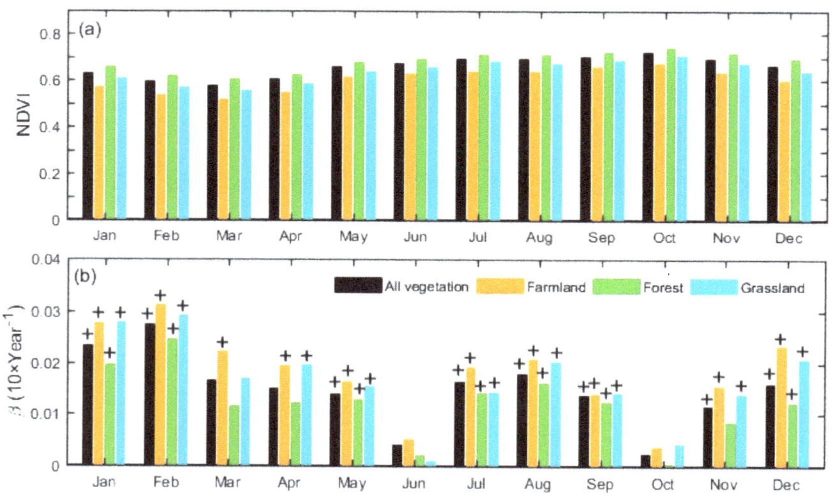

Figure 4. The regional averaged monthly NDVI (**a**), and its trend rates (unit: 10 × year^{-1}) in each month during 1982–2015 (**b**). Note: '+' significant at $p < 0.05$.

3.2. Correlation between NDVI and Climate Extremes on a Monthly Scale

Figure 5 investigates the spatial distributions of the correlations between the monthly NDVI and each extreme index. The correlations between the monthly NDVI and different temperature indices showed a similar pattern, and roughly 80–95% of the vegetation area passed the significance test for the positive correlations (Figure 5a–f). In some areas, the significant coefficient can be as high as 0.8. Moreover, Wang et al. [55] pointed out that the temperature extremes generally had high rates of increase in most of Guangdong during the past few decades. This indicates that the NDVI increased with the extreme temperature rising in this region.

The spatial patterns of the correlations for the different precipitation indices were similar (Figure 5g–j), but different from those for the temperature indices. The precipitation extremes showed much weaker correlations with the NDVI than the temperature extremes. Approximately 60–75% of the vegetation area passed the significance test for the correlations with the precipitation indices. The significant coefficients showed north–south patterns. The precipitation indices were significantly positively correlated with the NDVI in a few areas of southern Guangdong, while they were strongly negatively correlated with the NDVI in the parts of the northern and northeastern regions. Therefore, the effects of precipitation extremes on vegetation dynamics had an obvious north–south difference, and the northern and northeastern regions were mainly characterized by the inhibitory effect of precipitation extremes on plant growth.

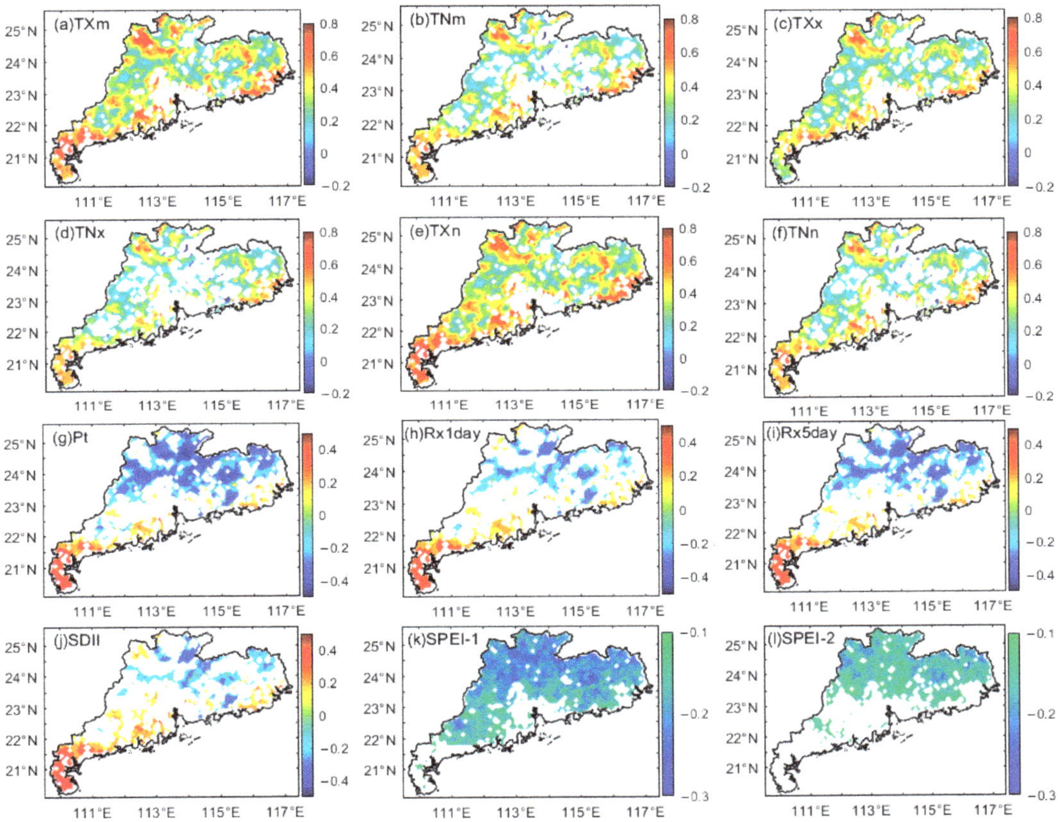

Figure 5. Spatial distributions of the correlation coefficients between the monthly NDVI and each extreme index during 1982–2015: (**a**) TXm (maximum temperature); (**b**) TNm (minimum temperature); (**c**) TXx (max TX); (**d**) TNx (max TN); (**e**) TXn (min TX); (**f**) TNn (min TN); (**g**) Pt (total precipitation); (**h**) Rx1day (max 1-day precipitation amount); (**i**) Rx5day (max 5-day precipitation amount); (**j**) SDII (simple daily intensity index); (**k**) SPEI-1 (the standardized precipitation evapotranspiration index on the 1-month timescale); and (**l**) SPEI-2 (the standardized precipitation evapotranspiration index on the 2-month timescale). Note: only the vegetation areas with the correlation passing the 0.05 significance test are colored in the figures.

Furthermore, with respect to the drought effect (Figure 5k,l), 90% of the vegetation showed significant and negative correlations with the 1-month timescale SPEI, and approximately 69% of it with the 2-month timescale SPEI. The negative correlations with the 1-month timescale SPEI showed a similar spatial pattern, but stronger magnitudes compared with those of the 2-month scale SPEI. The stronger correlations mostly occurred in northern and northeastern Guangdong. As the timescale of the SPEI increased, the correlations between the NDVI and the SPEI became much weaker. Only half of the vegetation significantly correlated with the 3-month timescale SPEI, around 30% of it with the 4- or 5-month timescale SPEI, and less than 12% with the SPEI on the timescale longer than six months (results not shown). Thus, the influence of the SPEI on vegetation mainly occurred on the short timescale (1–2 months), especially the 1-month timescale. In terms of the correlation between the 1-month timescale SPEI and the NDVI on the regional average, the significant coefficient was approximately −0.21 for all vegetation, with a value of −0.22 for forests and −0.19 for farmlands. The responses of the different vegetation to drought did not show much difference.

Then, based on the regional and monthly average, the correlations between vegetation and different climate extremes were further explored to identify the potentially different effects of climate extremes in different months (Figure 6). For any certain month, once one of the extreme temperature (precipitation) indices showed a significant correlation with the NDVI, this research considers that the temperature (precipitation) extremes were significantly correlated with the NDVI in that month. For each type of vegetation, the correlations varied greatly from January to December. The NDVI was significantly positively correlated with temperature extremes in most months (Figure 6a–f). The significant and positive correlations occurred in February, March, April, June, August, September, October, and November. The coefficient was often higher than 0.4. However, some negative correlations also existed between the NDVI and temperature extremes. In the forest area, the NDVI was significantly and negatively correlated with TNm and TNx in October, and with TNm in December.

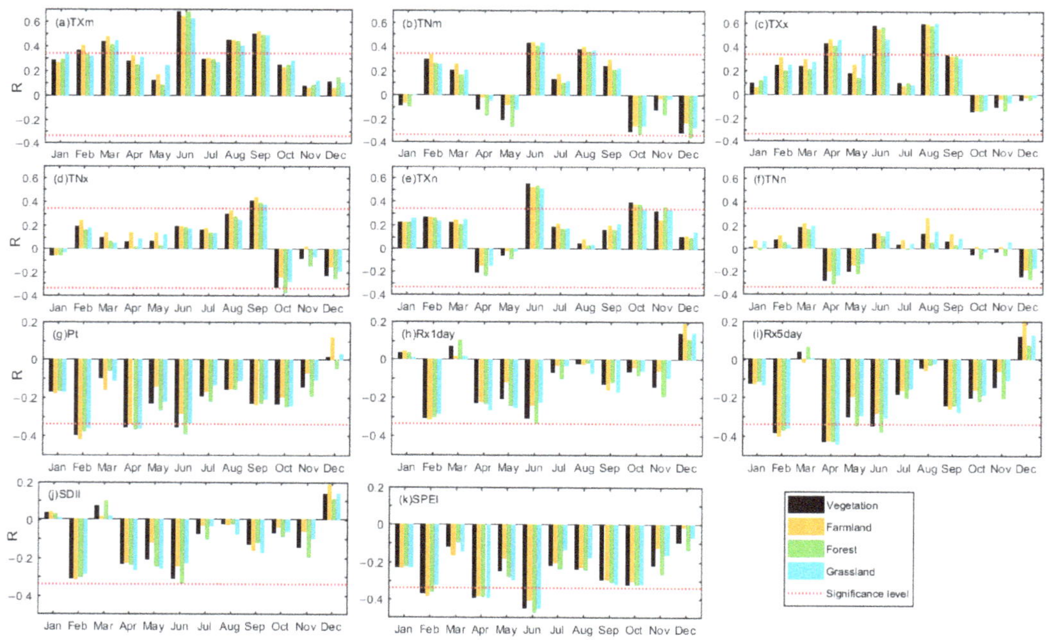

Figure 6. Correlation coefficients between the regional averaged NDVI and the extreme indices during 1982–2015 in each month. Note: the SPEI on the 1-month timescale in (**k**).

Nevertheless, the NDVI only displayed significant and negative correlations with the extreme precipitation indices (Pt, Rx5day) in February, April, and June (Figure 6g–j). Compared with farmlands and grasslands, the forest area was more affected by precipitation extremes in June. With regard to drought, the NDVI was significantly negatively correlated with the 1-month timescale SPEI in February, April, and June (Figure 6k). In June, the correlation was stronger, and the forest area was more affected by drought, with a correlation coefficient close to −0.5. In May and September, the NDVI showed significant and negative correlations with the SPEI on the timescale longer than one month, and the correlation magnitudes were a little weaker compared with those of the 1-month timescale SPEI (results not shown). In addition, there was a significant and positive correlation between the NDVI and the long-timescale SPEI in November. Generally, the magnitudes of correlations between vegetation and climate extremes did not show much difference among the three types of vegetation.

3.3. Sensitivity of Vegetation Responses to Drought

To further explore the strength of drought impact on vegetation, Figure 7a illustrates the spatial distribution of maximum correlation coefficients (R_{max}) between the NDVI and the SPEI during 1982–2015. The R_{max} was statistically significant in 94.47% of the vegetation area. A total of 73.02% of the vegetation area showed a negative correlation coefficient, whereas only 21.45% of it displayed a positive correlation coefficient. The R_{max} value was −0.6 in most vegetation areas, indicating that vegetation was significantly inhibited by drought in Guangdong. The areas with the greater negative correlations were mainly located in northern and northeastern Guangdong, which is similar to the result from Figure 5.

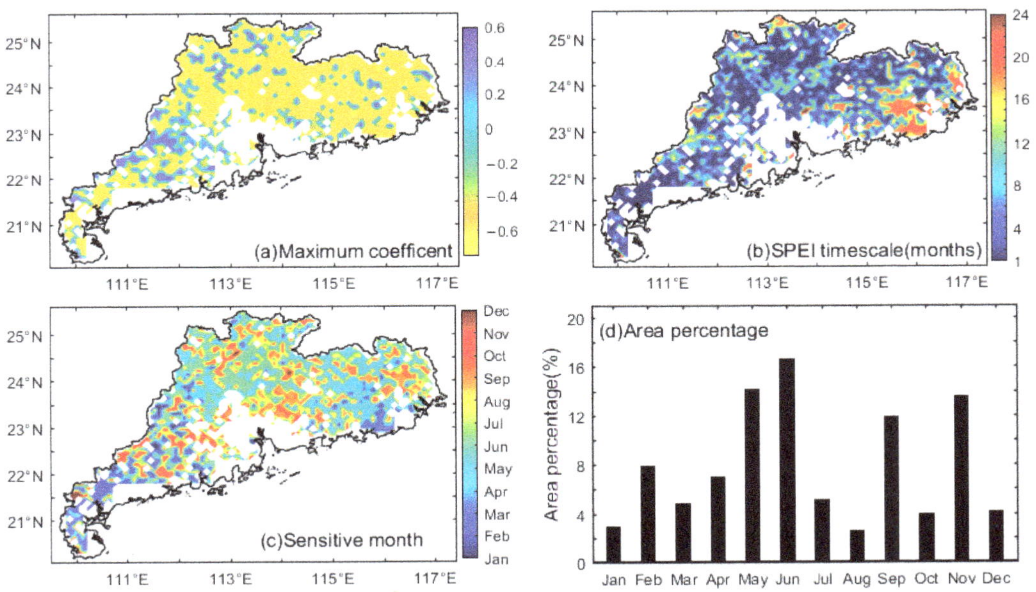

Figure 7. Spatial distribution of maximum correlation coefficients (R_{max}) between the NDVI and the SPEI during 1982–2015 (**a**); spatial distribution of SPEI timescales for maximum correlation coefficients (**b**); spatial distribution of the most sensitive month for vegetation responses to drought (**c**); the area percentage for each sensitive month (**d**). Note: only the vegetation areas with the correlation passing the 0.05 significance test are colored in the figures.

Figure 7b analyzes the dominant timescale of vegetation responses to drought. The vegetation in Guangdong was more sensitive to the short-timescale drought. The proportion of the 1-month timescale was the highest (30.57% of the vegetation area), followed by that of the 2-month timescale (11.27%). The timescales of three months and four months respectively accounted for 6.50% and 4.97%, whereas those of the other longer months (5–24 months) rarely appeared. Thus, the dominant timescale of vegetation responses to drought was 1–2 months.

In addition, the most sensitive month in which vegetation responded to drought had great spatial heterogeneity in Guangdong (Figure 7c,d). The most sensitive month was June which accounted for 16.64% of the vegetation area, and this was followed by May, November, and September respectively making up 14.13%, 13.57%, and 11.83%. However, February and April respectively constituted 7.8% and 7.5%, while the other six months respectively occupied less than 5.0%. Additionally, by comparing the spatial distribution of R_{max} with that of the most sensitive months, it can be found that drought in November had a positive impact on vegetation rather than a negative one. Thus, in Guangdong, the most sensitive month of vegetation responses to drought was June, followed by May and September.

3.4. Time Lags of NDVI Responses to Climate Extremes

Vegetation dynamics usually has delay responses to climate extremes [35–37,54,56]. Figure 8 investigates the time-lagged response of the NDVI to the different climate extremes based on the regional mean. The coefficients with the time lag longer than five months were not shown as they were much weaker than those with the shorter time lag. For all vegetation, the coefficients were as high as 0.70 when the NDVI lagged the extreme temperature indices by 1–2 months. Except for TXm and TXn, the time lags of vegetation responses to the temperature indices were mostly two months. The responses of farmlands and grasslands to TXm and TXn displayed a 1-month time lag. With regard to the NDVI responses to precipitation extremes, the coefficients were around 0.60 when the NDVI lagged precipitation extremes by 2–3 months. The time lag for forests and grasslands was mostly three months, whereas it was approximately two months for farmlands. Thus, the response time lag of farmlands to precipitation extremes was one month shorter than that of forests and grasslands. However, the NDVI responses to the SPEI did not show any time lags, indicating that the influence of drought on vegetation occurred quickly in Guangdong.

Figure 9 illustrates the spatial heterogeneity of the time lag to further explore the vegetation's delayed responses. For the temperature extremes (Figure 9a–f), the vegetation responses with a 2-month lag accounted for the largest part (45.0–63.0% of the vegetation area), and the vegetation responses with a 1-month lag mainly occurred in the parts of the farmland area (20.0–36.0% of the vegetation area). Only 8.0–25.0% of the vegetation showed a time lag of three months. The delayed responses of vegetation to precipitation extremes (Figure 9g–j) were spatially different from those to temperature extremes. The unequal lagged effects of temperature and precipitation extremes particularly occurred in some northern and northeastern regions.

By comparing the distribution of the lag results (Figure 9g–j) with that of land use and land cover classes (Figure 1e), we can vaguely find that the lag time of farmlands to precipitation extremes was shorter than that of the other two vegetation areas. The grids covered by a 3-month delay accounted for 48.0–53.0% of the vegetation area, whereas those with a 2-month delay constituted 15.0–28.0%. The vegetation responses with 4-month and 1-month lags were respectively scattered in a few areas of the central-northern and south-western regions. However, the vegetation rarely showed delayed responses to drought; in other words, the grids with no time lag accounted for 88.0% of the vegetation area. The asymmetric responses of vegetation to the different climate extremes were distinguished in Guangdong.

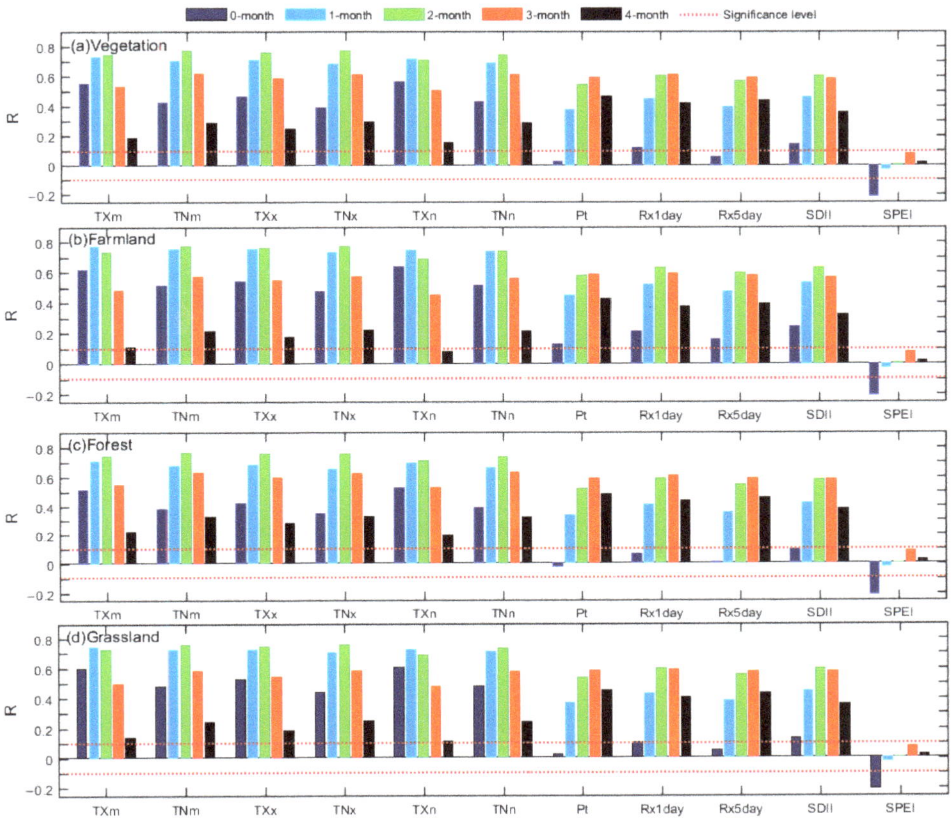

Figure 8. Correlation coefficients between the regional averaged monthly NDVI and the extreme indices for different time lags during 1982–2015. Note: '0-month' NDVI and extreme indices were collected over the same period; '1-month', '2-month', '3-month', '4-month' NDVI lagged the extreme indices by 1–4 months.

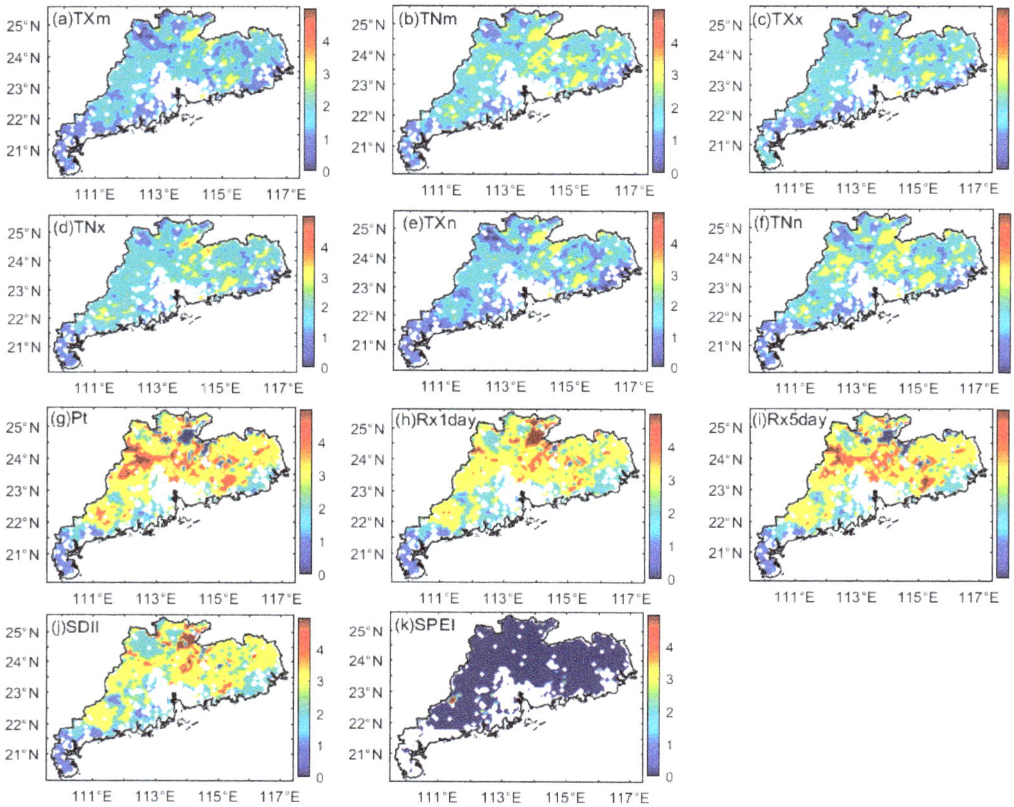

Figure 9. Spatial distributions of the time lag of vegetation responses to the different climate extremes. Note: this study conducted seven correlation analyses (namely, 0–6 months), and then superimposed their results to take the maximum value; the numbers 0, 1, 2, 3, and 4 represent the number of lag months; only the vegetation areas with the correlation passing the 0.05 significance test are colored in the figures.

4. Discussion

4.1. Variations in Climate Extremes

The SPEI is a good indicator for identifying drought and assessing the intensity and duration of drought [33,42,44]. The multiyear mean SPEI approximately decreased from southwest to northeast (results not shown). According to Figure 3, the trend rates of each climate index were strongly temporal heterogeneous, which highly proves that it is of great significance to carry out the research on a monthly scale. Both the increased rates of extreme temperatures and drought were the highest in February. The SPEI showed a significant drought trend in most months, and the extreme temperatures significantly increased in most corresponding months. The high temperature can enhance evaporation and increase soil drought conditions [57]. Thus, drought in Guangdong is mainly attributed to the increasing extreme temperature. In addition to the great warming rate of extreme temperatures, the significant decrease in precipitation would also contribute to the strong drought trend in February and April.

In order to further explore the spatial distributions of drought variations, Figure 10 illustrates the trend rate of the spatial SPEI, which was calculated by selecting the maximum rates of the SPEI among 24-timescale series (from 1-month to 24-month) in each pixel. The northeastern region and parts of northern Guangdong mostly witnessed a significant decreasing SPEI (in other words, a significant drought trend). In those regions, the trend rate of drought was mostly higher than 0.04 year^{-1} in many months, with the highest value around 0.06 year^{-1}. The grids with intensified drought accounted for 73.02% of the vegetation area in April, respectively followed by 65.23% in February, 63.0% in May, 53.51% in March, and 46.80% in June. Around 41.0% of the vegetation area showed drought trends in July, August, September, October, and November, but only 35.63% in December and 25.96% in January. By comparing the spatial patterns of the variations of drought and temperature extremes [55], we can find that the vegetation area experiencing drought intensification mostly corresponded well with that suffering from the high increases of temperature extremes. This further clarifies the contribution that the temperature rise made to drought intensification in Guangdong.

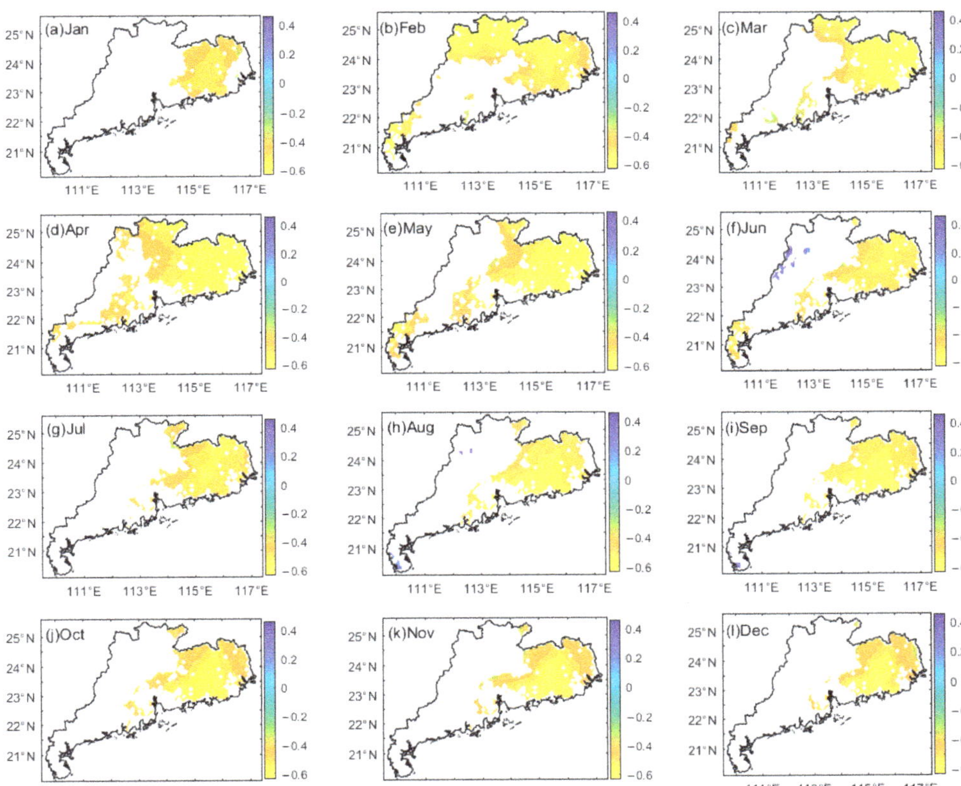

Figure 10. Spatial distributions of the maximum trend rates of the SPEI time series (from 1-month to 24-month) in each month during 1982–2015 (unit: 10 × year^{-1}). Note: only the vegetation areas with the rate passing the 0.05 significance test are colored in the figures.

4.2. Impact of Climate Extremes on Vegetation Change

Knowledge of vegetation responses to climate extremes is helpful for evaluating the vegetation vulnerability and resilience to climate extremes [24]. Over the past few decades, some parts of Guangdong have suffered from huge anthropogenic activities, such as rapid urbanization, economic development, and population growth [39,40]. The vegetation coverage in Guangdong was affected by local social changes and anthropogenic disturbances, which can weaken the accuracy of the relationship between climate extremes and the NDVI. In order to avoid the impact caused by the above disturbances, this study only considered the vegetation areas with unchanged vegetation types when analyzing the vegetation variation and the effects of climate extremes on vegetation. Thus, the results related to vegetation are reliable.

The vegetation greenness in China presents great northwest–southeast differences (Figure 1a). Highly different from northwestern China, most areas in Guangdong have a high vegetation cover rate. In the vegetation area of Guangdong, the vegetation activity strengthened in most months except June and October during 1982–2015 (Figure 4b), and it positively responded to the increased temperature indices in most months (Figure 6). Extreme temperature has a complex effect on vegetation [25,26]. Although high temperature usually precludes vegetation growth in many areas [26], sufficient precipitation would mitigate the negative effect of the increased high temperatures in South China [37]. Located in southern China, Guangdong had annual precipitation as high as 1780.9 mm in the past 34 years. Therefore, the adverse effect of the rise in extreme temperature rarely occurred in Guangdong, and the increasing extreme temperature promoted plant growth in most months.

The impact of precipitation extremes on vegetation was weaker than that of temperature extremes in Guangdong (Figures 5 and 6), which is similar to that in Guangxi [37]. The vegetation in Guangdong has sufficient water for growth and extreme precipitation is a less important factor affecting vegetation, which differs significantly from that in arid and semi-arid regions where vegetation is sensitive to limited precipitation [26,29,34]. The negative effect of precipitation extremes mainly occurred in the northern and northeastern parts of Guangdong which possesses less precipitation. The decreased extreme precipitation in February and April tended to prevent vegetation growth.

Drought implies a deficit in water availability which differs from the normal level [42]. Vegetation recovery is usually slow after severe damage from drought [8]. Owing to the regulation of precipitation, the correlations between vegetation and drought in humid regions are relatively weaker than those in arid regions of China [29]. However, this study finds that vegetation growth was widely inhibited by drought in Guangdong (Figure 5k,l and Figure 6k). February, April, and June generally witnessed increased extreme temperatures and the adverse effect of precipitation extremes on vegetation, which can well explain the negative effect of drought on vegetation in those months. In addition, on account of the significant drought trend in February and April, the vegetation growth was further greatly restrained in those three months.

Vegetation type can influence the effect of climate extremes on vegetation dynamics [13,58]. For the effect of temperature extremes on vegetation, the correlation magnitudes among the three types of vegetation were slightly different in each month, but the differences were not significant. Except for June, the correlation between precipitation extremes and vegetation activities in each month was basically the same or insignificant for the three types of vegetation. In June, the forest area was more sensitive to extreme precipitation than the other two vegetation areas. In terms of drought impact, the forest area was relatively more affected in June, whereas the farmland area was more influenced in February. Thus, the analysis on a monthly scale clearly reflects the influences of climate extremes on vegetation in different growing periods.

4.3. Sensitivity Analysis of Vegetation Responses to Drought

The physiological response of vegetation to drought determines the resistance and resilience of vegetation to water deficit [8]. Drought conditions may be different when being calculated during different periods. Vegetation responds differently to drought on different timescales [8], and the timescale of drought can highly affect the magnitude of vegetation responses [20,59]. It is necessary to consider drought on different timescales when the effect of drought on vegetation is assessed [29,45].

The timescale of vegetation responses to drought in Guangdong was thoroughly investigated in this study. Figure 5 points out that the influence of drought on vegetation mainly happened on the short timescales (1–3 months), with drought on the 1-month timescale showing the greatest impact on vegetation. According to Figures 5 and 7a, vegetation was significantly inhibited by drought in most vegetation areas of Guangdong, with the greater negative correlations mainly located in northern and northeastern Guangdong. The effect of drought on vegetation in Guangdong highly differs from that in many other regions where drought was positively related to vegetation dynamics [19,20,29,59].

Meanwhile, although the vegetation in humid regions was mostly inclined to respond to long-term drought [8], the vegetation in Guangdong was the most sensitive to drought on timescales of 1–2 months (Figure 7b). The dominant timescale in Guangdong was much shorter than that in many other regions, such as Qiling [20], and the Continental United States [17]. These imply that the vegetation in Guangdong had a rapid response and weak resistance to drought. In addition, although the forest area was generally sensitive to longer-timescale droughts, as the plant roots are mostly long and can reach groundwater [60], this research did not find a tremendously clear difference in the drought timescales among farmlands, forests, and grasslands.

According to Figure 6, February, April, and June witnessed a negative vegetation response to 1-month timescale drought. May and September experienced a similar but weaker vegetation response to drought when the SPEI timescale was longer than one month or even 12 months. Those partly indicate that the vegetation in February, April, and June was more sensitive to short-timescale drought, while the vegetation in May and September was more sensitive to longer-timescale drought. Furthermore, the most sensitive month of vegetation responses to drought in Guangdong was June, closely followed by May and September (Figure 7c,d). According to Qi et al. [20], the vegetation in the Qinling Mountains of China was also the most sensitive to drought in May and June. June and May are the critical periods for vegetation growth in Guangdong, especially for crops, such as rice, vegetables, peanuts, tuber crops, sugar cane, corn, and soybeans. For instance, the rice in June is mostly at the heading and flowering stage, often suffering from high temperatures [61]. Although irrigation can mitigate the adverse effects of drought and other extreme climatic events on rice to some extent, drought often tends to amplify the negative effects of extreme heat on crop production [14]. Thus, the crop growth in those periods will be greatly inhibited.

4.4. Lagged Responses of Vegetation to Different Climate Extremes

The lag of vegetation responses to climate extremes was thoroughly investigated in this study (Figures 8 and 9). The farmland, forest, and grassland areas generally responded to temperature extremes with two months' time-lag effects, but they showed different lag responses to precipitation extremes. The farmland area responded to extreme precipitation with a lag of two months, one month earlier than the forest and grassland areas. Therefore, the response of forests and grasslands to temperature extremes was one month earlier than that to precipitation extremes, but the farmland area showed a similar time lag to the temperature and precipitation extremes. The time lags in vegetation responses to temperature and precipitation extremes in Guangdong were longer than those in Guangxi and the whole Central Asia [35,37]. The vegetation response to soil moisture typically showed a 1-month time lag [62], which partly explains the lag in vegetation responses to temperature extremes.

Although both the temperature and precipitation extremes showed obvious time-lag effects on vegetation, drought did not have any time-lag effects on vegetation growth in Guangdong. This implies that the responses of vegetation to drought were faster than those to other climate extremes in Guangdong. The farmland, grassland, and forest vegetation responded to drought immediately, without any difference among the three types of vegetation. However, drought has lag effects on vegetation in some other places, such as the Continental United States [17]. The lack of lag in vegetation responses to drought further indicates that drought will generate a great threat to vegetation growth in Guangdong with the increase in extreme temperature.

Overall, the temperature extremes showed the strongest correlation with vegetation dynamics in Guangdong, followed by drought. The increasing extreme temperature promoted plant growth in most months. Nevertheless, drought exerted a negative effect on plant growth in February, April, May, June, and September, and the precipitation extremes made a negative effect in February, April, and June. Although the inhibiting effects of drought and precipitation extremes on vegetation were relatively weaker than the promoting effects of temperature extremes, the inhibiting effects of drought and precipitation extremes should be paid more attention. Based on the categorization of extreme climate effects on vegetation dynamics [26], Guangdong belongs to the heat-promoting and drought-inhibiting compound type.

The monthly scale analysis shows that the responses of vegetation to climate extremes displayed great temporal heterogeneity in different months. Thus, it is an effective approach to understand how vegetation responds to climate extremes in different growing periods. This study can help find effective preventive measures for the protection of the Guangdong ecosystem. Drought can easily reduce vegetation activity and slow plant growth [8]. Appropriate measures, for example, introducing drought-resistant vegetation, should be formulated to reduce the damage caused by drought in February, April, May, June, and September. People can also lessen drought effects on agriculture by using irrigation and film-mulching, especially in June and May. Moreover, this work is particularly important under the background of ongoing global warming. The effect of future drought will almost certainly be worsened by rising temperatures [28]. The future temperature may frequently go beyond the vegetation's optimum and enhance drought degrees. Prolonged exposure to climate extremes may result in fragile vegetation. People should take precautions against heat threats and related threats in Guangdong.

4.5. Limitations and Uncertainties

Although the GIMMS NDVI data have been proven to be reliable for this kind of study [47], there were some limitations and uncertainties in the study. Firstly, the GIMMS NDVI data are indirect remote sensing data, and their accuracy is limited to the satellite sensor sensitivity [63,64]. There may be some errors in the modeled NDVI because of the influence of the fragmentized terrain and saturated problem [64]. In addition, the differences in the data resolution may lead to weaknesses in the result. There were uncertainties when the climate data were resampled to the spatial resolution of the GIMMS NDVI data. Furthermore, changes in vegetation have feedback to the climate system [3,5], and the vegetation–climate interactions are highly heterogeneous [8,58]. This study did not strictly consider the feedback of vegetation to climate extremes. More information is required to analyze the vegetation–climate interactions in the future. Future work should address these uncertainties to obtain more valuable conclusions.

5. Conclusions

This paper analyzed the impacts of different climate extremes (temperature extremes, precipitation extremes, and drought) on vegetation dynamics on a monthly scale in Guangdong from 1982 to 2015, mainly focusing on vegetation response to drought. The major conclusions are summarized as follows.

(1) Drought showed an increasing trend in most months, corresponding to the increase in extreme temperatures. The stronger trend rates of drought happened in February and April. Except for June and October, the vegetation in other months showed a significant greening trend.

(2) The vegetation dynamics displayed strong and positive correlations with the enchanted temperature extremes in most months. However, it showed great negative correlations with precipitation extremes and 1-month timescale drought in February, April, and June, and with long-timescale drought in May and September. The response of vegetation to drought was the most sensitive in June.

(3) For various timescales, the most significant negative correlation between vegetation and drought mainly occurred on the timescales of one to two months. With a rapid response and weak resistance to drought, the vegetation did not have any time lag. However, the vegetation responded to temperature and precipitation extremes with a time lag of at least two months.

The results promote our insight into the interaction between vegetation and climate and help improve the sustainable use of natural resources and facilitate the implementation of reasonable mitigation measures.

Author Contributions: Conceptualization, L.W. and F.H.; methodology, M.L., L.W. and F.H.; validation, C.Z. and Y.M.; formal analysis, H.C. and K.Z.; investigation, M.L. and L.W.; data curation, L.W. and F.H.; writing—original draft preparation, L.W., F.H. and M.L.; writing—review and editing, L.W., C.Z., Y.M., H.C. and K.Z.; visualization, L.W. and F.H.; supervision, M.L.; project administration, L.W. and F.H.; funding acquisition, L.W. and F.H. All authors have read and agreed to the published version of the manuscript.

Funding: This work was supported by the National Natural Science Foundation of China (Grant No. 42005142) and Compilation of the second South China Regional Climate Change Assessment Report (Grant No. CCSF202012).

Data Availability Statement: All the data are available in the public domain at the links provided in the texts.

Acknowledgments: Thanks to the China Meteorological Administration (CMA) for providing the meteorological data. Thanks to all editors and commenters.

Conflicts of Interest: The authors declare no conflict of interest.

References

1. Braswell, B.H.; Schimel, D.S.; Linder, E.; Moore, B. The response of global terrestrial ecosystems to interannual temperature variability. *Science* **1997**, *278*, 870–872. [CrossRef]
2. Nolan, C.; Overpeck, J.T.; Allen, J.R.; Anderson, P.M.; Betancourt, J.L.; Binney, H.A.; Brewer, S.; Bush, M.B.; Chase, B.M.; Cheddadi, R.; et al. Past and future global transformation of terrestrial ecosystems under climate change. *Science* **2018**, *361*, 920–923. [CrossRef] [PubMed]
3. Schlesinger, W.H.; Jasechko, S. Transpiration in the global water cycle. *Agric. For. Meteorol.* **2014**, *189*, 115–117. [CrossRef]
4. Jiang, L.; Bao, A.; Guo, H.; Ndayisaba, F. Vegetation dynamics and responses to climate change and human activities in Central Asia. *Sci. Total Environ.* **2017**, *599–600*, 967–980. [CrossRef]
5. Cui, L.; Wang, L.; Qu, S.; Singh, R.P.; Lai, Z.; Yao, R. Spatiotemporal extremes of temperature and precipitation during 1960–2015 in the Yangtze River Basin (China) and impacts on vegetation dynamics. *Theor. Appl. Climatol.* **2019**, *136*, 675–692. [CrossRef]
6. Silberstein, R.P.; Vertessy, R.A.; Morris, J.; Feikema, P.M. Modelling the effects of soil moisture and solute conditions on long-term tree growth and water use: A case study from the Shepparton irrigation area, Australia. *Agric. Water Manag.* **1999**, *39*, 283–315. [CrossRef]
7. Ciais, P.; Reichstein, M.; Viovy, N.; Granier, A.; Ogée, J.; Allard, V.; Aubinet, M.; Buchmann, N.; Bernhofer, C.; Carrara, A.; et al. Europe-wide reduction in primary productivity caused by the heat and drought in 2003. *Nature* **2005**, *437*, 529–533. [CrossRef]
8. Vicente-Serrano, S.M.; Gouveia, C.; Camarero, J.J.; Beguería, S.; Trigo, R.; López-Moreno, J.I.; Azorín-Molina, C.; Pasho, E.; Lorenzo-Lacruz, J.; Revuelto, J.; et al. Response of vegetation to drought time-scales across global land biomes. *Proc. Natl. Acad. Sci. USA* **2013**, *110*, 52–57. [CrossRef]
9. Rammig, A.; Wiedermann, M.; Donges, J.F.; Babst, F.; Von Bloh, W.; Frank, D.; Thonicke, K.; Mahecha, M.D. Coincidences of climate extremes and anomalous vegetation responses: Comparing tree ring patterns to simulated productivity. *Biogeosciences* **2015**, *12*, 373–385. [CrossRef]
10. Choat, B.; Brodribb, T.J.; Brodersen, C.R.; Duursma, R.A.; López, R.; Medlyn, B.E. Triggers of tree mortality under drought. *Nature* **2018**, *558*, 531–539. [CrossRef]
11. Pan, N.; Feng, X.; Fu, B.; Wang, S.; Ji, F.; Pan, S. Increasing global vegetation browning hidden in overall vegetation greening: Insights from time-varying trends. *Remote Sens. Environ.* **2018**, *214*, 59–72. [CrossRef]
12. Abera, T.A.; Heiskanen, J.; Pellikka, P.; Maeda, E.E. Impact of rainfall extremes on energy exchange and surface temperature anomalies across biomes in the Horn of Africa. *Agric. For. Meteorol.* **2020**, *280*, 107779. [CrossRef]
13. Zhang, W.; Wang, L.; Xiang, F.; Qin, W.; Jiang, W. Vegetation dynamics and the relations with climate change at multiple time scales in the Yangtze River and Yellow River Basin, China. *Ecol. Indic.* **2020**, *110*, 105892. [CrossRef]
14. Lesk, C.; Coffel, E.; Winter, J.; Ray, D.; Zscheischler, J.; Seneviratne, S.I.; Horton, R. Stronger temperature–moisture couplings exacerbate the impact of climate warming on global crop yields. *Nat. Food* **2021**, *2*, 683–691. [CrossRef]

15. Wu, C.; Peng, J.; Ciais, P.; Peñuelas, J.; Wang, H.; Beguería, S.; Andrew Black, T.; Jassal, R.S.; Zhang, X.; Yuan, W.; et al. Increased drought effects on the phenology of autumn leaf senescence. *Nat. Clim. Chang.* **2022**, *12*, 943–949. [CrossRef]
16. Sohoulande Djebou, D.C. Bridging drought and climate aridity. *J. Arid. Environ.* **2017**, *144*, 170–180. [CrossRef]
17. Zhong, S.; Sun, Z.; Di, L. Characteristics of vegetation response to drought in the CONUS based on long-term remote sensing and meteorological data. *Ecol. Indic.* **2021**, *127*, 107767. [CrossRef]
18. Reichstein, M.; Bahn, M.; Ciais, P.; Frank, D.; Mahecha, M.D.; Seneviratne, S.I.; Zscheischler, J.; Beer, C.; Buchmann, N.; Frank, D.C.; et al. Climate extremes and the carbon cycle. *Nature* **2013**, *500*, 287–295. [CrossRef]
19. Deng, H.; Yin, Y.; Xiang, H. Vulnerability of vegetation activities to drought in Central Asia. *Environ. Res. Lett.* **2020**, *15*, 084005. [CrossRef]
20. Qi, G.; Song, J.; Qi, L.; Bai, H.; Sun, H.; Zhang, S.; Cheng, D. Response of vegetation to multi-timescales drought in the Qinling Mountains of China. *Ecol. Indic.* **2022**, *135*, 108539. [CrossRef]
21. You, Q.; Kang, S.; Aguilar, E.; Pepin, N.; Flügel, W.A.; Yan, Y.; Xu, Y.; Zhang, Y.; Huang, J. Changes in daily climate extremes in China and their connection to the large scale atmospheric circulation during 1961–2003. *Clim. Dyn.* **2011**, *36*, 2399–2417. [CrossRef]
22. Yu, M.; Li, Q.; Hayes, M.J.; Svoboda, M.D.; Heim, R.R. Are droughts becoming more frequent or severe in China based on the standardized precipitation evapotranspiration index: 1951–2010? *Int. J. Climatol.* **2014**, *34*, 545–558. [CrossRef]
23. Xu, X.; Piao, S.; Wang, X.; Chen, A.; Ciais, P.; Myneni, R.B. Spatio-temporal patterns of the area experiencing negative vegetation growth anomalies in China over the last three decades. *Environ. Res. Lett.* **2012**, *7*, 035701. [CrossRef]
24. Mulder, C.P.H.; Iles, D.T.; Rockwell, R.F. Increased variance in temperature and lag effects alter phenological responses to rapid warming in a subarctic plant community. *Glob. Chang. Biol.* **2016**, *23*, 801–814. [CrossRef] [PubMed]
25. Peng, S.; Piao, S.; Ciais, P.; Myneni, R.B.; Chen, A.; Chevallier, F.; Dolman, A.J.; Janssens, I.A.; Penuelas, J.; Zhang, G.; et al. Asymmetric effects of daytime and night-time warming on Northern Hemisphere vegetation. *Nature* **2013**, *501*, 88–92. [CrossRef]
26. Li, S.; Wei, F.; Wang, Z.; Shen, J.; Liang, Z.; Wang, H.; Li, S. Spatial heterogeneity and complexity of the impact of extreme climate on vegetation in China. *Sustainability* **2021**, *13*, 5748. [CrossRef]
27. Xu, G.; Zhang, H.; Chen, B.; Zhang, H.; Innes, J.L.; Wang, G.; Yan, J.; Zheng, Y.; Zhu, Z.; Myneni, R.B. Changes in vegetation growth dynamics and relations with climate over China landmass from 1982 to 2011. *Remote Sens.* **2014**, *6*, 3263–3283. [CrossRef]
28. Trenberth, K.E.; Dai, A.; Van Der Schrier, G.; Jones, P.D.; Barichivich, J.; Briffa, K.R.; Sheffield, J. Global warming and changes in drought. *Nat. Clim. Chang.* **2014**, *4*, 17–22. [CrossRef]
29. Zhang, Q.; Kong, D.D.; Singh, V.P.; Shi, P.J. Response of vegetation to different time-scales drought across China: Spatiotemporal patterns, causes and implications. *Glob. Planet. Chang.* **2017**, *152*, 1–11. [CrossRef]
30. Barichivich, J.; Briffa, K.R.; Myneni, R.B.; Osborn, T.J.; Melvin, T.M.; Ciais, P.; Piao, S.; Tucker, C. Large-scale variations in the vegetation growing season and annual cycle of atmospheric CO_2 at high northern latitudes from 1950 to 2011. *Glob. Chang. Biol.* **2013**, *19*, 3167–3183. [CrossRef]
31. Piao, S.L.; Nan, H.J.; Huntingford, C. Evidence for a weakening relationship between interannual temperature variability and northern vegetation activity. *Nat. Commun.* **2014**, *5*, 5058. [CrossRef] [PubMed]
32. Dubovyk, O.; Landmann, T.; Dietz, A.; Menz, G. Quantifying the impacts of environmental factors on vegetation dynamics over climatic and management gradients of Central Asia. *Remote Sens.* **2016**, *8*, 600. [CrossRef]
33. Schuldt, B.; Buras, A.; Arend, M.; Vitasse, Y.; Beierkuhnlein, C.; Damm, A.; Gharun, M.; Grams, T.E.; Hauck, M.; Hajek, P.; et al. A first assessment of the impact of the extreme 2018 summer drought on Central European forests. *Basic Appl. Ecol.* **2020**, *45*, 86–103. [CrossRef]
34. Liu, Y.; Lei, H. Responses of natural vegetation dynamics to climate drivers in China from 1982 to 2011. *Remote Sens.* **2015**, *7*, 10243–10268. [CrossRef]
35. Luo, M.; Sa, C.; Meng, F.; Duan, Y.; Liu, T.; Bao, Y. Assessing extreme climatic changes on a monthly scale and their implications for vegetation in Central Asia. *J. Clean. Prod.* **2020**, *271*, 122396. [CrossRef]
36. Wen, Y.; Liu, X.; Yang, J.; Lin, K.; Du, G. NDVI indicated inter-seasonal non-uniform time-lag responses of terrestrial vegetation growth to daily maximum and minimum temperature. *Glob. Planet. Chang.* **2019**, *177*, 27–38. [CrossRef]
37. Wang, L.; Hu, F.; Miao, Y.; Zhang, C.; Zhang, L.; Luo, M. Changes in Vegetation Dynamics and Relations with Extreme Climate on Multiple Time Scales in Guangxi, China. *Remote Sens.* **2022**, *14*, 2013. [CrossRef]
38. Wu, X.; Hao, Z.; Hao, F.; Zhang, X. Variations of compound precipitation and temperature extremes in China during 1961–2014. *Sci. Total Environ.* **2019**, *663*, 731–737. [CrossRef]
39. Zhao, X.; Zhang, Z.; Wang, X.; Zuo, L.; Liu, B.; Yi, L.; Xu, J.; Wen, Q. Analysis of Chinese cultivated land's spatial temporal changes and causes in recent 30 years. *Trans. Chin. Soc. Agric. Eng.* **2014**, *30*, 1–11. (In Chinese)
40. Abbas, S.; Nichol, J.E.; Wong, M.S. Trends in vegetation productivity related to climate change in China's Pearl River Delta. *PLoS ONE* **2021**, *16*, e0245467. [CrossRef]
41. Easterling, D.R.; Alexander, L.V.; Mokssit, A.; Detemmerman, V. CCI/CLIVAR workshop to develop priority climate indices. *Bull. Am. Meteor. Soc.* **2003**, *84*, 1403–1407.
42. Slette, I.J.; Smith, M.D.; Knapp, A.K.; Vicente-Serrano, S.M.; Camarero, J.J.; Beguería, S. Standardized metrics are key for assessing drought severity. *Glob. Chang. Biol.* **2020**, *26*, e1–e3. [CrossRef] [PubMed]

43. Vicente-Serrano, S.M.; Beguería, S.; Lopez-Moreno, J.I. A Multiscalar Drought Index Sensitive to Global Warming: The Standardized Precipitation Evapotranspiration Index. *J. Clim.* **2010**, *23*, 1696–1718. [CrossRef]
44. Schwalm, C.R.; Anderegg, W.R.; Michalak, A.M.; Fisher, J.B.; Biondi, F.; Koch, G.; Litvak, M.; Ogle, K.; Shaw, J.D.; Wolf, A.; et al. Global patterns of drought recovery. *Nature* **2017**, *548*, 202–205. [CrossRef] [PubMed]
45. Deng, Y.; Wu, D.; Wang, X.; Xie, Z. Responding time scales of vegetation production to extreme droughts over China. *Ecol. Indic.* **2022**, *136*, 108630. [CrossRef]
46. Chua, Z.-W.; Kuleshov, Y.; Watkins, A.B. Drought detection over Papua New Guinea using satellite-derived products. *Remote Sens.* **2020**, *12*, 3859. [CrossRef]
47. Zhu, Z.; Bi, J.; Pan, Y.; Ganguly, S.; Anav, A.; Xu, L.; Samanta, A.; Piao, S.; Nemani, R.R.; Myneni, R.B. Global Data Sets of Vegetation Leaf Area Index (LAI)3g and Fraction of Photosynthetically Active Radiation (FPAR)3g Derived from Global Inventory Modeling and Mapping Studies (GIMMS) Normalized Difference Vegetation Index (NDVI3g) for the Period 1981 to 2011. *Remote Sens.* **2013**, *5*, 927–948.
48. Tucker, C.J.; Pinzon, J.E.; Brown, M.E.; Slayback, D.A.; Pak, E.W.; Mahoney, R.; Vermote, E.; Saleous, N.E. An extended AVHRR 8-km NDVDI dataset compatible with MODIS and SPOT vegetation NDVI data. *Int. J. Remote Sens.* **2005**, *26*, 4485–4498. [CrossRef]
49. Fensholt, R.; Proud, S.R. Evaluation of Earth Observation based global long term vegetation Trends-Comparing GIMMS and MODIS global NDVI time series. *Remote Sens. Environ.* **2012**, *119*, 131–147. [CrossRef]
50. Sarmah, S.; Jia, G.; Zhang, A.; Singha, M. Assessing seasonal trends and variability of vegetation growth from NDVI3g, MODIS NDVI and EVI over South Asia. *Remote Sens. Lett.* **2018**, *9*, 1195–1204. [CrossRef]
51. Zeng, F.W.; Collatz, G.J.; Pinzon, J.E.; Ivanoff, A. Evaluating and quantifying the climate-driven interannual variability in Global Inventory Modeling and Mapping Studies (GIMMS) Normalized Difference Vegetation Index (NDVI3g) at Global Scales. *Remote Sens.* **2013**, *5*, 3918–3950. [CrossRef]
52. Sen, P.K. Estimates of the regression coefficient based on Kendall's tau. *J. Am. Stat. Assoc.* **1968**, *63*, 1379–1389. [CrossRef]
53. Mann, H.B. Nonparametric tests against trend. *Econom. J. Econ. Soc.* **1945**, *13*, 245–259. [CrossRef]
54. Wu, D.; Zhao, X.; Liang, S.; Zhou, T.; Huang, K.; Tang, B.; Zhao, W. Time-lag effects of global vegetation responses to climate change. *Glob. Chang. Biol.* **2015**, *21*, 3520–3531. [CrossRef] [PubMed]
55. Wang, L.; Hu, F.; Hu, J.; Chen, C.; Liu, X.; Zhang, D.; Chen, T.; Miao, Y.; Zhang, L. Multistage spatiotemporal variability of temperature extremes over South China from 1961 to 2018. *Theor. Appl. Climatol.* **2021**, *146*, 243–256. [CrossRef]
56. Wen, Z.; Wu, S.; Chen, J.; Lü, M. NDVI indicated long-term interannual changes in vegetation activities and their responses to climatic and anthropogenic factors in the Three Gorges Reservoir Region, China. *Sci. Total Environ.* **2017**, *574*, 947–959. [CrossRef]
57. Yan, H.; Yu, Q.; Zhu, Z.; Myneni, R.B.; Yan, H.; Wang, S.; Shugart, H.H. Diagnostic analysis of interannual variation of global land evapotranspiration over 1982–2011: Assessing the impact of ENSO. *J. Geophys. Res. Atmos.* **2013**, *118*, 8969–8983. [CrossRef]
58. Hua, W.; Chen, H.; Zhou, L.; Xie, Z.; Qin, M.; Li, X.; Ma, H.; Huang, Q.; Sun, S. Observational Quantification of Climatic and Human Influences on Vegetation Greening in China. *Remote Sens.* **2017**, *9*, 425. [CrossRef]
59. Vicente-Serrano, S.M.; Azorin-Molina, C.; Peña-Gallardo, M.; Tomas-Burguera, M.; Domínguez-Castro, F.; Martín-Hernández, N.; Beguería, S.; El Kenawy, A.; Noguera, I.; García, M. A high-resolution spatial assessment of the impacts of drought variability on vegetation activity in Spain from 1981 to 2015. *Nat. Hazards Earth Syst. Sci.* **2019**, *19*, 1189–1213. [CrossRef]
60. Jiang, P.; Ding, W.; Yuan, Y.; Ye, W. Diverse response of vegetation growth to multi-time-scale drought under different soil textures in China's pastoral areas. *J. Environ. Manag.* **2020**, *274*, 110992. [CrossRef]
61. Sun, W.; Huang, Y. Global warming over the period 1961–2008 did not increase high-temperature stress but did reduce low-temperature stress in irrigated rice across China. *Agric. For. Meteorol.* **2011**, *151*, 1193–1201. [CrossRef]
62. Chen, T.; De Jeu, R.; Liu, Y.; Van der Werf, G.; Dolman, A. Using satellite based soil moisture to quantify the water driven variability in NDVI: A case study over mainland Australia. *Remote Sens. Environ.* **2014**, *140*, 330–338. [CrossRef]
63. Julien, Y.; Sobrino, J.A. Comparison of cloud-reconstruction methods for time series of composite NDVI data. *Remote Sens. Environ.* **2010**, *114*, 618–625. [CrossRef]
64. Brantley, S.T.; Zinnert, J.C.; Young, D.R. Application of hyperspectral vegetation indices to detect variations in high leaf area index temperate shrub thicket canopies. *Remote Sens. Environ.* **2011**, *115*, 514–523. [CrossRef]

remote sensing

MDPI

Article

Effect of Vegetation Carryover and Climate Variability on the Seasonal Growth of Vegetation in the Upper and Middle Reaches of the Yellow River Basin

Xinru Zhang [1], Qian Cao [2], Hao Chen [1], Quan Quan [3], Changchao Li [1], Junyu Dong [1], Mengjie Chang [1], Shuwan Yan [1] and Jian Liu [1,*]

1 Environment Research Institute, Shandong University, Qingdao 266237, China
2 School of Geography and Information Engineering, China University of Geosciences, Wuhan 430074, China
3 State Key Laboratory of Eco-hydraulics in Northwest Arid Region of China, Xi'an University of Technology, Xi'an 710054, China
* Correspondence: ecology@sdu.edu.cn

Abstract: Vegetation dynamics are often affected by climate variability, but the past state of vegetation has a non-negligible impact on current vegetation growth. However, seasonal differences in the effects of these drivers on vegetation growth remain unclear, particularly in ecologically fragile areas. We used the normalized difference vegetation index (NDVI), gross primary productivity (GPP), and leaf area index (LAI) to describe the vegetation dynamic in the upper and middle reaches of the Yellow River basin (YRB). Three active vegetation growing seasons (early, peak, and late) were defined based on phenological metrics. In light of three vegetation indicators and the climatic data, we identified the correlation between the inter-annual variation of vegetation growth in the three sub-seasons. Then, we quantified the contributions of climate variability and the vegetation growth carryover (VGC) effect on seasonal vegetation greening between 2000–2019. Results showed that both the vegetation coverage and productivity in the study area increased over a 20-year period. The VGC effect dominated vegetation growth during the three active growing seasons, and the effect increased from early to late growing season. Vegetation in drought regions was found to generally have a stronger vegetation carryover ability, implying that negative disturbances might have severer effects on vegetation in these areas. The concurrent seasonal precipitation was another positive driving factor of vegetation greening. However, sunshine duration, including its immediate and lagged impacts, had a negative effect on vegetation growth. In addition, the VGC effect can sustain into the second year. The VGC effect showed that initial ecological restoration and sustainable conservation would promote vegetation growth and increase vegetation productivity. This study provides a comprehensive perspective on understanding the climate–vegetation interactions on a seasonal scale, which helps to accurately predict future vegetation dynamics over time in ecologically fragile areas.

Keywords: climate variability; seasonality; vegetation dynamics; vegetation growth carryover; Yellow River basin

1. Introduction

In terrestrial ecosystems, vegetation controls the cycling of water and energy between the soil and the atmosphere and provides ecosystem services on which humans depend [1]. The structure and function of vegetation are influenced by climate factors (e.g., temperature, precipitation, and sunshine) and non-climate factors (vegetation itself and land use changes) [2–4]. To maintain the health and function of the ecosystem, we need to determine what factors affect vegetation growth [5,6].

Many studies have focused on the vegetation dynamics, resulting from climate variation [7,8]. For instance, temperature controls the high-latitude vegetation growth in the northern hemisphere [9]. Vegetation greening is primarily affected by solar radiation in tropical rainforest areas [10], while in arid and semiarid areas, the limiting factor of vegetation growth is precipitation [11,12]. However, the response of vegetation to climate factors is complex and delayed [13]. For example, there is about a 3-month lag between temperature change and vegetation activity in China [14]. The lag effect of the same type of vegetation to different climate factors is different, and the lag effect of different types of vegetation to the same climate factors is also different [15]. In drought areas, the vegetation dynamic is most sensitive to climate factors in the current and previous month [16]. These observations suggest that vegetation dynamics may be influenced not only by the current climate conditions but also by earlier climate conditions. Therefore, the time-lag effects of climate should be considered when exploring the mechanism of vegetation–climate interaction.

Besides climate factors, vegetation dynamics are also influenced by its past states [17,18]. The biological cycles of vegetation consist of many successive growth periods that are tightly connected to the past and present [19], implying that the present growth state of plants will influence subsequent growth. This phenomenon has been defined as vegetation growth carryover [17]. For example, vegetation carryover has an important impact on the daily net carbon exchange in dry-land ecosystems [20], which allows plants to send greening signals by increasing carbon uptake, resulting in more substantial leaves and roots [1]. Such structural changes in vegetation may enhance its resistance to external disturbance unless the change in external environmental conditions exceeds the tolerance of plant growth [17]. The key question is how substantial the role of the carryover effect is, especially when compared with the immediate and lagged impacts of climate factors. Previous researchers analyzed the VGC effect on vegetation growth at hemispheric or global scales [17,18]. However, processes that dominate at large scales are not necessarily dominant at small scales. Another important question to answer is how does the VGC affect the vegetation growth in ecologically fragile areas?

Here, we focus on the Yellow River basin (YRB). The YRB spans the arid, semiarid, and semi-humid areas in China [4]. Droughts occur frequently in the YRB with increasing duration and intensity [21]. Moreover, problems, such as soil erosion and reduced biodiversity, are also prominent in the basin [22], which makes the YRB vulnerable to environmental change [23]. The upper and middle reaches of the YRB account for 91% of the total area of the YRB [4]. There are a large number of tributaries flowing into the main stream in the upper reaches of the Yellow River, which is the main source of runoff in the basin [24]. The loss of the soil and water is serious in the middle reaches, which is the main source area of sediment in the basin [25]. Therefore, our research area is the upper and middle reaches of the YRB, which is very important to the development of the lower reaches and the whole basin. Previous studies on vegetation dynamics in the upper and middle reaches of the YRB have focused on the growing season or annual time scale [4,6,21]. However, the long-term climate changes often differ between seasons, so the vegetation responses to climate also vary significantly between seasons. In recent years, some studies have explored the characteristic patterns and driving factors of vegetation growth from the perspective of seasonal dynamics. For example, a study has shown that the seasonal change of temperature can regulate the spatial heterogeneity of deciduous broadleaf forests greening [26]. A previous study reported that the decrease in vegetation greenness in the Northern hemisphere after 1997 was mainly due to the decrease in vegetation cover in the early parts of the growing season and the slowdown increase in vegetation cover in the peak of the growing season [27]. Since the seasonal dynamics of vegetation in different regions and ecosystems are different, it is not accurate to study the seasonal changes of vegetation in a certain region by simply using meteorological seasons based on a constant calendar (for example, spring is defined as March–May). Therefore, it is more universal to define seasonality based on local vegetation phenology information. Analyzing the interseasonal relationship of vegetation growth will help to better understand the climate-vegetation interaction.

In this study, we first assessed the vegetation indexes change from three different datasets to detect and characterize vegetation growth trends in the upper and middle of the YRB from 2000–2019. The satellite-derived NDVI, GPP, and LAI were regarded as indicators to reflect vegetation dynamics and productivity. Secondly, we divided the growing season into three active seasons based on phenological metrics [17,28]. Then, we assessed the effect of vegetation carryover on vegetation growth across the phenology-based season using partial correlation analysis and compared the size of this effect against that of the immediate and lagged impacts of climate factors. Temperature (TEM), precipitation (PRE), and sunshine duration (SSD) were used to represent the climate conditions. Finally, the individual contributions of climate variability and the vegetation carryover on vegetation greening were calculated. Our research is expected to increase understanding of vegetation–climate interactions at the seasonal scale and provide insight on how to better manage vegetation under global climate change in the future.

2. Materials and Methods

2.1. Study Area

The Yellow River is the second-longest river in China. It has a watershed area of 79.5 km^2, 91% of which (72.3 × 10^4 km^2) is considered the upper and middle reaches [4]. Our study area covered the upper and middle reaches of the YRB, and its undulating terrain varies in elevation between 84 and 6119 m (Figure 1a). The upper and middle reaches of the YRB span many climatic zones (Figure 1b). The upper reaches of the YRB are dominated by polar climate (ET and EF). The Ningxia Plain and the western part of Hetao Plain have cold desert climate (BWk). The eastern part of Hetao Plain and Ordos Plain have cold steppe climate (BSk). The southern part of the study area is dominated by hot, warm-summer continental climate (Dwa and Dwb) [29]. The main land-cover types (>10^4 km) are grasslands, croplands, deciduous broadleaf forests, barrens, savannas, urban and built-up lands, and mixed forests (Figure 1d and Table 1). The quality of the ecological environment in the upper and middle reaches directly determines the environmental conditions in the lower reaches of the YRB, so this area has crucial strategic significance in terms of geographical location and ecological protection [4].

Table 1. The main land cover types (>10^4 km) and proportion of the upper and middle reaches of the YRB in 2019.

Land Type.	Area (10^4 km^2)	Area Proportion (%)
Grasslands	52.33	67.77
Croplands	13.71	17.75
Deciduous Broadleaf Forests	3.60	3.54
Barrens	2.41	3.12
Savannas	2.04	2.64
Mixed Forests	1.35	1.54
Urban and built-up lands	1.16	1.50

Note: Land cover types with an area of less than 10^4 km^2 are not listed.

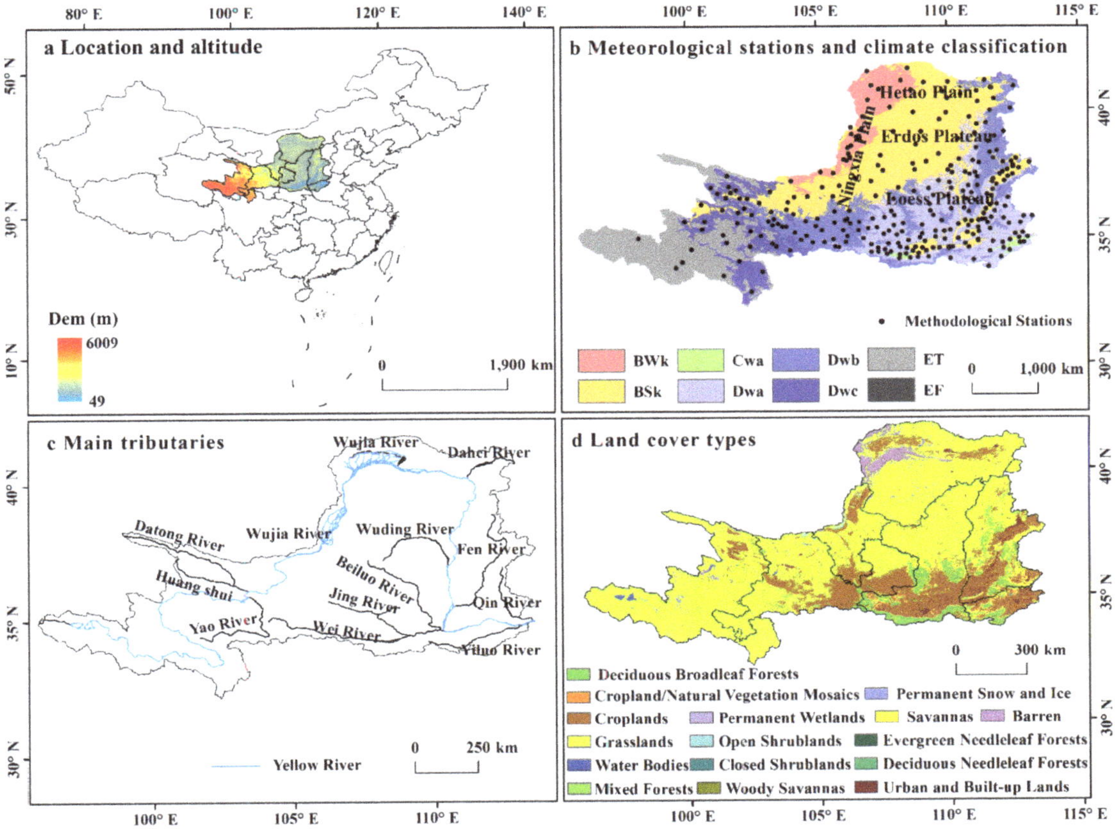

Figure 1. Location and altitude (**a**), stations and climate classification (BWk: cold desert; BSk: cold steppe climate; Cwa: dry-winter humid subtropical climate; Dwa: hot-summer continental climate; Dwb: warm-summer continental climate; Dwc: boreal climate; ET: tundra climate; EF: ice-cap climate) (**b**), main tributaries (**c**), and land cover types in 2019 (**d**) of the upper and middle reaches of the Yellow River Basin.

2.2. Data Sources and Processing

Normalized difference vegetation index (NDVI) reflects vegetation growth trend and is widely used to describe the physiological condition of vegetation [17]. Here, NDVI data were provided by the Moderate Resolution Imaging Spectroradiometer (MODIS) products (https://lpdaac.usgs.gov/) at a spatial resolution of 500 m × 500 m and temporal resolution of 16 days. The time-series data covers 2000–2019. We selected MOD13A1 NDVI data of four imagery data sets, including h25v05, h26v04, h26v05, and h27v05. NASA's MRT (MODIS Reprojection Tools) software was used for data processing, including image mosaic, projection, format transformation, and resampling. Then, vector boundaries of the upper and middle reaches of the YLB were used to extract the study area in ArcMap 10.2.2. The synthetic NDVI data may have outliers or missing values. Therefore, the MODIS NDVI datasets were smoothed and denoised by the Savitzky–Golay (S-G) filter [30,31]. The basic idea is first to filter the data after reprojection, and then the pixels with good quality remain unchanged, while those with poor quality are filtered to form a new NDVI sequence. Considering that NDVI is easily saturated in dense vegetation areas, we also introduced two data products, leaf area index (LAI) and gross primary productivity (GPP) to independently verify the robustness of the NDVI results. LAI and GPP data (MOD15A2H) are also derived from the MODIS products (https://lpdaac.usgs.gov/). The LAI and GPP dataset are available at a spatial resolution of 500 m × 500m and 8-day temporal resolution. All pretreatments were the same as for NDVI.

Temperature affects many plant processes, such as photosynthesis, respiration, and transpiration [26]. In arid and semi-arid areas, vegetation is very sensitive to precipitation [21]. Sunshine is the source of energy for plant growth [32]. Therefore, precipitation, temperature and sunshine duration were selected as meteorological indicators in this study. Monthly mean temperature and monthly total precipitation with a spatial resolution of 0.00833° were obtained from the Institute of Tibetan Plateau Research, Chinese Academy of Sciences (https://data.tpdc.ac.cn/). Daily sunshine duration (meteorological station data) was acquired from the Chinese National Bureau (https://data.cma.cn/) (Figure 1b). The period for time-series meteorological data is 2000–2019. To maintain the same spatial resolution, vegetation indexes and shine duration were resampled or interpolated into a grid cell with a spatial resolution of 0.00833°.

We obtained the land cover data (MCD12Q1) from the United States Geological Survey (USGS) platform (https://lpdaac.usgs.gov/). The MCD12Q1 supplies global maps of land cover at annual time steps and 500 m spatial resolution since 2001 and was classified according to the International Geosphere-Biosphere Program (IGBP) land use classification scheme. The spatial distribution of the altitude (DEM) data was obtained from the Resources and Environment Science and Data Center, Chinese Academy of Sciences (https://www.resdc.cn/Default.aspx).

Details of data used in this research are shown in Table 2.

Table 2. Main data used in this study.

Data Type	Data Sources	Spatial Resolution	Temporal Resolution	Temporal Span	Acquisition Time
NDVI	https://lpdaac.usgs.gov/	500 m	16-day	2000–2019	2021.10.11
GPP	https://lpdaac.usgs.gov/	500 m	8-day	2000–2019	2021.10.19
LAI	https://lpdaac.usgs.gov/	500 m	8-day	2000–2019	2021.10.23
TEM	https://data.tpdc.ac.cn/	1 km	monthly	2000–2019	2021.10.15
PRE	https://data.tpdc.ac.cn/	1 km	monthly	2000–2019	2021.10.05
SSD	https://data.cma.cn/	2400 stations	daily	2000–2019	2021.12.23
Land Cover (Type 1)	https://www.resdc.cn/ Default.aspx	500 m	yearly	2000–2019	2022.4.3

2.3. Trend Analysis and Significance Test

Linear regression analysis was done to simulate the trend of vegetation index variation on the pixel scale from 2000 to 2019. The formula is as follows:

$$slope = \frac{n\sum_{i}^{n}(i \times Var_i) - \sum_{i=1}^{n} i \times \sum_{i=1}^{n} Var_i}{n\sum_{i=1}^{n} i^2 - \left(\sum_{i=1}^{n} i\right)^2} \tag{1}$$

where the slope is the inter-annual variation rate of relevant variables i, Var_i the value of the relevant variable in the ith year (season), and n represents the length of the dataset (n = 20). Slope > 0 indicates an increasing trend of the Var_i during the study period, and slope < 0 indicates a negative trend of the Var_i during the study period.

$$\sum_{i=2}^{n}\sum_{j=1}^{i=1} sign(X_i - X_j) \tag{2}$$

$$sign(X_i - X_j) = \begin{cases} -1, X_j - X_i < 0 \\ 0, X_j - X_i = 0 \\ +1, X_j - X_i > 0 \end{cases} \tag{3}$$

$$Var(s) = n(n-1)(2n+5)/18 \tag{4}$$

$$\begin{cases} Z = \frac{S-1}{\sqrt{Var(S)}}, S > 0 \\ Z = 0, S = 0 \\ Z = \frac{S+1}{\sqrt{Var(S)}}, S < 0 \end{cases} \tag{5}$$

At the same time, the Mann–Kendall test by Equations. (2)–(5) was used to check the significance level of the vegetation trend.

2.4. Identifying Phenology-based Seasons

The length and period of the growing season for each grid cell are not directly defined by a constant calendar-based meteorological season, for example, the growing season is April–October [33] or the spring season is April–May [34,35]. Instead, we defined growth seasons based on vegetation phenology [17,18,36]. We extracted the start of the growing season (SOS) and the end of the growing season (EOS) based on phenological events [17,28], and then divided the growing season into three active seasons: early growing season (EGS), peak growing season (PGS), and late growing season (LGS). The specific extraction process was as follows. The 16-day NDVI time series datasets were input into TIMESAT 3.3 software [32] to extract SOS and EOS for determining the length of the growing season. Dynamic thresholds of vegetation SOS and vegetation EOS were set at 20% and 80%, respectively [37]. Since the TIMESAT 3.3 software can only extract phenological parameters of n-1 year from n-year data, we obtained 2001–2019 phenological parameters based on 20 years of NDVI data. SOS and EOS are the 19-year averages. The two consecutive months with the largest NDVI value were defined as PGS (not before April or after October). EGS is defined as the intermediate month from the month of SOS to the month of PGS, and LGS is defined as the intermediate month from the month of PGS to the month of EOS. The preceding-season NDVI for EGS is the NDVI of the last year's LGS [17].

2.5. Correlation between the Vegetation Indicators and Their Driving Factors

Partial correlation analysis was applied to assess the relationship between seasonal vegetation growth and productivity (NDVI, GPP, and LAI) and their driving factors between 2000–2019. We considered the concurrent climate factors (temperature, TEM, precipitation, PRE, and sunshine duration, SSD) and climate factors (TEM_{ps}, PRE_{ps}, and SSD_{ps}) and NDVI ($NDVI_{ps}$) of the preceding season as driving factors of seasonal NDVI (GPP and LAI) variations.

2.6. Contribution of Climate Variability and the VGC Effect on Vegetation Greening

We quantified the individual contributions of climate factors (TEM, PRE, and SSD of the present and preceding season) and VGC effect (preceding-season NDVI) to the observed NDVI trend of each season during 2000–2019, so as to decompose the linear trend of NDVI (dNDVI/dt) for each season over a 20-year period into the additive contribution of eight components:

$$\begin{aligned} \frac{dNDVI}{dt} &= \frac{\partial NDVI}{\partial NDVI_{ps}}\frac{dNDVI_{ps}}{dt} + \frac{\partial NDVI}{\partial TEM_{ps}}\frac{dTEM_{ps}}{dt} + \frac{\partial NDVI}{\partial PRE_{ps}}\frac{dPRE_{ps}}{dt} \\ &+ \frac{\partial NDVI}{\partial SSD_{ps}}\frac{dSSD_{ps}}{dt} + \frac{\partial NDVI}{\partial TEM}\frac{dTEM}{dt} + \frac{\partial NDVI}{\partial PRE}\frac{dPRE}{dt} + \frac{\partial NDVI}{\partial SSD}\frac{dSSD}{dt} + \sigma \\ &= \Delta NDVI_{ps} + \Delta TEM_{ps} + \Delta PRE_{ps} + \Delta SSD_{ps} + \Delta TEM + \Delta PRE + \Delta SSD + \sigma \\ &= \Delta NDVI_{ps} + \Delta climate_{ps} + \Delta climate + \sigma \end{aligned} \tag{6}$$

where $\Delta NDVI_{ps}$, $\Delta climate_{ps}$, ΔTEM_{ps}, ΔPRE_{ps}, ΔSSD_{ps}, $\Delta climate$, ΔTEM, ΔPRE, and ΔSSD represent the individual contributions to the interannual NDVI changes, respectively. $\Delta climate_{ps}$ (preceding-season climate) is the sum of ΔTEM_{ps}, ΔPRE_{ps}, and ΔSSD_{ps}. $\Delta climate$ is the sum of ΔTEM, ΔPRE, and ΔSSD. Moreover, $\partial NDVI/\partial NDVI_{ps}$ is equal to the partial correlation coefficient between NDVI and $NDVI_{ps}$ when removing the disturbances of the concurrent and preceding season for temperature, precipitation, and sunshine duration; $dNDVI_{ps}/dt$ is the inter-annual variation rate (slope) of $NDVI_{ps}$; $\Delta NDVI_{ps}$ is calculated as the product of $\partial NDVI/\partial NDVI_{ps}$ and $dNDVI_{ps}/dt$. A similar method is also suitable for other variables. Before quantification analysis, we detrended all variables to transform the time series to Z-score standardized anomalies:

$$Z = \frac{x - \overline{x}}{\omega} \tag{7}$$

where $\overline{x}$ and ω are the mean and standard deviation of study variables in each active season, respectively. x is the value of the study variable.

3. Results

3.1. Spatiotemporal Variations of Vegetation Growth

We obtained the vegetation coverage levels of the upper and middle reaches of the YRB during the growing season from 2000 to 2019 (Table 3). The medium vegetation cover areas were the largest (33.54%), followed by medium-high vegetation cover areas (30.92%, Table 3). Overall, the spatial distribution of mean NDVI during the growing season declined trend from southeast to northwest (Figure 2a), which was consistent with the spatial pattern of precipitation (Figure S1a) but opposite to the spatial distribution of sunshine duration (Figure S1c). NDVI had no obvious relationship with the spatial pattern of temperature (Figure S1b). The NDVI of the forests and savannas is high, while the grasslands have a low NDVI. Although the upper reach is primarily occupied by grasslands (Figure 1d), rivers are widely distributed (Figure 1c), which is conducive to vegetation growth, showing a high NDVI value. Compared with other study areas, the southeast has more precipitation and has extensive forests, croplands, and savannas (Figure 1d), so the NDVI is relatively high.

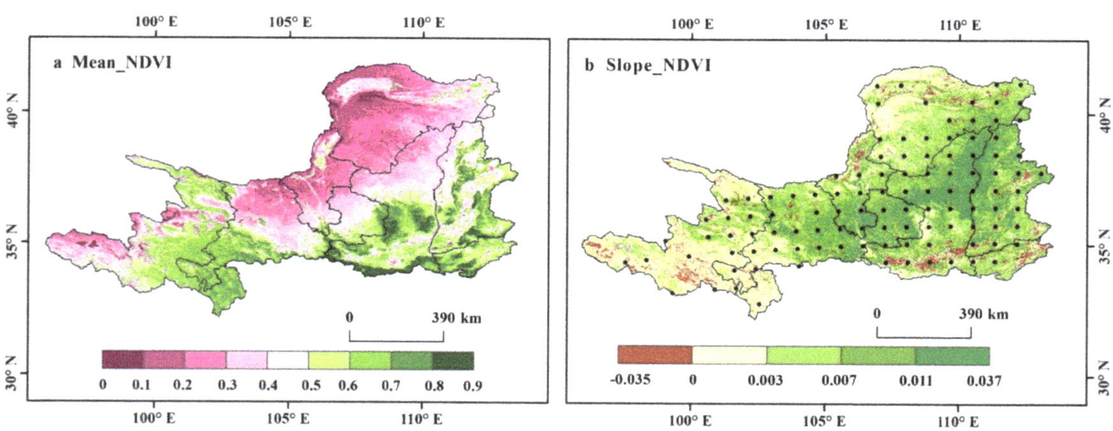

Figure 2. Spatial distribution of the mean NDVI (**a**) and inter-annual variation rate (slope) (**b**) during the growing season in the upper and middle reaches of the Yellow River basin from 2000 to 2019. Black dots indicate statistically significant correlations at the 95% confidence level. Same below.

Table 3. Vegetation coverage levels in the upper and middle reaches of the YRB during the growing season from 2000–2019.

Mean NDVI	Level	Proportion (%)
<0.1	Low vegetation cover area	0.88
0.1–0.3	Medium-low vegetation cover area	24.50
0.3–0.5	Medium vegetation cover area	33.54
0.5–0.7	Medium-high vegetation cover area	30.92
>0.7	High vegetation cover area	10.16

From 2000 to 2019, the mean NDVI during the growing season of the upper and middle reaches of the YRB exhibited a significant growth trend with a rate of 0.0056 per year ($p < 0.001$) (Figure 2b). Similarly, the mean GPP and LAI also showed a positive and significant trend ($p < 0.001$) at rates of 0.3718 per year and 0.0231 per year (Figure S2), respectively. The inter-annual variation of these vegetation indicators suggested that the vegetation coverage and productivity significantly increased in the study area over the two decades. In addition, we calculated the inter-annual variation rates of vegetation NDVI in each active season over the study period (Figure S3). The spatial distribution of inter-annual variation rates of NDVI in the early, peak, and late growing season (EGS, PGS, and LGS) was similar, with the rates equal to 0.0062 per year ($p < 0.001$), 0.0059 per year ($p < 0.001$), and 0.0046 per year ($p < 0.001$), respectively. EGS showed the strongest trend of vegetation greening, followed by PGS, although PGS showed a higher vegetation degradation trend than the other two sub-seasons (Figure S3b). Only LGS had a lower vegetation variation trend than the 20-year growing season mean trend (0.0056 per year). For climate variables, the precipitation in the growing season in the upper and middle reaches of the YRB increased significantly over the 20-year period, at a rate of 2.467 mm/year ($p < 0.05$) (Figure S1d). However, the interannual variation rate of sunshine duration decreased significantly, by –6.0620 h/year ($p < 0.05$) (Figure S1f). We found that the temperature increased at a rate of 0.012 °C/year ($p > 0.05$), which did not pass the significance test (Figure S1e).

3.2. Inter-Seasonal VGC Dominates Seasonal Vegetation Growth

At the seasonal scale, the relationships among the three vegetation indicators (NDVI, GPP, and LAI) and each driving factor were analyzed from 2000–2019. The results showed that the preceding-season NDVI significantly ($p < 0.05$) controls the current season vegetation NDVI in the three active seasons (Figure 3). The VGC effect was also positive for seasonal vegetation growth across the majority (91%, 93%, and 95% for EGS, PGS, and LGS, respectively) of the upper and middle reaches of the YRB (Figure 4). The partial correlation coefficients between the concurrent seasonal NDVI and the preceding-season NDVI in EGS, PGS, and LGS were 0.5168, 0.5743, and 0.6146, respectively (Figure 3), indicating that the strength of the VGC effect on vegetation growth increased from early to late growing season. In addition, the positive effect of VGC controlling vegetation growth in the three active seasons was also verified by examining the other vegetation growth and productivity indicators GPP and LAI (Figures S4 and S5). The matrix diagram and spatial distribution diagram of partial correlation coefficients of NDVI, GPP, and LAI showed a strong similarity. The vegetation growth and productivity of the preceding season were the dominant factors predicting vegetation greenness in each active growing season, implying that our results are reliable and the VGC cannot be ignored in the research of vegetation dynamics.

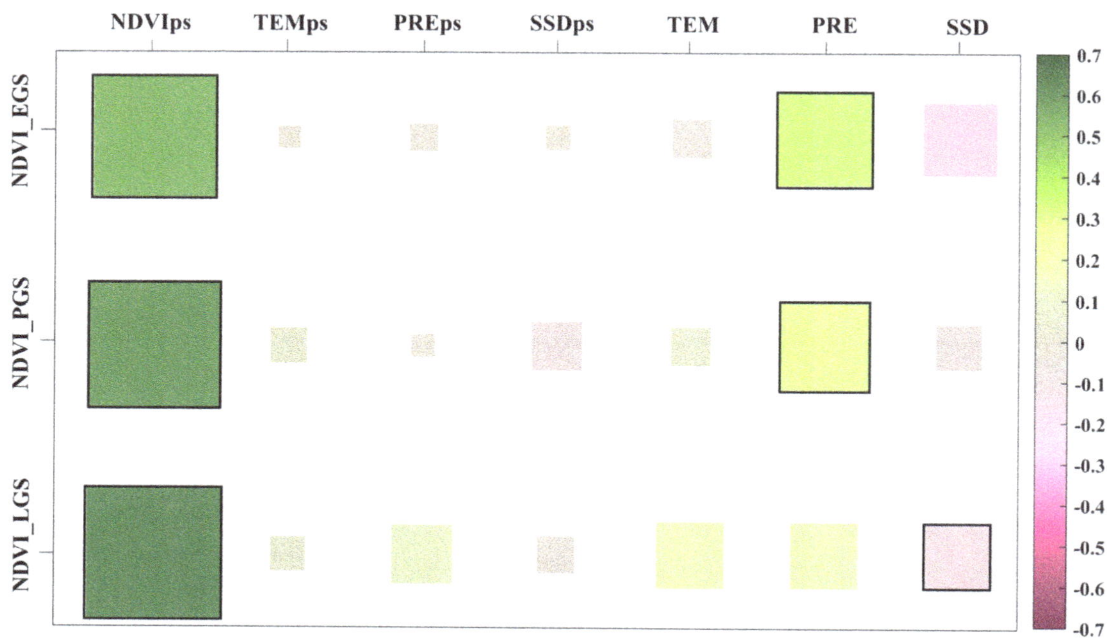

Figure 3. Partial correlation coefficients between 20-year seasonal NDVI time series and concurrent climate factors (temperature, TEM, precipitation, PRE, and sunshine duration, SSD), and climate factors (TEM_{ps}, PRE_{ps}, and SSD_{ps}) and NDVI ($NDVI_{ps}$) of the preceding season. The color block size represents the element value size. Squares with black outlines show statistically significant correlations at the 95% confidence level.

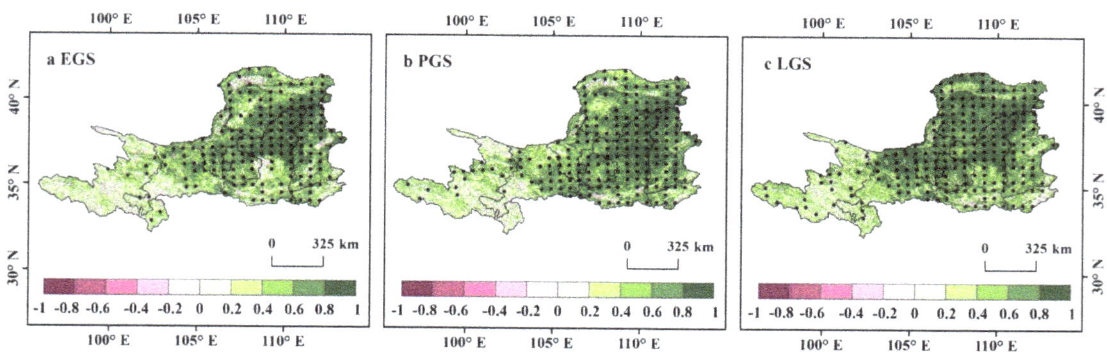

Figure 4. Spatial distribution of the partial correlations between NDVI of each season and that of the preceding season. Correlationpatterns are shown for EGS (**a**), PGS (**b**), and LGS (**c**).

Compared with the positive effect of the VGC, the intensity and direction of climate factors, including the immediate and lagged impacts of climate, determining the inter-annual variation of seasonal NDVI (GPP and LAI) varied greatly between seasons and across regions (Figures 5, 6 and S6–S9). The concurrent precipitation is the primary climate factor determining the inter-annual variation of EGS and PGS NDVI of the upper and middle reaches of the YRB over the 20-year period (partial correlation: 0.3122, $p < 0.05$; 0.2686, $p < 0.05$, respectively) (Figures 3 and 5d–e). The positive impact of concurrent precipitation on vegetation growth in EGS and PGS was also consistent in the analyses of other vegetation indicators GPP (Figures S4 and S6d–e) and LAI (Figures S4 and S8d–e). However, the concurrent precipitation had a much weaker impact on LGS vegetation, in comparison to that of EGS and PGS (Figures 3 and 5f). The immediate impact of precipitation diminished over the season (Figures 3 and 5d–f), contrary to the VGC effect. The precipitation of the preceding season had no obvious positive effect on vegetation growth in EGS and PGS, yet the positive effect in LGS increased (Figures 6d–f, S7d–f and S9d–f). The temperature of the current and preceding season on vegetation growth was very weak and statistically insignificant during the three active seasons. The weak temperature influence may be due to a spatial canceling-out of the positive effects and the negative effects (Figures 5a–c and 6a–c). Both the current- and preceding-season sunshine duration had negative effects on vegetation growth (Figures 5, 6 and S6–S9). The same results were also observed for GPP and LAI (Figures S6–S9).

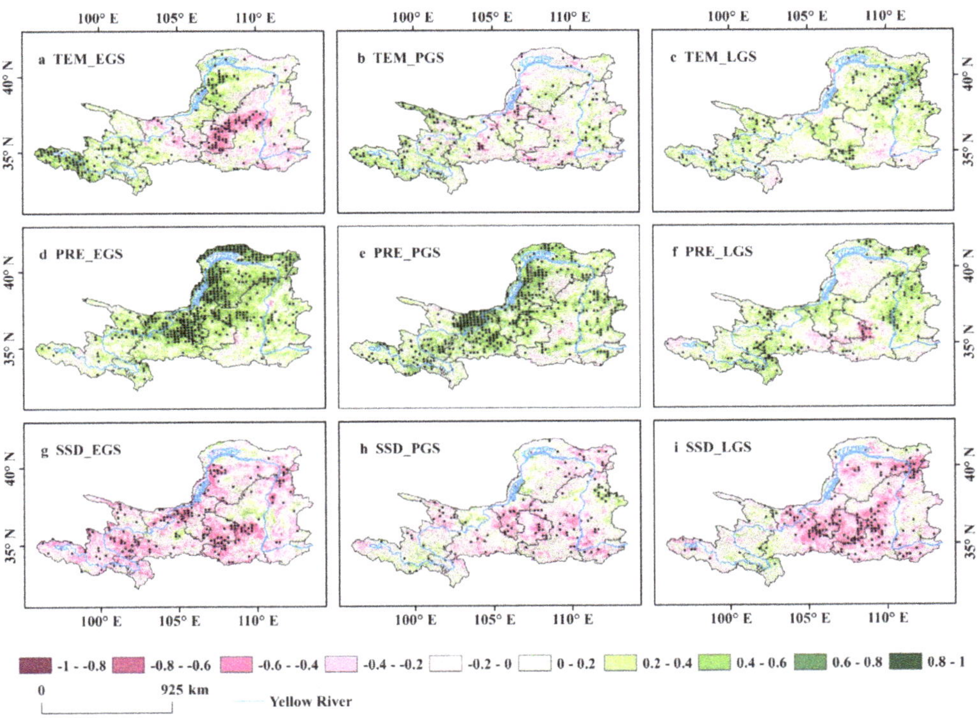

Figure 5. Spatial distribution of the partial correlations between NDVI of each season and the concurrent climate factors. Correlation patterns are shown for EGS (**a**,**d**,**g**), PGS (**b**,**e**,**h**), and LGS (**c**,**f**,**i**), and for correlation with concurrent TEM (**a**–**c**), PRE (**d**–**f**), and SSD (**g**–**i**).

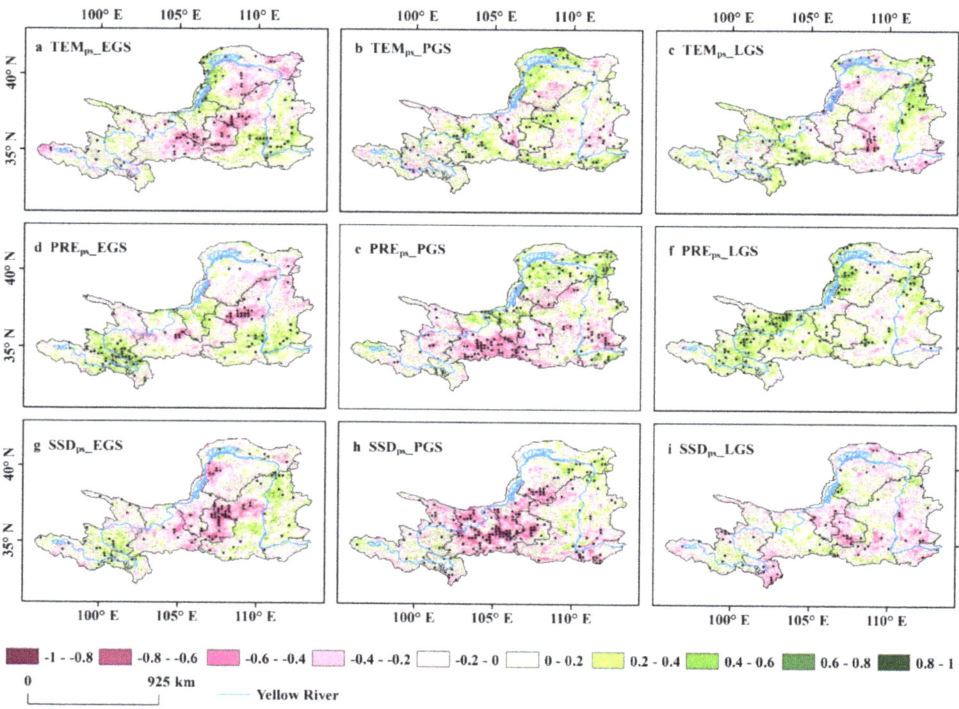

Figure 6. Spatial distribution of the partial correlations between NDVI of each season and climate factors of the preceding season. Correlation patterns are shown for EGS (**a**,**d**,**g**), PGS (**b**,**e**,**h**), and LGS (**c**,**f**,**i**), and for correlation with TEM_{ps} (**a–c**), PRE_{ps} (**d–f**), and SSD_{ps} (**g–i**).

3.3. The Persistence of the Vegetation Carryover into the Next Year

At the annual scale, we performed a partial correlation analysis between the current-year NDVI and the NDVI of the lag years. For a time lag of one year, 90.7% of study areas showed positive lagged correlations (Figure 7a). Among them, 52.5% passed the significance test ($p < 0.05$). The positive inter-annual vegetation carryover was most significant in eastern areas, particularly over the Shaanxi and Shanxi provinces. By contrast, only a few areas, including the upper reaches of the YRB and the northern desert, had a negative but generally insignificant ($p > 0.05$) vegetation carryover effect. This positive inter-annual VGC effect indicates that the vegetation productivity of the current year can carry over to the next year. If time lags are two years, 80.5% of study areas showed a positive relationship between the current-year NDVI and the NDVI of two years earlier (Figure 7b), but only 30.0% of the areas passed the significance test ($p < 0.05$). Compared to a one-year time lag, the area of positive correlation passing the significance test is significantly reduced (22.5%). If time lags are three and four years, the positive lagged correlation areas are 73.4% and 67.6%, respectively (Figure 7c–d), among which 29.5% and 23.1% passed the significance test ($p < 0.05$), respectively. From the third year of time lags, the positive VGC effect areas that passed the significance test remained at about 30%, indicating that the VGC effect can be carried over to the subsequent year yet weakly to years after that.

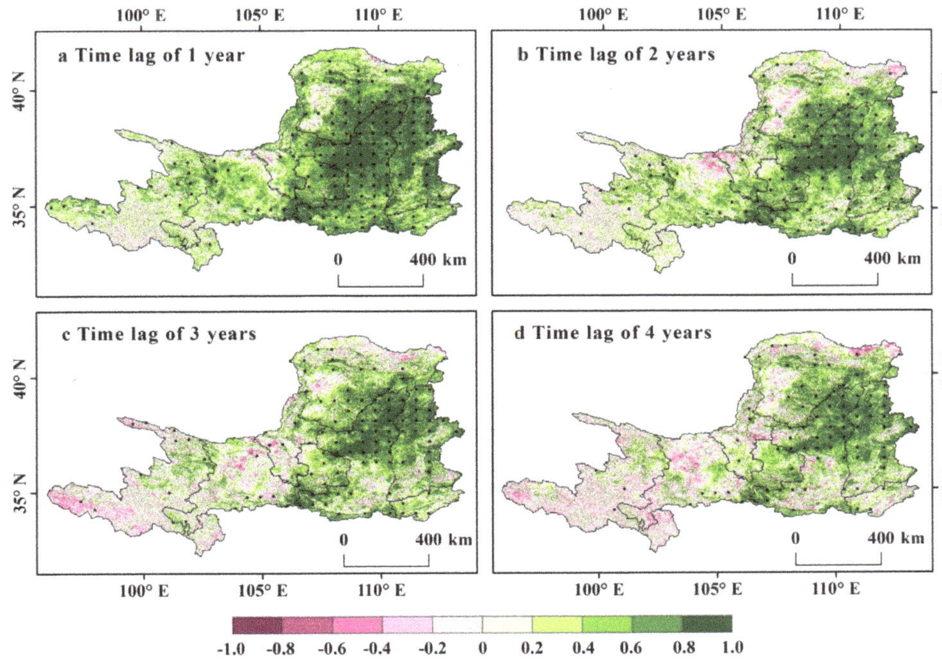

Figure 7. Spatial distribution of the partial correlations between NDVI of each year and that of the lag time, after controlling for the climate variable of both years.

3.4. The Individual Contributions of Climate Effects and the VGC Effect

For each active season, we further quantified the individual contributions of the VGC effect, climate effects of the concurrent season, and climate effects of the preceding season on the 20-year vegetation growth (Figure 8). We combined the impacts of temperature, precipitation, and sunshine duration into the single variable of climate effects. The results showed that the greening trends in the three active growing seasons were predominantly attributed to the VGC (the contributions are 50%, 67%, and 58%, respectively) (Figure 8), confirming the positive carryover effect of vegetation seasonal growth. It is interesting to note that the VGC effect had the strongest correlation with NDVI during the LGS (Figure 3) but the VGC effect contributed the most to vegetation growth in PGS (Figure 8). We also found that PGS was the season whose inter-annual variation trend (slope) of vegetation growth was most similar to that of the growing-season mean (gray dashed line, Figure 8). Concurrent seasonal climate factors also played a positive role in the three active growing seasons of vegetation growth (the contributions are 14.24% (EGS), 19.38% (PGS), and 13.80% (LGS), Figure 8). In comparison, the climate of the preceding season played a much smaller role in the vegetation greening for the three active growing seasons (the contributions are 0.64% (EGS), 5.58% (PGS), and 10.03% (LGS), Figure 8). Residuals is the contributions of other driving factors, including human activities and natural disturbances, that could influence vegetation growth but were not analyzed in our research. However, the residuals had a positive impact on vegetation greening, especially in the EGS (Figure 8). From Figure 3, VGC- and humid-induced vegetation greening from the early growth season sustained to the late growing season (ignoring residuals), which was the primary factor dominating vegetation growth in the upper and middle reaches of the Yellow River in the 20 years period.

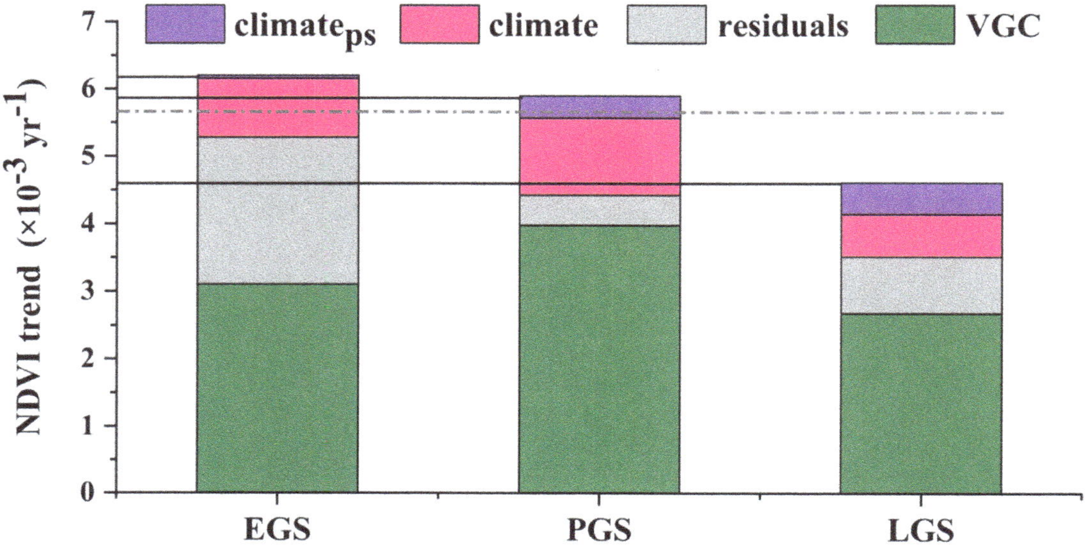

Figure 8. Individual contributions of the vegetation growth carryover (VGC) and the immediate (climate) and lagged (climate$_{ps}$) climate effects to seasonal NDVI trends over a 20-year period (2000–2019). The gray dashed line indicates the 20-year growing-season mean NDVI trend. The black solid lines indicate the observed NDVI trends of each season.

4. Discussion

4.1. The Effects of Climate Variability on Vegetation Growth

Regarding climate variability, vegetation growth in dry-land ecosystems is more sensitive to external precipitation changes (Figures 3, 5 and 6). The concurrent seasonal precipitation in the EGS and PGS had a significant positive effect on vegetation growth in the upper and middle reaches of the YRB, which was consistent with prior work [38]. The main reason for this sensitivity to precipitation may be that our study area is dominated by arid and semiarid areas [39]. Another possible explanation involves the widely distributed grasslands in the area (Figure 1d). Because herbs absorb the water from the upper part of the soil, they respond quickly to changes in precipitation [40]. Compared with the EGS and PGS seasons, the effect of concurrent precipitation on vegetation growth was weak in the LGS season. This phenomenon may be attributed to vegetation growth needing more moisture during green-up and maturity (EGS and PGS) than in senescence (LGS) [41]. The preceding-season precipitation in EGS and PGS had a weak effect on concurrent vegetation seasonal growth, indicating that the lagged effect of precipitation is not significant in the area with grassland as the main vegetation type [21]. This result is reasonable because the xylem in the stem of herbaceous plants is underdeveloped, so their water storage capacity is low [42]. In addition, the root system of plants matures in the late growing season, and the roots deeper in the soil dry out more slowly, so the precipitation of the preceding season has a positive effect on vegetation growth in LGS [21]. The sensitivity of precipitation clearly states that water availability is very important for vegetation growth in water-controlled ecosystems [43]. The immediate and lagged impacts of temperature had strong spatial heterogeneity and did not exhibit significant temporal effects on vegetation growth, which may be because the positive and negative effects cancel each other out. Especially in the EGS, the concurrent season temperature had a positive effect on vegetation growth in the western and northern parts of the study area, while it had a negative effect on vegetation growth in the central part of the study area (Figure 5a). This stemmed from the melting of snow and ice in early spring in the western alpine region, which promoted vegetation growth and improved vegetation productivity [44]. The upper and middle reaches of the YRB have plenty of sunshine, with an average annual sunshine duration of 2605 h in the past 30 years [4]. The sunshine duration, including the immediate and lagged impacts, was negatively correlated with seasonal vegetation growth. This may be because high sunshine duration increases evapotranspiration and thus drought, further inhibiting vegetation growth [41].

4.2. The Effects of Vegetation Growth Carryover on Vegetation Growth

Most people believe that vegetation growth depends mainly on external climate factors and human activity [1,45,46]. Our findings suggest that vegetation carryover has a significant positive impact on vegetation growth (Figures 3 and 4). That is to say, the VGC effect is the primary positive factor controlling the seasonal vegetation growth, also shown by others [17,18]. The positive effect of VGC in the three active growth seasons makes it easier to carry over the vegetation state in each season, and then affects vegetation growth in subsequent seasons. For instance, although sunshine duration had a negative effect on seasonal vegetation growth by increasing evapotranspiration [41,47], the positive and strong VGC effect can override this adverse effect. The same phenomenon was observed previously for high temperature [17,18]. In addition, the VGC effect in the central and northern part of our study area was stronger than in other regions, confirming that vegetation in drought regions may have a stronger VGC effect [10,18]. This strong vegetation carryover effect in drought regions implies that disturbances might have severer effects on vegetation in these areas. Initial ecological restoration and sustainable conservation would promote vegetation growth and increase vegetation productivity, which is of great significance to arid areas. Our results also exhibited that the persistence of the vegetation carryover can be sustained to the next year. So, a year with better vegetation growth is often followed by another year with better vegetation growth. Prior work has reported that drought legacies mostly last for one year, with significantly fewer legacies in the second and third years [48], as did our study of the VGC effect (Figure 7). Correspondingly, if vegetation does not overcome the influence of negative disturbances (e.g., wildfires, insect pests, and extreme climate) in time, the vegetation will slow growth in the coming years [18]. All the above evidence confirms the importance of the VGC effect on seasonal vegetation growth. Vegetation carryover is a critical pathway that should not be ignored.

4.3. Contributions of Climate Variability and Vegetation Carryover to Vegetation Greening

This study revealed that the strong observed greening trend of vegetation in each active season was mainly contributed by the VGC effect (Figure 8). Although the VGC effect was weaker during the early growing season, it increased from early to late growing season. High-quality vegetation growth in the early growing season determines vegetation growth in the subsequent seasons. The current-season climate factor, especially precipitation, is another important contributor, followed by the climate of the preceding season. Concerning the effect of current-season precipitation, vegetation was more sensitive in the early growing season. These phenomena may be attributed to the characteristics of vegetation—a low tolerance for lack of water during the early growth stage [18,49]. From this perspective, it is most effective to intervene in the vegetation to promote vegetation growth in spring through implementing ecological conservation and improving water conditions. There is still a proportion of unexplained variance (residuals) in vegetation growth in the three active seasons. These unexplained variances reflected the contributions of other drivers of vegetation growth, such as carbon dioxide fertilization and nitrogen deposition [1,50,51], human activities, [24,52,53], natural disturbances [7,15,54], and precipitation pattern shift [55]. The higher growth rate (black solid line) observed in the early parts of the growing season was primarily inherited from the vegetation carryover and residuals. However, this study focused on the impacts of climate factors and VGC effects at a seasonal scale and, therefore, did not quantitatively analyze other factors. As climate change is a continuous process, it may accumulate over time and then affect vegetation growth. Studies have shown that vegetation is significantly correlated with 2–6 months of cumulative precipitation [7], and precipitation on the Loess Plateau has a positive cumulative effect, especially in April and May [52]. This means that early spring precipitation may affect vegetation greening. Thus, the residuals in EGS may be contributed by the cumulative effect of precipitation. In addition, the melting of alpine snow and ice in early spring also provides moisture for vegetation growth in EGS [52]. The large-scale ecological restoration project, Grain to Green Project (GTGP), which was implemented to improve the ecological environment and maintain the ecological balance in the study area, has reduced arable land and increased forest and grassland. A prior study demonstrated that vegetation restoration is faster for lands converted to woodlands and grasslands than land with no conversion under similar dry conditions [38]. This is also the reason why the VGC effect in the study area is stronger than other factors. In the PGS, it showed a higher contribution of vegetation growth carryover effect, which is primarily inherited from the preceding-season vegetation greening (67%) rather than from the contributions of the climate factors (19%). Without considering the vegetation growth carryover effect, numerous studies suggested that climate factors and human activities are the driving factors for vegetation greening [4,21,45]. Yet, our study shows that the VGC effect contributes much more to vegetation greening than other factors. Though the positive VGC effect of vegetation growth can override negative climate impacts under the present climate and carry over into the subsequent year, it is still an open question whether this positive vegetation carryover will continue with future climate change.

4.4. Limitations and Implications

Although we studied the responses of vegetation dynamics to various drivers on the phenology-based seasonal scale, due to the complex driving factors, there are still some limitations and uncertainties in our study. Firstly, due to differences in sensor performance and soil background color, satellite-based NDVI data have inherent limitations [18]. Secondly, other factors, such as natural disturbance and human activities have a non-negligible impact on vegetation growth [1,50n51]. Taking all these factors into account is the most likely to yield more comprehensive and accurate outcomes. Finally, we noticed that climate factors have both a time delay effect and a cumulative effect on global vegetation growth. Our research only focused on the immediate and lagged impacts of climate. Therefore, in future work, we should discuss the influence of other variables in detail.

5. Conclusions

Based on the phenological seasonal scale, we exhibit vegetation dynamics in the upper and middle reaches of the Yellow River basin from 2000 to 2019, as well as immediate and lagged impacts of climate and vegetation growth carryover effects. The following conclusions are drawn: (1) From 2000 to 2019, the spatial distribution of mean NDVI, GPP, and LAI in the growing season exhibited a declining trend from southeast to northwest in the upper and middle reaches of the YRB. The vegetation coverage and productivity experienced a significant increase throughout the study area. (2) Significant positive correlations in the statistical analyses suggest that the vegetation carryover effect dominated the vegetation greening during the three active growing seasons, and this effect increased from early to late growing season. The concurrent season climate factors, especially precipitation, are also important drivers, followed by the climate of the preceding season. Sunshine duration, including its immediate and lagged impacts, always played a negative effect on vegetation growth. (3) As expected, the strongly observed greening trends of the three growing seasons were predominately attributed to the vegetation growth state of the preceding season. In addition, the vegetation growth carryover can be carried over to the subsequent year.

This carryover in seasonal vegetation growth is conducive to the next stage of vegetation growth. Although the vegetation growth carryover effect can counteract the negative effects of climate factors, whether this growth carryover effect will still exist under future climate change still needs to be addressed in future research.

Supplementary Materials: The following supporting information can be downloaded at: https://www.mdpi.com/article/10.3390/rs14195011/s1, Figure S1: Spatial distribution (a–c) and inter-annual variation rates (slope) (d–f) of mean precipitation (PRE), mean temperature (TEM), and mean sunshine duration (SSD) in growing season in the upper and middle reaches of the YRB over a 20-year period (2000–2019); Figure S2: The inter-annual variation rates (slope) of GPP (a) and LAI (b) in the growing season in the middle and upper of the YRB over a 20-year period (2000–2019). Black dots indicate statistically significant correlations at the 95% confidence level. Same below; Figure S3: The inter-annual variation rates (slope) of NDVI in three active seasons in the upper and middle reaches of the YRB over a 20-year period (2000–2019). The slopes patterns are shown for EGS (a), PGS (b) and LGS (c); Figure S4: Partial correlations between vegetation growth and that of its driving factors. This is same as Figure 3, except that vegetation growth is here based on GPP (a) and LAI (b); Figure S5: Spatial distribution of the partial correlations between GPP (a–c) or LAI (d–f) of each season and that of the preceding season. Correlation patterns are shown for EGS (a,d), PGS (b,e), and LGS (c,f); Figure S6: Spatial distribution of the partial correlations between GPP of each season and the concurrent climate factors. Correlation patterns are shown for EGS (a,d,g), PGS (b,e,h) and LGS (c,f,i), and for correlation with concurrent temperature (TEM) (a–c), precipitation (PRE) (d–f) and sunshine duration (SSD) (g–i); Figure S7: Spatial distribution of the partial correlations between GPP of each season and climate factors of the preceding season. Correlation patterns are shown for EGS (a,d,g), PGS (b,e,h) and LGS (c,f,i), and for correlation with preceding season temperature (TEM) (a–c), precipitation (PRE) (d–f), and sunshine duration (SSD) (g–i); Figure S8: Spatial distribution of the partial correlations between LAI of each season and the concurrent climate factors. Correlation patterns are shown for EGS (a,d,g), PGS (b,e,h) and LGS (c,f,i), and for correlation with concurrent temperature (TEM) (a–c), precipitation (PRE) (d–f), and sunshine duration (SSD) (g–i); Figure S9: Spatial distribution of the partial correlations between LAI of each season and climate factors of the preceding season. Correlation patterns are shown for EGS (a,d,g), PGS (b,e,h), and LGS (c,f,i), and for correlation with preceding season temperature (TEM) (a–c), precipitation (PRE) (d–f) and sunshine duration (SSD) (g–i).

Author Contributions: Conceptualization, X.Z. and J.L.; Funding acquisition, J.L.; Methodology, X.Z., Q.C., H.C. and J.L.; Resources, J.L.; Software, X.Z.; Visualization, X.Z.; Writing—original draft, X.Z.; Writing—review and editing, Q.C., H.C., Q.Q., C.L., J.D., M.C., S.Y. and J.L. All authors have read and agreed to the published version of the manuscript.

Funding: This research was funded by the National Natural Science Foundation of China (No. 32071523); the Scientific Project of Jinan Environmental Research Academy (No. SDGP37010020210-2000636); and the Hydraulic Science and Technology Plan Foundation of Shaanxi Province (No. 2019slkj-B1).

Institutional Review Board Statement: Not applicable.

Informed Consent Statement: Not applicable.

Data Availability Statement: Publicly available datasets were analyzed in this study. The data source and access links are indicated in the text. The code for the analysis of the data in this article is available from the corresponding author upon reasonable request.

Acknowledgments: The authors thank Pamela Holt (Shandong University) for improving the English language of our manuscript.

Conflicts of Interest: The authors declare no conflict of interest.

References

1. Piao, S.; Wang, X.; Wang, K.; Li, X.; Bastos, A.; Canadell, J.G.; Ciais, P.; Friedlingstein, P.; Sitch, S. Interannual variation of terrestrial carbon cycle: Issues and perspectives. *Glob. Chang. Biol.* **2020**, *26*, 300–318. [CrossRef] [PubMed]
2. Gemitzi, A.; Albarakat, R.; Kratouna, F.; Lakshmi, V. Land cover and vegetation carbon stock changes in Greece: A 29-year assessment based on CORINE and Landsat land cover data. *Sci. Total Environ.* **2021**, *786*, 147408. [CrossRef] [PubMed]
3. Konapala, G.; Mishra, A.K.; Wada, Y.; Mann, M.E. Climate change will affect global water availability through compounding changes in seasonal precipitation and evaporation. *Nat. Commun.* **2020**, *11*, 3044. [CrossRef] [PubMed]

4. Lu, C.; Hou, M.; Liu, Z.; Li, H.; Lu, C. Variation Characteristic of NDVI and its Response to Climate Change in the Middle and Upper Reaches of Yellow River Basin, China. *IEEE J. Sel. Top. Appl. Earth Obs. Remote Sens.* **2021**, *14*, 8484–8496. [CrossRef]
5. Duan, H.; Qi, Y.; Kang, W.; Zhang, J.; Wang, H.; Jiang, X. Seasonal Variation of Vegetation and Its Spatiotemporal Response to Climatic Factors in the Qilian Mountains, China. *Sustainability* **2022**, *14*, 4926. [CrossRef]
6. Zhang, W.; Wang, L.; Xiang, F.; Qin, W.; Jiang, W. Vegetation dynamics and the relations with climate change at multiple time scales in the Yangtze River and Yellow River Basin, China. *Ecol. Indic.* **2020**, *110*, 105892. [CrossRef]
7. Chen, Z.; Wang, W.; Fu, J. Vegetation response to precipitation anomalies under different climatic and biogeographical conditions in China. *Sci. Rep.* **2020**, *10*, 830. [CrossRef]
8. Yan, Y.; Liu, X.; Wen, Y.; Ou, J. Quantitative analysis of the contributions of climatic and human factors to grassland productivity in northern China. *Ecol. Indic.* **2019**, *103*, 542–553. [CrossRef]
9. Lucht, W.; Prentice, I.C.; Myneni, R.B.; Sitch, S.; Friedlingstein, P.; Cramer, W.; Bousquet, P.; Buermann, W.; Smith, B. Climate control of the high-latitude vegetation greening trend and Pinatubo effect. *Science* **2002**, *296*, 1687–1689. [CrossRef]
10. Seddon, A.W.R.; Macias-Fauria, M.; Long, P.R.; Benz, D.; Willis, K.J. Sensitivity of global terrestrial ecosystems to climate variability. *Nature* **2016**, *531*, 229–232. [CrossRef]
11. Ayanlade, A.; Jeje, O.D.; Nwaezeigwe, J.O.; Orimoogunje, O.O.I.; Olokeogun, O.S. Rainfall seasonality effects on vegetation greenness in different ecological zones. *Environ. Challenges.* **2021**, *4*, 100144. [CrossRef]
12. Soergel, B.; Kriegler, E.; Weindl, I.; Rauner, S.; Dirnaichner, A.; Ruhe, C.; Hofmann, M.; Bauer, N.; Bertram, C.; Bodirsky, B.L.; et al. A sustainable development pathway for climate action within the UN 2030 Agenda. *Nat. Clim. Chang.* **2021**, *11*, 656–664. [CrossRef]
13. Anderegg, W.R.L.; Schwalm, C.; Biondi, F.; Camarero, J.J.; Koch, G.; Litvak, M.; Ogle, K.; Shaw, J.D.; Shevliakova, E.; Williams, A.P.; et al. Pervasive drought legacies in forest ecosystems and their implications for carbon cycle models. *Science* **2015**, *349*, 528–532. [CrossRef]
14. Piao, S.; Fang, J.; Zhou, L.; Guo, Q.; Henderson, M.; Ji, W.; Li, Y.; Tao, S. Interannual variations of monthly and seasonal normalized difference vegetation index (NDVI) in China from 1982 to 1999. *J. Geophys. Res. Atmos.* **2003**, *108*, 4401. [CrossRef]
15. Wu, D.; Zhao, X.; Liang, S.; Zhou, T.; Huang, K.; Tang, B.; Zhao, W. Time-lag effects of global vegetation responses to climate change. *Glob. Change Biol.* **2015**, *21*, 3520–3531. [CrossRef]
16. Sun, Q.; Liu, C.; Chen, T.; Zhang, A. A Weighted-Time-Lag Method to Detect Lag Vegetation Response to Climate Variation: A Case Study in Loess Plateau, China, 1982–2013. *Remote Sens.* **2021**, *13*, 923. [CrossRef]
17. Lian, X.; Piao, S.; Chen, A.; Wang, K.; Li, X.; Buermann, W.; Huntingford, C.; Peñuelas, J.; Xu, H.; Myneni, R.B. Seasonal biological carryover dominates northern vegetation growth. *Nat. Commun.* **2021**, *12*, 983. [CrossRef]
18. Yuan, Y.; Bao, A.; Jiapaer, G.; Jiang, L.; De Maeyer, P. Phenology-based seasonal terrestrial vegetation growth response to climate variability with consideration of cumulative effect and biological carryover. *Sci. Total Environ.* **2022**, *817*, 152805. [CrossRef]
19. O'Connor, C.M.; Norris, D.R.; Crossin, G.T.; Cooke, S.J. Biological carryover effects: Linking common concepts and mechanisms in ecology and evolution. *Ecosphere* **2014**, *5*, art28. [CrossRef]
20. Liu, Y.; Schwalm, C.R.; Samuels-Crow, K.E.; Ogle, K. Ecological memory of daily carbon exchange across the globe and its importance in drylands. *Ecol. Lett.* **2019**, *22*, 1806–1816. [CrossRef]
21. Jiang, W.; Niu, Z.; Wang, L.; Yao, R.; Gui, X.; Xiang, F.; Ji, Y. Impacts of Drought and Climatic Factors on Vegetation Dynamics in the Yellow River Basin and Yangtze River Basin, China. *Remote Sens.* **2022**, *14*, 930. [CrossRef]
22. Fang, M.; Si, G.; Yu, Q.; Huang, H.; Huang, Y.; Liu, W.; Guo, H. Study on the Relationship between Topological Characteristics of Vegetation Ecospatial Network and Carbon Sequestration Capacity in the Yellow River Basin, China. *Remote Sens.* **2021**, *13*, 4926. [CrossRef]
23. Ma, L.; Xia, H.; Meng, Q. Spatiotemporal Variability of Asymmetric Daytime and Night-Time Warming and Its Effects on Vegetation in the Yellow River Basin from 1982 to 2015. *Sensors* **2019**, *19*, 1832. [CrossRef]
24. Liu, W.; Shi, C.; Zhou, Y. Trends and attribution of runoff changes in the upper and middle reaches of the Yellow River in China. *J. Hydro-Enviro. Res.* **2021**, *37*, 57–66. [CrossRef]
25. Zhao, Y.; Hu, C.; Zhang, X.; Lv, X.; Yin, X.; Wang, Z. Response of sediment discharge to soil erosion control in the middle reaches of the Yellow River. *Catena* **2021**, *203*, 105330. [CrossRef]
26. Yu, M.; Gao, Q. Increasing Summer Rainfall and Asymmetrical Diurnal and Seasonal Warming Enhanced Vegetation Greenness in Temperate Deciduous Forests and Grasslands of Northern China. *Remote. Sens.* **2020**, *12*, 2569. [CrossRef]
27. Kong, D.; Zhang, Q.; Singh, V.P.; Shi, P. Seasonal vegetation response to climate change in the Northern Hemisphere (1982–2013). *Glob. Planet. Change.* **2017**, *148*, 1–8. [CrossRef]
28. Gao, M.; Piao, S.; Chen, A.; Yang, H.; Liu, Q.; Fu, Y.H.; Janssens, I.A. Divergent changes in the elevational gradient of vegetation activities over the last 30 years. *Nat. Commun.* **2019**, *10*, 2970. [CrossRef]
29. Beck, H.E.; Zimmermann, N.E.; McVicar, T.R.; Vergopolan, N.; Berg, A.; Wood, E.F. Present and future Köppen-Geiger climate classification maps at 1-km resolution. *Sci. Data* **2018**, *5*, 180214. [CrossRef]
30. Chen, J.; Jönsson, P.; Tamura, M.; Gu, Z.; Matsushita, B.; Eklundh, L. A simple method for reconstructing a high-quality NDVI time-series data set based on the Savitzky–Golay filter. *Remote Sens. Environ.* **2004**, *91*, 332–344. [CrossRef]
31. Jönsson, P.; Eklundh, L. TIMESAT—A program for analyzing time-series of satellite sensor data. *Comput. Geosci.* **2004**, *30*, 833–845. [CrossRef]

32. Moss, D.N. Solar energy in photosynthesis. *Sol. Energy.* **1967**, *11*, 173–179. [CrossRef]
33. Peng, S.; Piao, S.; Ciais, P.; Myneni, R.B.; Chen, A.; Chevallier, F.; Dolman, A.J.; Janssens, I.A.; Peñuelas, J.; Zhang, G.; et al. Asymmetric effects of daytime and night-time warming on Northern Hemisphere vegetation. *Nature* **2013**, *501*, 88–92. [CrossRef] [PubMed]
34. Lian, X.; Piao, S.; Li, L.Z.X.; Li, Y.; Huntingford, C.; Ciais, P.; Cescatti, A.; Janssens, I.A.; Peñuelas, J.; Buermann, W.; et al. Summer soil drying exacerbated by earlier spring greening of northern vegetation. *Sci. Adv.* **2020**, *6*, eaax0255. [CrossRef] [PubMed]
35. Zhu, Z.; Piao, S.; Lian, X.; Myneni, R.B.; Peng, S.; Yang, H. Attribution of seasonal leaf area index trends in the northern latitudes with "optimally" integrated ecosystem models. *Glob. Chang. Biol.* **2017**, *23*, 4798–4813. [CrossRef] [PubMed]
36. Wang, L.; Tian, F.; Huang, K.; Wang, Y.; Wu, Z.; Fensholt, R. Asymmetric patterns and temporal changes in phenology-based seasonal gross carbon uptake of global terrestrial ecosystems. *Glob. Ecol. Biogeogr.* **2020**, *29*, 1020–1033. [CrossRef]
37. Pei, T.; Ji, Z.; Chen, Y.; Wu, H.; Hou, Q.; Qin, G.; Xie, B. The Sensitivity of Vegetation Phenology to Extreme Climate Indices in the Loess Plateau, China. *Sustainability* **2021**, *13*, 7623. [CrossRef]
38. Ren, Y.; Liu, J.; Liu, S.; Wang, Z.; Liu, T.; Shalamzari, M.J. Effects of Climate Change on Vegetation Growth in the Yellow River Basin from 2000 to 2019. *Remote Sens.* **2022**, *14*, 687. [CrossRef]
39. Wang, W.; Sun, L.; Luo, Y. Changes in Vegetation Greenness in the Upper and Middle Reaches of the Yellow River Basin over 2000–2015. *Sustainability* **2019**, *11*, 2176. [CrossRef]
40. Knapp, A.K.; Carroll, C.J.W.; Denton, E.M.; La Pierre, K.J.; Collins, S.L.; Smith, M.D. Differential sensitivity to regional-scale drought in six central US grasslands. *Oecologia* **2015**, *177*, 949–957. [CrossRef]
41. Ding, Y.; Li, Z.; Peng, S. Global analysis of time-lag and -accumulation effects of climate on vegetation growth. *Int. J. Appl. Earth Obs. Geoinf.* **2020**, *92*, 102679. [CrossRef]
42. Craine, J.M.; Ocheltree, T.W.; Nippert, J.B.; Towne, E.G.; Skibbe, A.M.; Kembel, S.W.; Fargione, J.E. Global diversity of drought tolerance and grassland climate-change resilience. *Nat. Clim. Chang.* **2012**, *3*, 63–67. [CrossRef]
43. Costa, G.B.; Mendes, K.R.; Viana, L.B.; Almeida, G.V.; Mutti, P.R.; e Silva, C.M.S.; Bezerra, B.G.; Marques, T.V.; Ferreira, R.R.; Oliveira, C.P.; et al. Seasonal Ecosystem Productivity in a Seasonally Dry Tropical Forest (Caatinga) Using Flux Tower Measurements and Remote Sensing Data. *Remote Sens.* **2022**, *14*, 3955. [CrossRef]
44. Wang, T.; Peng, S.; Lin, X.; Chang, J. Declining snow cover may affect spring phenological trend on the Tibetan Plateau. *Proc. Natl. Acad. Sci. USA* **2013**, *110*, E2854–E2855. [CrossRef]
45. Ji, Y.; Zhou, G.; Wang, S.; Wang, L. Prominent vegetation greening and its correlation with climatic variables in northern China. *Environ. Monit. Assess.* **2020**, *192*, 636. [CrossRef]
46. Omer, A.; Elagib, N.A.; Zhuguo, M.; Saleem, F.; Mohammed, A. Water scarcity in the Yellow River Basin under future climate change and human activities. *Sci. Total Environ.* **2020**, *749*, 141446. [CrossRef]
47. Vicente-Serrano, S.M.; Gouveia, C.; Camarero, J.J.; Begueria, S.; Trigo, R.; Lopez-Moreno, J.I.; Azorín-Molina, C.; Pasho, E.; Lorenzo-Lacruz, J.; Revuelto, J.; et al. Response of vegetation to drought time-scales across global land biomes. *Proc. Natl. Acad. Sci. USA* **2013**, *110*, 52–57. [CrossRef]
48. Gazol, A.; Camarero, J.J.; Sánchez-Salguero, R.; Vicente-Serrano, S.M.; Serra-Maluquer, X.; Gutiérrez, E.; de Luis, M.; Sangüesa-Barreda, G.; Novak, K.; Rozas, V.; et al. Drought legacies are short, prevail in dry conifer forests and depend on growth variability. *J. Ecol.* **2020**, *108*, 2473–2484. [CrossRef]
49. Zhang, J.; Zhang, Y.; Qin, S.; Wu, B.; Wu, X.; Zhu, Y.; Shao, Y.; Gao, Y.; Jin, Q.; Lai, Z. Effects of seasonal variability of climate factors on vegetation coverage across drylands in northern China. *Land Degrad. Dev.* **2018**, *29*, 1782–1791. [CrossRef]
50. Zhu, Z.; Piao, S.; Myneni, R.B.; Huang, M.; Zeng, Z.; Canadell, J.G.; Ciais, P.; Sitch, S.; Friedlingstein, P.; Arneth, A.; et al. Greening of the Earth and its drivers. *Nat. Clim. Chang.* **2016**, *6*, 791–795. [CrossRef]
51. Nemani, R.R.; Keeling, C.D.; Hashimoto, H.; Jolly, W.M.; Piper, S.C.; Tucker, C.J.; Myneni, R.B.; Running, S.W. Climate-driven increases in global terrestrial net primary production from 1982 to 1999. *Science* **2003**, *300*, 1560–1563. [CrossRef]
52. Guo, E.; Wang, Y.; Wang, C.; Sun, Z.; Bao, Y.; Mandula, N.; Jirigala, B.; Bao, Y.; Li, H. NDVI Indicates Long-Term Dynamics of Vegetation and Its Driving Forces from Climatic and Anthropogenic Factors in Mongolian Plateau. *Remote Sens.* **2021**, *13*, 688. [CrossRef]
53. Omer, A.; Zhuguo, M.; Zheng, Z.; Saleem, F. Natural and anthropogenic influences on the recent droughts in Yellow River Basin, China. *Sci. Total Environ.* **2020**, *704*, 135428. [CrossRef]
54. Ayanlade, A.; Oluwaranti, A.; Ayanlade, O.S.; Borderon, M.; Sterly, H.; Sakdapolrak, P.; Jegede, M.O.; Weldemariam, L.F.; Ayinde, A.F.O. Extreme climate events in sub-Saharan Africa: A call for improving agricultural technology transfer to enhance adaptive capacity. *Clim. Serv.* **2022**, *27*, 100311. [CrossRef]
55. Fang, J.; Piao, S.; Zhou, L.; He, J.; Wei, F.; Myneni, R.B.; Tucker, C.J.; Tan, K. Precipitation patterns alter growth of temperate vegetation. *Geophys. Res. Lett.* **2005**, *32*, L2144. [CrossRef]

remote sensing

MDPI

Article

Linear and Non-Linear Vegetation Trend Analysis throughout Iran Using Two Decades of MODIS NDVI Imagery

Arsalan Ghorbanian [1,2], Ali Mohammadzadeh [1] and Sadegh Jamali [2,*]

[1] Department of Photogrammetry and Remote Sensing, Faculty of Geodesy and Geomatics Engineering, K. N. Toosi University of Technology, Tehran 19967-15433, Iran; a.ghorbanian@email.kntu.ac.ir (A.G.); a_mohammadzadeh@kntu.ac.ir (A.M.)

[2] Department of Technology and Society, Faculty of Engineering, Lund University, P.O. Box 118, 221 00 Lund, Sweden

* Correspondence: sadegh.jamali@tft.lth.se

Abstract: Vegetation is the main component of the terrestrial Earth, and it plays an imperative role in carbon cycle regulation and surface water/energy exchange/balance. The coupled effects of climate change and anthropogenic forcing have undoubtfully impacted the vegetation cover in linear/non-linear manners. Considering the essential benefits of vegetation to the environment, it is vital to investigate the vegetation dynamics through spatially and temporally consistent workflows. In this regard, remote sensing, especially Normalized Difference Vegetation Index (NDVI), has offered a reliable data source for vegetation monitoring and trend analysis. In this paper, two decades (2000 to 2020) of Moderate Resolution Imaging Spectroradiometer (MODIS) NDVI datasets (MOD13Q1) were used for vegetation trend analysis throughout Iran. First, the per-pixel annual NDVI dataset was prepared using the Google Earth Engine (GEE) by averaging all available NDVI values within the growing season and was then fed into the PolyTrend algorithm for linear/non-linear trend identification. In total, nearly 14 million pixels (44% of Iran) were subjected to trend analysis, and the results indicated a higher rate of greening than browning across the country. Regarding the trend types, linear was the dominant trend type with 14%, followed by concealed (11%), cubic (8%), and quadratic (2%), while 9% of the vegetation area remained stable (no trend). Both positive and negative directions were observed in all trend types, with the slope magnitudes ranging between −0.048 and 0.047 (NDVI units) per year. Later, precipitation and land cover datasets were employed to further investigate the vegetation dynamics. The correlation coefficient between precipitation and vegetation (NDVI) was 0.54 based on all corresponding observations (n = 1785). The comparison between vegetation and precipitation trends revealed matched trend directions in 60% of cases, suggesting the potential impact of precipitation dynamics on vegetation covers. Further incorporation of land cover data showed that grassland areas experienced significant dynamics with the highest proportion compared to other vegetation land cover types. Moreover, forest and cropland had the highest positive and negative trend direction proportions. Finally, independent (from trend analysis) sources were used to examine the vegetation dynamics (greening/browning) from other perspectives, confirming Iran's greening process and agreeing with the trend analysis results. It is believed that the results could support achieving Sustainable Development Goals (SDGs) by serving as an initial stage study for establishing conservation and restoration practices.

Keywords: Iran; vegetation cover; normalized difference vegetation index (NDVI); vegetation trend; MODIS; linear trend; non-linear trend

1. Introduction

Vegetation is a key constituent of the Earth's biosphere, which serves as the inter-linkage substance between the soil, water, and atmosphere [1,2]. Vegetation significantly influences the terrestrial carbon cycle, surface water/energy exchange and balance, and ecosystem service values from regional to global scales [3–7]. For instance, vegetation is responsible for regulating carbon-cycle balance, mitigating global warming and climate change effects, alleviating greenhouse gasses, and reducing thermal heat stress. The unequivocal climate change is an undeniable challenge of the current era that notably has affected human life and the environment [8]. Climate change can significantly influence vegetation activities and its associated environmental benefits [9]. Besides, coupled anthropogenic activities, such as land cover conversion and natural resource exploitation, can also impact the vegetation statutes [10]. Accordingly, monitoring vegetation dynamics is a fundamental requirement to enrich our understanding of how the terrestrial Earth is responding to natural and anthropogenic forcings [11]. Additionally, vegetation greening and browning are the primary factors in biodiversity measures and ecosystem functioning, emphasizing the necessity of monitoring vegetation dynamics for conservation and restoration practices and management legislation [12–14].

The availability of accurate and consistent datasets for vegetation monitoring and trend analysis is a principle to guarantee reliable results [15]. Over the past decades, advances in remote sensing created an unprecedented opportunity to monitor vegetation dynamics from regional to global scales. Remote sensing has provided spatially and temporally consistent datasets that have been widely employed for vegetation monitoring and trend analysis [16–19]. Accordingly, remote sensing could offer researchers and policymakers profound insights into vegetation monitoring that could be used propose and adopt effective workflows and regulations. In this regard, many vegetation indices, such as Normalized Difference Vegetation Index (NDVI), Enhanced Vegetation Index (EVI), and Soil Adjusted Vegetation Index (SAVI), have been developed to target vegetation condition monitoring [20–22]. The development of these vegetation indices was mainly shaped by the spectral differences of vegetation cover in different electromagnetic spectrum domains (e.g., red and near-infrared) to focus on vegetation detection and state examination [21]. Among these vegetation indices, NDVI has proven to be a reliable index for vegetation studies and thus has been applied for different tasks, such as land cover mapping [23,24], vegetation health monitoring [25,26], and trend analysis [27–29]. The NDVI as a ratio index has the advantage of reducing specific types of band-related noises and influences, such as sun and sensor view angle, topography, atmospheric attenuation, and solar irradiance variations [30–33]. The NDVI has been recognized as a sufficiently stable index allowing meaningful comparison of inter-annual vegetation dynamics [30]. Additionally, due to its simplicity and applicability for various tasks, several NDVI datasets have been generated using remote sensing images acquired by different sensors with different spatial resolutions. For instance, the NDVI datasets of the Global Inventory Modeling and Mapping Studies (GIMMS) of the Advanced Very High Resolution Radiometer (AVHRR) and the Moderate Resolution Imaging Spectroradiometer (MODIS) onboard the Terra and Aqua platforms are among the most used and validated NDVI datasets [18]. The former provides a global NDVI dataset at a coarse spatial resolution of ~9 km from 1981 to 2013 and is primarily suitable for global and continental scales vegetation monitoring. The latter offers an NDVI dataset from 2000 to the present at different temporal (i.e., daily to 16-day) and spatial (250 m to 1 km) resolutions, making it an appealing choice for vegetation monitoring and trend analysis. Different spatial resolutions of MODIS NDVI datasets enable the adoption of an appropriate NDVI dataset based on the extent of the interested area.

The combination of NDVI datasets and trend analysis algorithms has been extensively employed for vegetation monitoring worldwide. For instance, Jamali et al. [34] developed an automated vegetation trend analysis algorithm using polynomials, Polynomial for Trend analysis (PolyTrend), to study the vegetation status in the Sahel, on the African Continent. In this regard, GIMMS NDVI datasets between 1982 and 2006 were employed for linear and non-linear trend analysis. Their findings suggested that over half of the study area exhibited statistically significant trends, and the linear trend with the positive direction (i.e., increase in NDVI value) had the highest proportion. Likewise, Wang et al. [13] employed 17 years of MODIS NDVI imagery to explore the vegetation dynamics in Shiyang River Basin,

China. They implemented a trend analysis workflow that considered linear, exponential, logarithmic, and logistics trend types. The results indicated that the increasing trends were dominant in all trend types, and the most and the least trend types were linear (36.53%) and logistics (15.27%). In another study, Liu et al. [35] applied non-parametric linear trend analysis, Theil-Sen regression [36], to NDVI datasets derived from Landsat Archive imagery. They employed 35 years of NDVI data from Landsat 4, 5, 7, and 8 over six regions in the Russian Tundra. They observed that, in general, the Russian Tundra was subjected to the greening process, of which 30.7% and 0.5% of the study area experienced NDVI gain and loss, respectively.

Similar to vegetation monitoring studies around the globe, several studies were carried out to investigate the vegetation condition in Iran. For instance, Emamian et al. [37] employed 12 years of NDVI data from MODIS imagery to explore the vegetation dynamics over Khorasan, a northeastern province of Iran. Their analysis revealed that most parts of the province exhibited no vegetation trends, and the vegetation status remained stable in total, while the most increasing and decreasing trends occurred in semi-humid and extra arid climate zones. Furthermore, Gholamnia et al. [38] examined the spatio-temporal patterns of vegetation conditions in the semi-arid region of Kurdistan province, Iran. To this end, 17 years of MODIS NDVI datasets were utilized to identify the abrupt changes using the Breaks For Additive Season and Trend (BFAST) algorithm [39]. They observed that 2007 and 2010 were the years with significant percentages of breakpoints (i.e., abrupt changes), and the change magnitudes were between −0.14 and 0.12 during the study period. Likewise, Sedighifar et al. [40] utilized 17 years of MODIS EVI data to analyze the spatio-temporal vegetation dynamics in the northern parts of Iran. They employed the linear regression approach to determine the trend of vegetation cover dynamics between 2000 and 2017 and reported a rising trend in vegetation cover. In another study, Kazemzadeh et al. [10] explored the vegetation trend in the Taleghan watershed using NDVI data from 2000 to 2019. For this purpose, the NDVI dataset was derived from Landsat Archive imagery, and the PolyTrend was implemented for trend investigations. Their findings indicated that nearly all study areas were subjected to vegetation gain (i.e., positive trend), dominated by linear trends and slope magnitudes between 0.01 and 0.2. Moreover, Eskandari Dameneh et al. [41] studied the desertification process and susceptibility in Iran using remote sensing data. In their study, the MODIS NDVI dataset from 2001 to 2015 was analyzed using the Theil-Sen algorithm for linear trend detection. In general, they observed an insignificant minor positive trend based on average NDVI values during 15 years of investigations. Their results indicated increasing trends of NDVI in northern and western (northwest to southwest) parts of the study area, while opposite trends occurred in the northeast and southwest.

Although several studies were conducted for vegetation monitoring in Iran, they have mainly investigated relatively small areas or linear trends. To the best of the author's knowledge, there is no country-wide vegetation monitoring in Iran, with consistent spatial characteristics and consideration of linear/non-linear trend patterns. The importance of considering non-linear trend types is that vegetation cover may experience non-linear changes due to external disturbances (e.g., fire, clear-cut, insect outbreak, land cover conversion), natural processes, and ecological shifts between equilibrium states [42–44]. Additionally, non-linear trends can manifest downward and upward trends (NDVI gain/loss after NDVI loss/gain) through a time series, providing more detailed information. Therefore, the obtained results could benefit interdisciplinary researchers to profoundly investigate the drivers of vegetation change.

Accordingly, this paper aims to provide the first country-wide vegetation trend analysis using MODIS NDVI imagery at 250 m spatial resolution over the past two decades (2000 to 2020), considering both linear/non-linear trend patterns. To this end, the mean annual NDVI values during the growing season were first calculated based on 16-day MODIS NDVI images and then were fed to PolyTrend to identify linear and non-linear changes. Meanwhile, the significance of trends, trend directions (positive or negative), and slope magnitudes were also computed for each vegetation pixel throughout Iran. MODIS land cover data were used to examine associations between vegetation trends and vegetation types. Furthermore, a land cover change map of the study area was produced based on MODIS land cover maps to investigate the relationship between land cover transitions and vegetation trend types. Additionally, in-situ precipitation observations for the same period were employed to investigate the relationship between vegetation and precipitation dynamics. Finally, vegetation trend types, in-situ precipitation, and MODIS land cover data were also used to examine associations between vegetation trends, vegetation types, and precipitation. The results would benefit both the scientific community and governors in studying the possible drivers of change and identifying hotspot zones (i.e., highly dynamic regions) for further monitoring. It is believed that the results could support achieving Sustainable Development Goals (SDGs) by serving as an initial stage study for establishing conservation and restoration practices.

2. Study Area

The study area, Iran, spanning between 24°–40°N (latitudes) and 44°–64°E (longitudes), is located in the Middle East, the western part of Asia (see Figure 1a). Iran, with an area of approximately 1.7 million km^2, is recognized as the seventeenth largest country in the world. The study area is a mountainous region with two main mountain ranges of Zagros and Alborz. The former begins in the northwestern parts of Iran and follows the western border of the country to the southern parts, and the latter is located along a belt from west to the east in the northern regions of the country that can be seen in the ALOS Global Digital Surface Elevation Model (see Figure 1b) [45]. These two mountain ranges and their surroundings are among the most vegetated regions in Iran, which include dense and sparse forest, rangeland, and farmland areas (see Figure 1c) [46]. The southern border of Iran is shaped by the Persian Gulf and the Oman Sea, and the Caspian Sea is located in the north. The immense latitudinal range of Iran and its geographical characteristics cause high climate variations from very-humid to hyper-arid climate zones and high precipitation variations [47,48]. These variations are responsible for diverse land cover types in Iran, from dense forests (between the Caspian Sea and the Alborz Mountains) to the uncovered plains and deserts, such as Dasht-e-Kavir and Dasht-e-Lut (central parts).

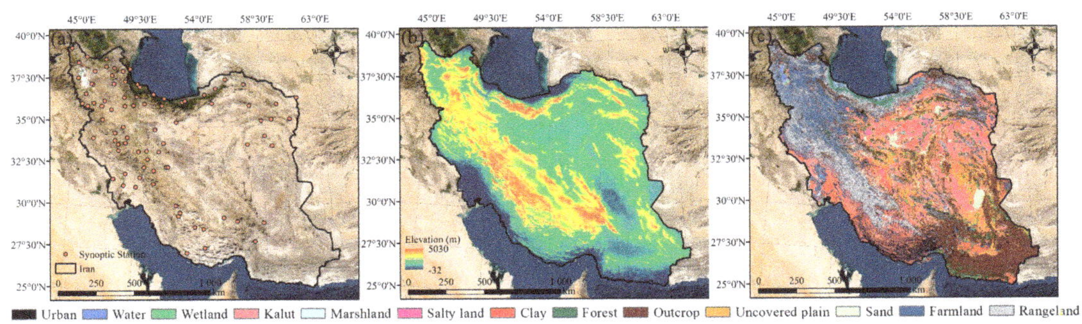

Figure 1. (**a**) Geographical location of Iran, along with spatial distribution of selected synoptic stations for precipitation monitoring, (**b**) ALOS Global Digital Surface Model [45], and (**c**) recent land cover map of Iran [46].

3. Material and Methods

In this section, first, the used datasets (see Table 1), including NDVI, in-situ precipitation, and land cover datasets, are introduced. Afterward, the implemented algorithm for vegetation trend analysis is explained.

Table 1. Summary of the specifications of the used datasets in this study.

Dataset		Sensor	Resolution Spatial	Time Period	Source
Type	Name				
Satellite	NDVI	MODIS (MOD13Q1)	250 m	2000–2020	Google Earth Engine (GEE)
Satellite	Land Cover	MODIS (MCD12Q1)	500 m	2001 and 2020	Google Earth Engine (GEE)
In-situ	Precipitation	Synoptic Station	-	2000–2020	Islamic Republic of Iran Meteorological Organization (IRIMO)

3.1. NDVI Dataset

The MODIS instrument onboard the Terra platform was launched on 18 December 1999 and became operational in early 2000. MODIS daily captures Earth's surface radiance in 36 spectral bands ranging from 0.4 μm to 14.4 μm of the electromagnetic spectrum. Due to its functional characteristics, MODIS imagery has been extensively used in different applications and for generating many global products [49]. Among various products, MODIS vegetation indices are well-approved products by the scientific community and have been widely incorporated into multiple tasks, such as vegetation dynamics monitoring [11,16,17,38,50]. Here, Collection 6 (version 6) of 16-day NDVI images (MOD13Q1) of the Terra-MODIS product was used. In addition to several advantages of NDVI, its lower sensitivity to topographical variations prioritized its utility over EVI [31,33] since the vegetation areas are mainly located near mountainous regions of the study area. Previous Collection of Terra-MODIS NDVI was adversely affected due to sensor degradation; however, the planned adjustments largely eliminated this effect for Collection 6 NDVI [51]. This NDVI dataset is a level 3 product generated every 16 days at the spatial resolution of 250 m. The MODIS algorithm specifies the best available pixel value within each 16-day interval based on the criteria of low clouds, low view angle (to minimize the BRDF effects), and the highest NDVI value, providing a high-quality NDVI dataset [32,52]. This product and its associated uncertainties have been well quantified through different approaches and achieved the stage 3 validation process [53]. In particular, this product has been validated using AERONET-corrected data, spaceborne and airborne sensors, as well as radiometric field measurements. Based on the validation analysis, the accuracy of NDVI values is within ± 0.025, demonstrating high-quality assurance of the MODIS 16-day NDVI dataset [54]. In total, 168 bi-monthly (16-day) NDVI images during the growing season (i.e., 21 April to 21 August) between 2000 and 2020 were employed in the present study. The growing season was determined based on the knowledge of the study area and the commonly considered months in previous relevant studies [10,55,56]. Later, all NDVI images within each year's growing season were averaged to produce yearly NDVI images for trend analysis. It is worth noting that the NDVI dataset was prepared using the Google Earth Engine (GEE) cloud computing platform [57,58] with the snippet of (*ee.ImageCollection("MODIS/006/MOD13Q1")*). Although the MODIS NDVI dataset includes high-quality NDVI values, the yearly aggregation procedure is expected to reduce the random changes in NDVI values, the possibility of outlier influence, and the temporal autocorrelation in time-series for trend analysis [10,34,37]. Figure 2 shows the per pixel vegetation condition of Iran based on the aggregated yearly NDVI images in 2000 (start), 2020 (end), multi-year NDVI mean, and the NDVI range (i.e., the difference between the largest and lowest NDVI values) over the last two decades. As evident in Figure 2a,b, vegetated areas are mainly located in northern and western parts, along the Zagros and Alborz Mountains, as well as in central southern parts of Iran.

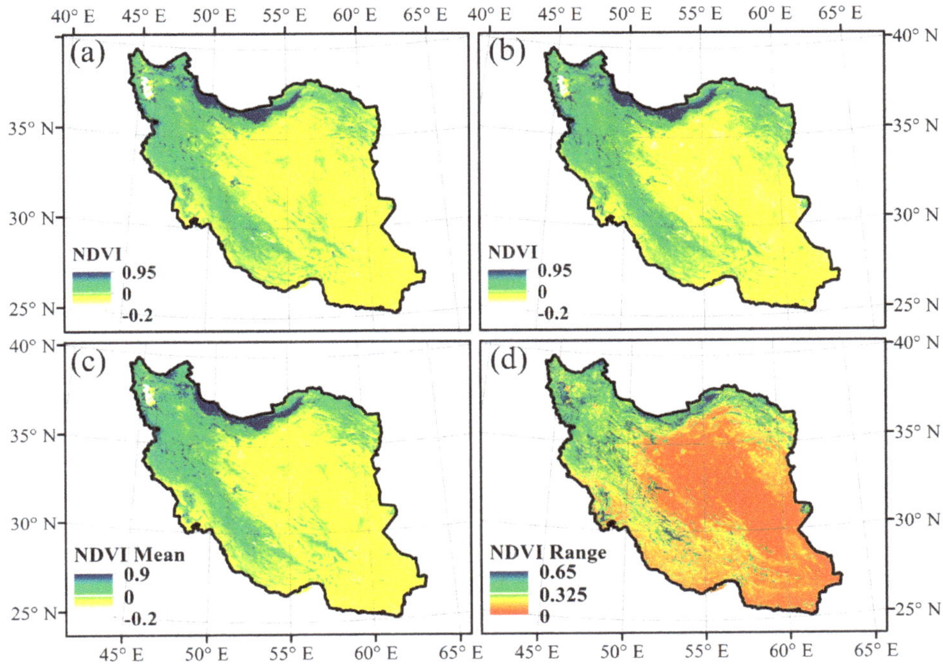

Figure 2. Normalized Difference Vegetation Index (NDVI) values at 250 m spatial resolution throughout Iran in (**a**) 2000 (start of the study period) and (**b**) 2020 (end of the study period), (**c**) per-pixel multi-year average NDVI values, and (**d**) per-pixel range (i.e., the difference between the largest and lowest values) of NDVI values over the 21 years.

3.2. Land Cover Dataset

Different vegetation types may behave multiformly through passing time and thus, might be considered as an impactful criterion for trend analysis. Accordingly, two MODIS land cover maps (MCD12Q1) of 2001 and 2020 with 500 m spatial resolution (relatively closer to the MODIS NDVI dataset) were employed for three purposes. The land cover dataset was downloaded from the GEE data catalog with the snippet of (*ee.ImageCollection ("MODIS/006/MCD12Q1")*). First, the 2001 land cover map was used to explore the dominant land cover type for each linear and non-linear trend pattern (see Section 3.4). Second, a land cover change map was generated by comparing these maps to identify land cover transitions, which then was incorporated to investigate the relationship between land cover transitions and trend patterns (see Section 3.4). Third, it was used to interpret the possible three-way relationship between vegetation trend direction (positive or negative), precipitation trend direction, and land cover types. Different land cover products have been developed using MODIS imagery, and in this study, the land cover product generated based on the International Geosphere-Biosphere Programme (IGBP) scheme was used. This product was generated using nadir Bidirectional Reflectance Distribution Function (BRDF) adjusted surface reflectance data and Random Forest classifier [59]. It contained 17 land cover classes, 12 of which classes are different vegetation covers, and the rest are permanent wetland, urban/built-up, permanent snow/ice, barren, and water bodies.

3.3. In-Situ Precipitation Dataset

Many studies have found supporting evidence for the impact of climate change on terrestrial vegetation cover [60–64]. Among climatic variables, precipitation, especially in semi-arid and closed climate zones, has been recognized as the most prominent factor in vegetation cover dynamics due to its dominant role in the water supply [65,66]. Accordingly, accurate precipitation observations from synoptic stations were employed to investigate the relationship between vegetation and precipitation inter-annual dynamics. In this regard, 85 synoptic stations (see Figure 1a) that were located on vegetation pixels (based on MODIS NDVI), selected from a total of 244 synoptic stations, were acquired from the Islamic Republic of Iran Meteorological Organization (IRIMO). These synoptic stations contained daily precipitation observations from 2000 to 2020. In particular, only synoptic stations located on vegetation pixels with 21 years of consistent precipitation observations were employed for further analysis. Furthermore, it can be seen in Figure 1a that synoptic stations are reasonably distributed across vegetation areas in Iran. The daily precipitation values during the growing season (i.e., similar to the NDVI dataset) were accumulated to generate a 21-year precipitation dataset for each synoptic station. Subsequently, they were utilized for trend analysis to explore their trend direction (i.e., positive or negative) for further comparisons with vegetation trend results.

3.4. Linear and Non-Linear Trend Analysis

It has been recognized that the vegetation covers do not always follow linear dynamics and may change in non-linear manners instead [34,42–44,67,68]. Accordingly, this paper implemented an automated vegetation trend analysis framework developed by Jamali et al. [34] that considers both linear and non-linear trend patterns. This algorithm, named PolyTrend, works with the per-pixel time-series NDVI values and assigns one of the (1) Cubic, (2) Quadratic, (3) Concealed, (4) Linear, (5) No Trend labels to the examined pixel. The concealed label is assigned to pixels with either cubic or quadratic trends that has insignificant net change during the study period (i.e., negligible NDVI difference between the start and end dates). PolyTrend is a 3-stage workflow (see Figure 3), which is initiated by checking the existence of a cubic ($ax^3 + bx^2 + cx + d$) trend. In this regard, first (stage 1), a cubic model is fitted to the input time-series NDVI values, followed by checking the statistical significance of the cubic coefficient (a) at the 95% confidence interval using the t-test statistics. In case of passing the significance test, the presence of both local maximum and minimum of the time-series in the fitted model is checked (comparing the roots of the fitted model with time-series NDVI values), and if approved, a linear model is fitted to the examined time-series NDVI values. The significance of the linear coefficient will lead to assigning the cubic label to the corresponding pixel; otherwise (i.e., the insignificance of the linear coefficient), the Concealed label is assigned. In case of rejecting both criteria, the algorithm will test the existence of a quadratic trend ($bx^2 + cx + d$) at the second stage. The procedure is advanced as in the cubic workflow, except for checking the presence of only one local maximum or minimum in the fitted model. If the considered criteria are not met, the possibility of a linear trend for the time-series NDVI values is explored (stage 3). To this end, a linear model is fitted to the time series, and if the model is found to be statistically insignificant, the No Trend label is assigned, and in case of being statistically significant, the corresponding pixel is identified with the Linear pattern. Further details of the PolyTrend are available at Jamali et al. [34]. Here, time-series NDVI values (see Section 3.1) of each pixel were inserted into the PolyTrend. It is worth noting that only pixels with at least one NDVI value of over 0.15 were considered as vegetation cover and employed in trend analysis [69–71]. In other words, non-vegetation and very sparsely vegetated pixels with low NDVI values < 0.15 during the study period (21 years) were excluded to reduce the computational burden of trend analysis due to the massive extent of Iran. In addition to trend types, the significance of trends, direction of trends, and the slope magnitude of trends were also computed for each vegetation pixel. These parameters would provide complementary information to the actual trend analysis results for profound investigations. For instance, vegetation pixels with statistically significant trends and high slope magnitude could present the hotspot locations that have been altered considerably.

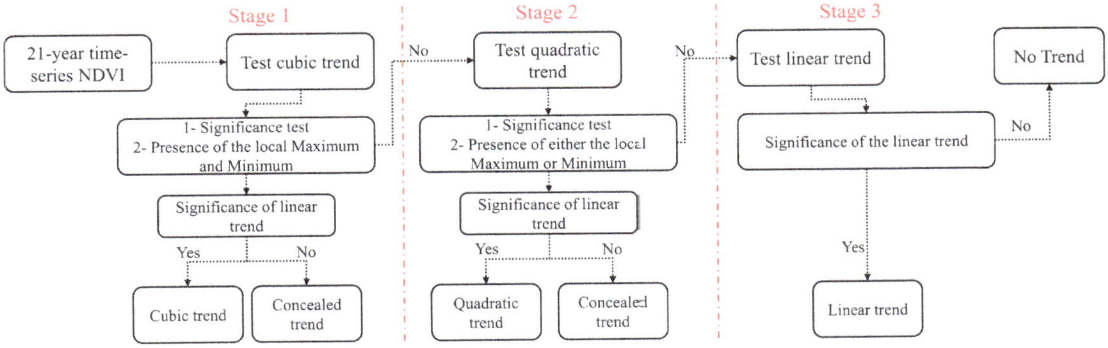

Figure 3. Schematic workflow of the PolyTrend for trend type identification using time-series NDVI values of MODIS imagery.

It is worth noting that PolyTrend was also applied to in-situ precipitation data to determine their trend directions (i.e., positive or negative). The derived trend directions were then used to compare the vegetation and precipitation mutual dynamics.

4. Results

This section covers the PolyTrend assessment, followed by providing trend analysis results in Iran over the last two decades. Finally, other data sources are included to investigate the associations between vegetation trends, precipitation trends, and land cover types.

4.1. Trend Linearity/Non-Linearity Assessment

The PolyTrend has been statistically tested using both synthesized and actual time-series NDVI data [34]. Furthermore, it has been employed for trend analysis in other studies and is claimed to be accurate in trend pattern identification [72–76]. However, in the first step, the results of the PolyTrend were evaluated through visual inspections for quality assurance of the detected vegetation trends in Iran. To this end, 100 candidate pixels were randomly (stratified random) selected throughout the study area, containing five different classes (see Section 3.4). Later, the trends detected by the PolyTrend for each individual pixel were visually and statistically examined. In general, the assessment step suggested the applicability of PolyTrend for vegetation trend analysis using two decades of MODIS NDVI imagery in Iran. Figure 4 shows the identified trends for 10 pixels derived from the candidate pixels for justification, including two samples for each trend class. Based on Figure 4, the PolyTrend correctly identified trend types with a 95% confidence interval. For instance, it properly identified the cubic and quadratic trends (Figure 4a–d), and assigned concealed trends to pixels with possible cubic and quadratic trends (Figure 4e,f), while their linear coefficients were insignificant (i.e., low net NDVI difference). Furthermore, the linear labels were given to pixels in which the required criteria for cubic and quadratic trends were not satisfied, but the linear coefficient was significant (Figure 4g,h). Finally, when no significant trend was observed for the input time series, the no trend label was assigned (Figure 4i,j). These pixels contained fluctuating NDVI values within very low boundaries (i.e., low variation) and ended without notable NDVI changes.

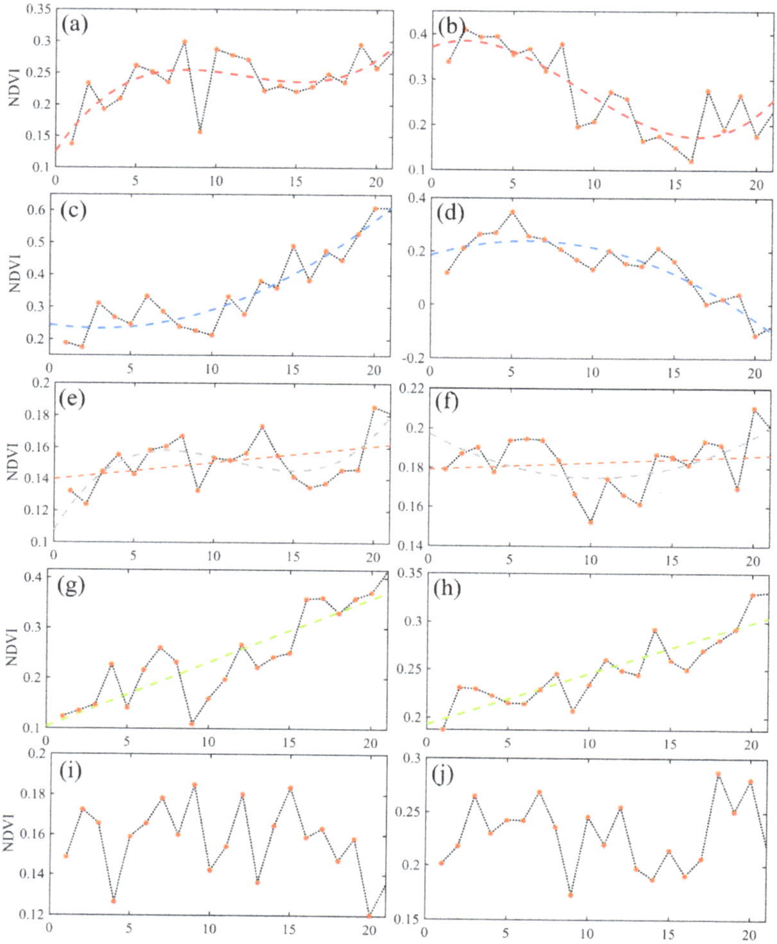

Figure 4. Ten candidate vegetation pixels for trend identification assessment of PolyTrend with (**a**,**b**) Cubic, (**c**,**d**) Quadratic, (**e**,**f**) Concealed, (**g**,**h**) Linear, and (**i**,**j**) No Trend patterns.

4.2. Trend Analysis

Figure 5 shows the per pixel trend map derived by using two decades of MODIS NDVI imagery and the PolyTrend algorithm. After applying the NDVI threshold value (see Section 3.4) for selecting vegetation pixels, nearly 14 million pixels were analyzed for trend identification, accounting for approximately 44% of Iran. Figure 5 illustrates that the vegetation covers were mainly distributed in the northern and western parts of the country. In general, no consistent spatial patterns in terms of different trend types were observed throughout Iran. The concealed class was observed primarily in northeastern and southwestern parts of the country. According to earlier descriptions (see Section 3.4), this class includes both cubic and quadratic trends without a significant net change in their NDVI values. In fact, the corresponding pixels were subjected to a substantial decrease and then increase (or increase and then decrease), which resulted in low NDVI changes between the start and end of the study period. Pixels identified as No Trend were mostly scattered from the west to the east in the northern parts of the country. Pixels with linear trends were mainly distributed from the northwestern to the southwestern parts of the country along the Zagros Mountains. However, several other regions of the country in the northeast and central-south had dominant linear trend pixels. Further visual inspections implied that the quadratic trends occurred in fewer locations and were seen in western and southern parts of Iran. Finally, as the dominant non-linear trend type, pixels with the cubic trend were detected in northern, southern, and northeastern parts of Iran. These pixels experienced significant downward and upward trends with one local maximum and minimum (see Figure 4a,b). As a sample, Figure 4b shows a pixel with a cubic trend, in which its time-series NDVI values started at 0.35 and had a downward tendency to nearly 0.12 in 2015, and then reached over 0.22 at the end of the study period.

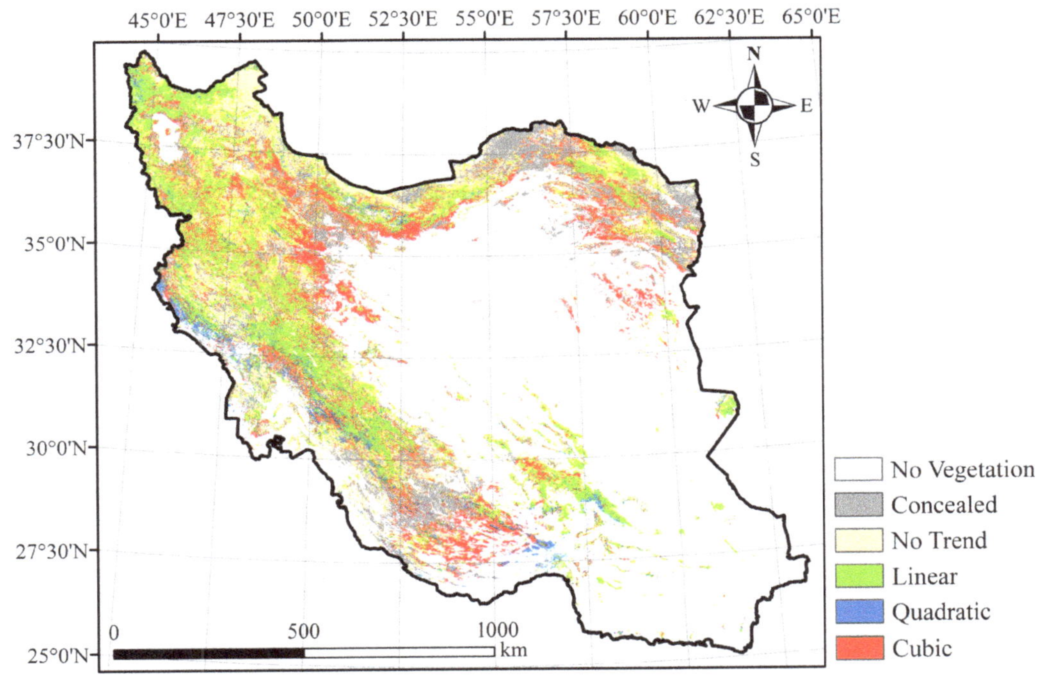

Figure 5. Vegetation trend map throughout Iran derived using two decades of MODIS NDVI imagery (2000–2020) and PolyTrend algorithm.

In addition to the trend types map (Figure 5), three other complementary maps, including trend directions, trend significance, and trend slopes, were also generated by PolyTrend, and are provided in Figure 6. Figure 6a shows that the positive trends are dominant over Iran, accounting for nearly 90% of vegetation pixels. This indicates that most parts of the country were subjected to the greening process. However, several hotspot locations can be seen with negative trend directions. Furthermore, Figure 6b illustrates that nearly 56% of vegetation pixels had significant trends over the past two decades that contained linear, quadratic, and cubic trend types. On the other hand, 46% of vegetation pixels were insignificantly changed and were assigned canceled and no trend labels. Figure 6c presents the slope magnitude (linear coefficient) of vegetation dynamics, derived by fitting an independent linear model to the time-series NDVI values, which are classified into eight classes for better illustrations. As anticipated, most vegetation pixels obtained positive slope values, and overall, slope magnitudes mainly fell in two classes with minor slope values between −0.002 and 0.002 (NDVI units) per year. However, visual inspections revealed that in some regions, homogeneous patches existed with notable (either positive or negative) slope values. These areas could be considered hotspot zones, being the candidates with higher priority for further examinations and management practices.

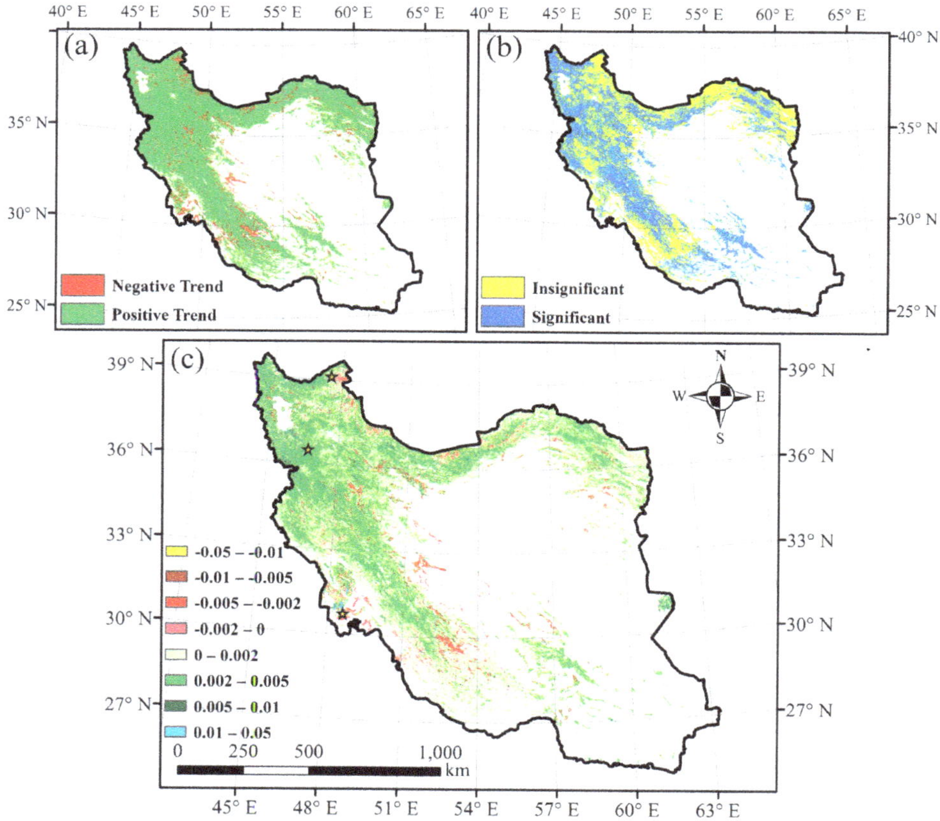

Figure 6. Complementary trend information provided by PolyTrend, including (**a**) trend direction, (**b**) statistical significance/insignificance of detected trends, and (**c**) and per-year slope magnitudes of vegetation dynamics over the past two decades (black-orange stars are candidate hotspot locations).

Further numerical results are provided in Tables 2 and 3. According to Table 1, the linear trend covering 14% of Iran was the most dominant trend pattern, followed by concealed (11%), cubic (8%), and quadratic (2%) trends, while 9% of the vegetation area remained stable (no trend). Slope magnitudes ranged between −0.048–0.047 (NDVI change per year), with the highest slope values for cubic and linear trends. Based on the numerical results, concealed, no trend, linear, quadratic, and cubic patterns covered positive (negative) areas of about 179,181 (33,608) km^2, 156,042 (28,273) km^2, 258,521 (6096) km^2, 39,849 (3228) km^2, and 143,756 (9575) km^2, respectively. Based on the obtained results, over 68% of vegetation pixels experienced minor dynamics over the past two decades, with slope magnitude ranges of −0.002 and 0.002. It was also observed that more pixels (30%) were subjected to vegetation gain with higher slope magnitudes (0.002–0.05), accounting for approximately 260,640 km^2, while nearly 3% (27,827 km^2) experienced higher negative slopes (−0.05––0.002).

Table 2. Numerical trend results obtained by analyzing two decades of MODIS NDVI imagery (2000–2020) and the PolyTrend algorithm.

Trend Type	Pixels (%)	Area (km^2)	Positive (km^2)	Negative (km^2)	Slope Value Range (NDVI Unit)
No Vegetation	56	1,081,447	-	-	-
Concealed	11	212,622	179,181	33,608	−0.016–0.017
No Trend	9	181,929	156,042	28,273	−0.013–0.016
Linear	14	270,890	258,521	6096	−0.043–0.041
Quadratic	2	38,587	39,849	3228	−0.041–0.037
Cubic	8	146,603	143,756	9575	−0.048–0.047

Table 3. Percent share of different slope magnitudes for vegetation pixels derived using the PolyTrend algorithm.

	Slope Magnitude Range							
	[−0.05–−0.01]	[−0.01–−0.005]	[−0.005–−0.002]	[−0.002–0]	[0–0.002]	[0.002–0.005]	[0.005–0.01]	[0.01–0.05]
Vegetation Pixels	0.15%	0.75%	2.34%	7.34%	59.15%	26.63%	3.17%	0.57%

The trend analysis results are valid at a 95% confidence interval, and the area values are computed by multiplying the pixel area (250 × 250 m^2) by the number of pixels in each trend type. The percentage values of Positive and Negative directions were computed as the ratio to the total number of vegetation pixels in the study area. The slope magnitudes are calculated by fitting a line to 21 years of NDVI values for each pixel, and the average standard error of slope magnitudes (linear coefficient) was 4.1×10^{-6}.

Further investigations using visual interpretation of satellite images (true-color Landsat Archive images with 30 m spatial resolution at the start and end of the study period) were carried out over three hotspot regions (see black-orange stars in Figure 6c) with high NDVI slope magnitudes. Figure 7a,b presents two satellite images captured in 2000 and 2020 over a region with high NDVI gain (lowermost star in Figure 6c). The trend analysis revealed that this region had a dominant positive trend with the highest slope magnitude range (class 7, 8). It was observed that high NDVI gains were mostly associated with agricultural expansion and vegetated wetland formation in this area. On the other hand, Figure 7c,d illustrates a location (uppermost star in Figure 6c) with a relatively intense negative slope magnitude, which was found to be linked to mountainous vegetation loss. Likewise, Figure 7e,f presents a region with agricultural expansion approximate to the waterbody areas, justifying the obtained greening process (positive trend) with high NDVI slope magnitudes using the PolyTrend.

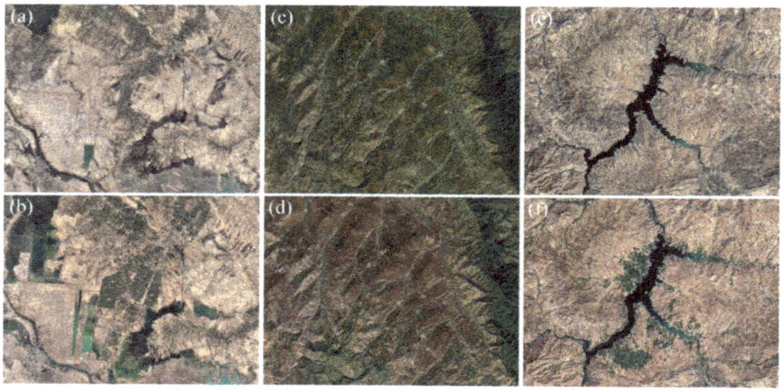

Figure 7. Satellite images (RGB) in 2000 (first row (**a**,**c**,**e**)) and 2020 (second row (**b**,**d**,**f**)) over three candidate regions (black-orange stars in Figure 6c) with high NDVI slope magnitude changes for further visual interpretations.

4.3. Vegetation Trend vs. Land Cover and Precipitation

Figure 8 shows the distribution of trend types and trend directions among different vegetation covers from the MODIS land cover dataset (see Section 3.2). Figure 8 shows the proportional amounts, meaning the number of pixels divided by the total number of pixels in the corresponding land cover class. To extract such information, first, the land cover dataset was resampled to 250 m spatial resolution to match the trend analysis results based on the nearest neighbor method. Afterward, a reclassification step was adopted to lower the land cover types into five more general vegetation land covers, namely (1) Forest, (2) Shrubland, (3) Savannas, (4) Grassland, and (5) Cropland. In addition to inherent misclassification errors in land cover maps, the spatial mismatch between original MODIS land cover and NDVI (used for trend analysis) datasets, as well as the employed threshold for vegetation pixel selection, several inconsistencies occurred. As a consequence, several vegetation pixels (determined through NDVI threshold > 0.15) in the trend analysis fell in non-vegetation classes in the MODIS land cover data. Therefore, they were excluded for further analysis, and only vegetation pixels identified using both datasets were used in this section. The results show the dominant trend types were no trend, concealed, no trend, linear, and no trend for land cover classes of forest, shrubland, savannas, grassland, and cropland, respectively, indicating the relative dominance of no trend for the considered vegetation types. Linear had the highest number and portion in all vegetation classes, considering the significant trend types, followed by the cubic trend pattern. The results also illustrate that the vegetation cover as grassland experienced the highest statistically significant dynamics, which were mainly located along the Zagros Mountains. Furthermore, Figure 8b shows that all considered vegetation covers were subjected to NDVI gain (i.e., positive trend directions), with the two highest portions for forest (99.63%) and grassland (97.65%). However, the amounts of negative trend direction for savannas (13.82%) and cropland (17.10%) were much higher than forest (0.37%), shrubland (3.45%), and grassland (2.35%).

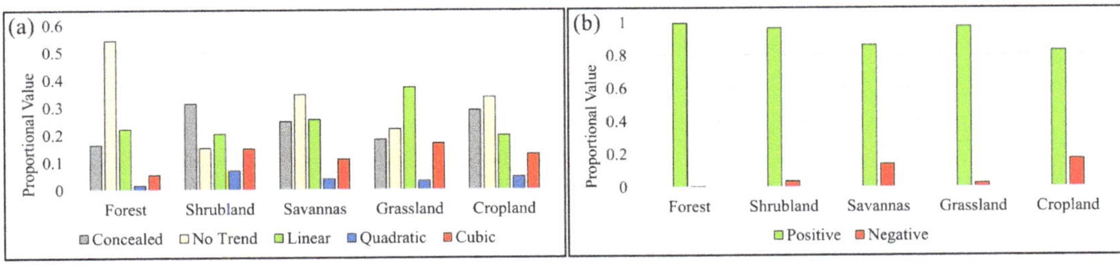

Figure 8. Distribution of (**a**) trend types and (**b**) trend directions in five dominant vegetation classes. The proportional value on the y-axis is the number of pixels for each trend type divided by the total number of pixels in the corresponding land cover class.

Afterward, two MODIS land cover maps of 2001 and 2020, which were initially resampled to 250 m spatial resolution and reclassified into nine classes of (1) Forest, (2) Shrubland, (3) Savannas, (4) Grassland, (5) Wetland, (6) Cropland, (7) Urban, (8) Barren, and (9) Water, were compared. The result was a change map containing the class-wise land cover transitions. Later, the generated change map and the trend types map were integrated to explore the relationship between land cover transitions and trend types. In this regard, the land cover transitions for linear, quadratic, and cubic patterns (statistically significant trends) were derived from the integrated map. The proportion of each land cover transition class to the total land cover transitions for each trend type was then calculated. The minor portions were excluded by focusing on land cover transitions with the five highest proportions that occurred more frequently > 5. Table 4 summarizes the calculated proportions for linear, quadratic, and cubic patterns. Accordingly, the land cover transition between Barren (2001) and Shrubland (2020) had the highest proportion for the three trend patterns. This was followed by grassland transition to shrubland for linear and cubic patterns, while the reverse change was observed for quadratic. The third-highest transition for linear, quadratic, and cubic patterns were shrubland to grassland, grassland to shrubland, and barren to grassland, respectively. Linear and quadratic patterns had the same land cover transition (i.e., barren to grassland), while the shrubland to grassland transition was found for cubic as the fourth-highest change. The grassland to cropland transition was also observed for all three trend patterns as the fifth-highest transition. In general, the results did not provide any specific relationship between land cover transitions and trend patterns. However, the results indicated the greening process in Iran since a high portion of Barren was changed to vegetation cover classes.

Table 4. The five highest land cover transitions between 2001 and 2020 for each statistically significant trend type were derived by integrating the land cover change map and trend type map of the study area. The trend analysis results are valid at a 95% confidence interval, and the area values are computed by multiplying the pixel area (250 × 250 m^2) by the number of pixels in each transition class. B: Barren, S: Shrubland, G: Grassland, C: Cropland.

	Highest Transitions between 2001 and 2020 (Area in km^2)				
	First	**Second**	**Third**	**Fourth**	**Fifth**
Linear	B → S (6130)	G → S (5147)	S → G (4651)	B → G (4272)	G → C (3318)
Quadratic	B → S (2443)	S → G (1098)	G → S (939)	B → G (779 km^2)	G → C (603)
Cubic	B → S (7150)	G → S (3821)	B → G (3389)	S → G (2119)	G → C (1474)

In the next step, the relationship between NDVI and precipitation values was investigated for each synoptic station. In this regard, the time-series NDVI values at each synoptic station location were extracted. Afterward, the correlation coefficient between NDVI and precipitation values was computed similarly to previous studies [33,56,77,78]. It is worth noting that the correlation coefficient values were calculated using all corresponding NDVI and precipitation values throughout Iran (n = 1785) and for each synoptic station individually (n = 21). Figure 9a presents the spatial distribution of obtained correlation coefficient values for each synoptic station, indicating higher values in northern and northwestern parts of the country. The correlation coefficient values ranged from 0.002 to 0.847, with an average value of 0.371. Figure 9b shows the precipitation and NDVI values scatterplot using all corresponding observations. The computed correlation coefficient was 0.54, indicating a relatively strong influence of precipitation on vegetation dynamics.

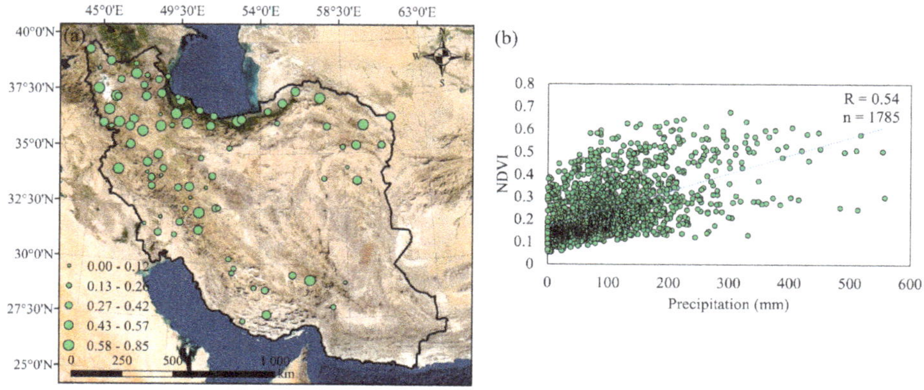

Figure 9. The correlation coefficient of time-series NDVI and precipitation (**a**) at each synoptic station and (**b**) using all corresponding observations over the study area.

Furthermore, the trend of time-series NDVI and precipitation observations were also investigated since they might provide complementary information to the previous analysis. Figure 10 presents the results of comparing vegetation and precipitation trend directions (derived by applying PolyTrend). In this regard, the trend matching procedure was implemented based on 4 cases: (1) positive directions in both (i.e., NDVI and precipitation increased), (2) positive/negative directions in vegetation/precipitation (i.e., NDVI increased and precipitation decreased), (3) negative directions in both, and (4) negative/positive directions in vegetation/precipitation (see Figure 10a). In general, the results implied that in nearly 60% of the cases (at 49 synoptic stations), both vegetation and precipitation had similar trend directions (either positive or negative), indicating the impact of precipitation dynamics on vegetation cover change. On the other hand, vegetation and precipitation had reverse trend directions in 36 cases, in 22 cases of which vegetation increased and precipitation decreased. Further investigations revealed that in these 22 cases, the synoptic stations are located in regions with a high amount of annual precipitation rates. This probably suggests that although the precipitation rates have declined, the required amount of water supply for vegetation growth still exists. Therefore, other influential factors may exist in these regions, as well as in other parts of Iran. In fact, although precipitation has been recognized as the most impactful criterion for vegetation dynamics, there are other impactful parameters when the water supply is enough to support biological processes [79,80]. Figure 10b illustrates that in the similar trend category (both positive/negative), the linear pattern was the dominant trend, while in the opposite trend, the concealed was dominant. This indicates that significant trend types were the dominant trend patterns in similar directions, meaning that vegetation with statistically significant trends had relatively more similar trend directions to precipitation dynamics.

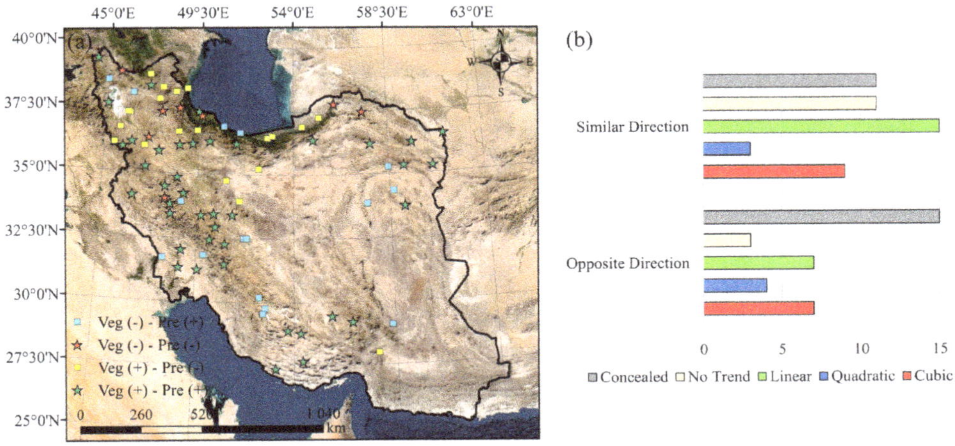

Figure 10. (**a**) Spatial distribution of the matching results between vegetation and precipitation trend directions (Veg: vegetation, Pre: precipitation, +: positive direction, -: negative direction), and (**b**) number of synoptic stations with similar and opposite trends with vegetation dynamics.

The final analysis included land cover data for five vegetation types to investigate the possible three-way associations. It is worth noting that no synoptic stations were located in Forest, and thus, only four vegetation covers were employed to derive subsequent information. In general, the findings were relatively discrepant when considering vegetation and precipitation trends over different vegetation land covers. In fact, the existence of both similar and opposite tend directions among four vegetation covers prevented a robust depiction of any association. In particular, 18(13), 7(4), 8(6), and 1(1) synoptic stations had similar (opposite) trend directions in grassland, shrubland, cropland, and savannas, respectively. Although dominant cases with statistically significant trends (vegetation) in each land cover type had similar trend directions, the high amount of opposite trend directions hindered an authentic conclusion.

5. Discussion

Long-term vegetation trend analysis is essential for vegetation condition monitoring and identifying hotspot locations for further studies. Such information could be used to identify areas prone to vegetation loss to avoid further land degradation [81]. Not only do areas with negative trends require further analysis, but regions with positive trends, especially with high slope magnitudes, should be candidate cases for further explorations. This is because such regions may be subjected to invasive vegetation growth or arbitrary agricultural expansion, which could bring adverse environmental issues [82,83]. Therefore, the trend analysis results could benefit scholars and governors in conducting research to determine the driving factors and adopting effective legislation. In this study, the long-term vegetation trend analysis was successfully performed using the PolyTrend algorithm that considers both linear and non-linear patterns, satisfying the main objective of this paper. The results indicated a higher rate of greening than browning across Iran, with a dominance of the linear trend pattern. Furthermore, the results of this study are also provided at province scales (see Figure 11), which could be of importance for sub-national decision-makers. This information, either at national or sub-national scales, could serve as initial information for restoration and conservation practices. Moreover, the relationships between vegetation dynamics and land cover, land cover transition, and precipitation were also investigated as other objectives of the current study. The grassland experienced the highest statistically significant dynamics between 2000 and 2020. Besides, the cropland and forest had the highest statistically significant negative and positive trends over the last two decades. The results suggested the relatively strong impact of precipitation on vegetation dynamics in general, with different correlation coefficient values across the country.

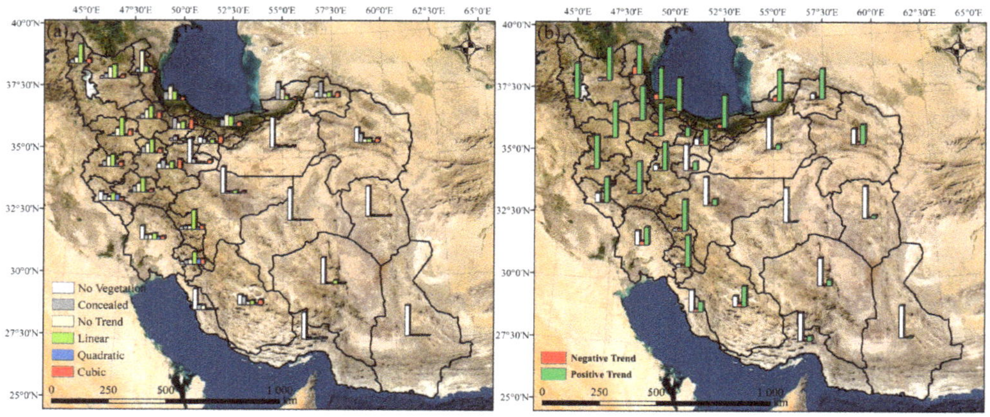

Figure 11. The proportional amount (divided by the area of each province) of (**a**) vegetation trend types and (**b**) trend directions at the province scale.

In this study, an automatic workflow was implemented for linear and non-linear trend analysis using the MODIS NDVI dataset. To the best of the authors' knowledge, this is the first long-term (two decades) time-series analysis of vegetation trends performed over Iran, which also considered both linear and non-linear trend patterns. The importance of considering non-linear trends is that vegetation may respond non-linearly to different disturbances [42–44]. In fact, external disturbances, such as fire, clear-cut, insect outbreaks, and land cover conversion, as well as natural processes and ecological shifts, could introduce non-linear alterations. Therefore, considering non-linear trend types could manifest more details and enable further investigations of vegetation dynamics in the surrounding environment. Comparing the obtained results with another previous study that employed linear trend analysis using the Theil-Sen algorithm [41], the results were relatively in agreement. Although the study periods, the implemented algorithms for trend detection, and the further assumptions differed, which could affect the results, the general depictions were in accordance. For instance, both studies (i.e., this study and [41]) observed a greening process (NDVI gain) over the western parts of the country, especially along the Zagros Mountains. Moreover, our results agreed with those reported by Kazemzadeh et al. [10], which performed a linear/non-linear trend analysis over the Taleghan watershed, Alborz, Iran. In particular, the dominant positive trends were observed in both studies.

The trend analysis results demonstrated a greening process in Iran (see Figure 6). Two other data sources were also employed to investigate the greening process in Iran. These datasets were independent of the trend analysis and thus, can affirm the greening process from other perspectives. The first dataset included two MODIS land cover maps at 500 m spatial resolution for two years, near the start (2001) and the end (2020) of the study period [59]. It is worth noting that the land cover map of 2001 was employed due to the unavailability of the 2000 land cover map. In this regard, original maps (with 17 classes) were initially reclassified into two broad classes: vegetation and non-vegetation, and then a change map was generated through image differencing. The results indicated that some regions were subjected to vegetation loss, while vegetation gain was observed in most areas. Considering the total transitions, nearly 36,240 km^2 of vegetation gain was recorded in Iran. It should be pointed out that this result suggested the greening process by having more vegetation pixels. Furthermore, the annual MODIS Net Primary Productivity (NPP) data at 500 m spatial resolution between 2001 and 2020 were used as the second independent data source. This is due to the fact that higher vegetation would lead to higher

NPP, and thus NPP changes can be employed to explore the greening process [84]. The findings of comparing the NPP data also confirmed the greening process in Iran, with an NPP increase of about $1.5 \times 10^5 \frac{Kg*C}{m^2}$ for the entire country of Iran. However, the spatial patterns depicted that some regions experienced NPP decrease as expected (i.e., negative direction in NDVI) but pixels with NPP increase were dominant. Therefore, the greening process in Iran was also confirmed with two other datasets that were not used for trend analysis. These could imply the robustness of the implemented workflow for vegetation trend analysis.

Currently, NDVI datasets can be obtained from AVHRR, Landsat−8, Himawari−8, and Sentinel−2. However, the utility of these data is limited due to their spatial resolutions or the unavailability of long-term NDVI for such study. For instance, AVHRR can provide long-term NDVI with a very coarse spatial resolution (~9 km). On the other hand, Landsat−8 could provide long-term NDVI data back to 1984, while its medium spatial resolution (30 m) makes it challenging for country-wide studies. This is due to the computational burden of employing numerous pixels for trend analysis, and thus it has been used mostly for local studies [10,82]. The other sources, e.g., Sentinel−2 and Himawari−8, lack long-term NDVI data since they were launched after 2014. Additionally, Landsat−8 and Sentinel−2 could not provide daily observations, and the existence of cloud cover might lead to a large NDVI data gap [85,86]. These points make MODIS NDVI an appealing choice for long-term and large-scale studies. Therefore, in this study, MODIS NDVI data at 250 m spatial resolution were employed for vegetation trend analysis throughout Iran. The results provided an excellent overview of vegetation dynamics in different parts of Iran. Additionally, it assisted in determining hotspot locations (i.e., high vegetation dynamics) for further studies. Subsequently, the identified hotspot zones could be studied with higher resolution satellite data, such as Landsat−8 and Sentinel−2. For instance, higher resolution NDVI data could be used to monitor vegetation growth in smaller-scale areas, providing detailed information [87]. Additionally, integrating NDVI datasets from different sources could alleviate the drawback of each, leading to a more appropriate NDVI dataset for the desired application [88]. In particular, multi-sensor NDVI fusion enables the production of spatially and temporally consistent NDVI datasets for vegetation cover monitoring [88,89]. This is because higher resolution satellite imagery could provide further details about the occurred changes. Meanwhile, further studies are required to investigate the exact time of abrupt changes in hotspot areas. This could help identify the associate drivers of abrupt changes, classifying them as natural or anthropogenic forcings [90–92].

This study employed MODIS land cover maps to investigate the relationship between vegetation trend types, land cover, and land cover transitions between 2001 and 2020. Although these results could be helpful in interpreting the vegetation cover changes for further studies, they included several uncertainties that should be considered. First, the spatial mismatch between original MODIS land cover maps (500 m) and trend analysis results (250 m) could be a source of uncertainty. This is due to the fact that spectral values of pixels with 250 m and 500 m spatial resolution in locations without homogenous land covers may differ, and they might be recognized as two different land covers. Second, the inherent misclassification error of land cover maps would propagate into the final results when producing change maps, which could also be considered another source of error when interpreting the results [24].

In this study, only precipitation data from synoptic stations were employed to study its relationship with vegetation trend directions. Despite their associate errors, these in-situ observations are the most accurate precipitation data due to their intuitive measurements. However, they are point observations, and their spatial inconsistency with satellite data (pixels) may introduce some uncertainties in the comparison procedure. Therefore, it is suggested to use long-term satellite data with higher-spatial resolutions for specified locations to minimize spatial inconsistencies. Although precipitation has been recognized as the main driving force of vegetation dynamics, which was also observed for 60% of cases in this study, and the correlation coefficient was 0.54, other climatic factors such as temperature [79] and solar radiation [80] affect the gradual vegetation dynamics. Therefore, utilizing different climatic parameters, along with topographical variables, could result in obtaining a more robust conclusion about the driving forces of vegetation dynamics in Iran. Consideration of other climatic parameters was beyond the scope of this study and shall be pursued in future studies.

6. Conclusions

In this study, a time series of MODIS NDVI data collected between 2000 and 2020 was employed to study the annual vegetation trends on a national scale in Iran. The PolyTrend, an automated trend analysis algorithm that considers both linear and non-linear trend patterns, was implemented. A visual assessment using candidate samples suggested the applicability of the PolyTrend for trend type identification. The results indicated a higher rate of greening in Iran with the dominance of the linear, followed by the cubic as the statistically significant trends. Further information illustrated some locations with more intensified vegetation change (positive or negative) that could be nominated regions in future studies. It is believed that the results could support achieving Sustainable Development Goals (SDGs) by serving as an initial stage study for establishing conservation and restoration practices. The correlation coefficient between precipitation and NDVI data was 0.54 considering all corresponding observations across the country. Furthermore, the vegetation and precipitation dynamics comparison suggested their mutual directions in 60% of cases with more statistically significant changes. However, some observed discrepancies suggested the consideration of other climatic and topographical variables to study the driving forces of vegetation cover more profoundly. Land cover types were also used to study two-way and three-way relationships between vegetation trend–land cover, and vegetation–precipitation–land cover. Besides, an attempt was made to investigate the relationship between trend types and land cover transitions. However, the observed inconsistencies hamper the driving of a robust relationship between these data sources. Accordingly, more accurate data, especially land cover maps, with more spatially consistent resolutions should be considered.

Author Contributions: Conceptualization, A.G., A.M. and S.J.; methodology, A.G.; software, A.G. and S.J.; validation, A.G.; formal analysis, A.G.; investigation, A.G. and S.J.; resources, A.G.; data curation, A.G.; writing—original draft preparation, A.G.; writing—review and editing, A.G., A.M. and S.J.; visualization, A.G.; supervision, A.M. and S.J.; project administration, A.G.; funding acquisition, S.J. All authors have read and agreed to the published version of the manuscript.

Funding: This research received no external funding.

Data Availability Statement: The NDVI and land cover datasets supporting the reported results of this paper are publicly available from the Google Earth Engine data catalog.

Acknowledgments: The authors would like to thank the Islamic Republic of Iran Meteorological Organization (IRIMO) for providing in-situ precipitation observations from synoptic stations. During this study, Arsalan Ghorbanian was funded by Erasmus + ICM programme for a 5-month stay at Lund University, Sweden.

Conflicts of Interest: The authors declare no conflict of interest.

References

1. Zhang, Y.; Song, C.; Band, L.E.; Sun, G.; Li, J. Reanalysis of Global Terrestrial Vegetation Trends from MODIS Products: Browning or Greening? *Remote Sens. Environ.* **2017**, *191*, 145–155. [CrossRef]
2. Ruan, Z.; Kuang, Y.; He, Y.; Zhen, W.; Ding, S. Detecting Vegetation Change in the Pearl River Delta Region Based on Time Series Segmentation and Residual Trend Analysis (TSS-RESTREND) and MODIS NDVI. *Remote Sens.* **2020**, *12*, 4049. [CrossRef]
3. Pakeman, R.J.; Fielding, D.A.; Everts, L.; Littlewood, N.A. Long-Term Impacts of Changed Grazing Regimes on the Vegetation of Heterogeneous Upland Grasslands. *J. Appl. Ecol.* **2019**, *56*, 1794–1805. [CrossRef]
4. Pan, N.; Feng, X.; Fu, B.; Wang, S.; Ji, F.; Pan, S. Increasing Global Vegetation Browning Hidden in Overall Vegetation Greening: Insights from Time-Varying Trends. *Remote Sens. Environ.* **2018**, *214*, 59–72. [CrossRef]

5. Chen, C.; Park, T.; Wang, X.; Piao, S.; Xu, B.; Chaturvedi, R.K.; Fuchs, R.; Brovkin, V.; Ciais, P.; Fensholt, R.; et al. China and India Lead in Greening of the World through Land-Use Management. *Nat. Sustain.* **2019**, *2*, 122–129. [CrossRef] [PubMed]
6. Chu, H.; Venevsky, S.; Wu, C.; Wang, M. NDVI-Based Vegetation Dynamics and Its Response to Climate Changes at Amur-Heilongjiang River Basin from 1982 to 2015. *Sci. Total Environ.* **2019**, *650*, 2051–2062. [CrossRef] [PubMed]
7. Yang, Y.; Wang, S.; Bai, X.; Tan, Q.; Li, Q.; Wu, L.; Tian, S.; Hu, Z.; Li, C.; Deng, Y. Factors Affecting Long-Term Trends in Global NDVI. *Forests* **2019**, *10*, 372. [CrossRef]
8. Sun, Q.; Miao, C.; Duan, Q.; Ashouri, H.; Sorooshian, S.; Hsu, K.-L. A Review of Global Precipitation Data Sets: Data Sources, Estimation, and Intercomparisons. *Rev. Geophys.* **2018**, *56*, 79–107. [CrossRef]
9. Gajewski, K. Impact of Holocene Climate Variability on Arctic Vegetation. *Glob. Planet. Chang.* **2015**, *133*, 272–287. [CrossRef]
10. Kazemzadeh, M.; Noori, Z.; Alipour, H.; Jamali, S.; Seyednasrollah, B. Natural and Anthropogenic Forcings Lead to Contrasting Vegetation Response in Long-Term vs. Short-Term Timeframes. *J. Environ. Manag.* **2021**, *286*, 112249. [CrossRef]
11. Dubovyk, O.; Landmann, T.; Erasmus, B.F.N.; Tewes, A.; Schellberg, J. Monitoring Vegetation Dynamics with Medium Resolution MODIS-EVI Time Series at Sub-Regional Scale in Southern Africa. *Int. J. Appl. Earth Obs. Geoinf.* **2015**, *38*, 175–183. [CrossRef]
12. Kim, J.Y.; Rastogi, G.; Do, Y.; Kim, D.-K.; Muduli, P.R.; Samal, R.N.; Pattnaik, A.K.; Joo, G.-J. Trends in a Satellite-Derived Vegetation Index and Environmental Variables in a Restored Brackish Lagoon. *Glob. Ecol. Conserv.* **2015**, *4*, 614–624. [CrossRef]
13. Wang, J.; Xie, Y.; Wang, X.; Dong, J.; Bie, Q. Detecting Patterns of Vegetation Gradual Changes (2001–2017) in Shiyang River Basin, Based on a Novel Framework. *Remote Sens.* **2019**, *11*, 2475. [CrossRef]
14. Jiang, W.; Yuan, L.; Wang, W.; Cao, R.; Zhang, Y.; Shen, W. Spatio-Temporal Analysis of Vegetation Variation in the Yellow River Basin. *Ecol. Indic.* **2015**, *51*, 117–126. [CrossRef]
15. Huang, S.; Tang, L.; Hupy, J.P.; Wang, Y.; Shao, G. A Commentary Review on the Use of Normalized Difference Vegetation Index (NDVI) in the Era of Popular Remote Sensing. *J. For. Res.* **2021**, *32*, 1–6. [CrossRef]
16. Eckert, S.; Hüsler, F.; Liniger, H.; Hodel, E. Trend Analysis of MODIS NDVI Time Series for Detecting Land Degradation and Regeneration in Mongolia. *J. Arid Environ.* **2015**, *113*, 16–28. [CrossRef]
17. Zhang, Y.; Ye, A. Spatial and Temporal Variations in Vegetation Coverage Observed Using AVHRR GIMMS and Terra MODIS Data in the Mainland of China. *Int. J. Remote Sens.* **2020**, *41*, 4238–4268. [CrossRef]
18. Fensholt, R.; Langanke, T.; Rasmussen, K.; Reenberg, A.; Prince, S.D.; Tucker, C.; Scholes, R.J.; Le, Q.B.; Bondeau, A.; Eastman, R.; et al. Greenness in Semi-Arid Areas across the Globe 1981–2007—An Earth Observing Satellite Based Analysis of Trends and Drivers. *Remote Sens. Environ.* **2012**, *121*, 144–158. [CrossRef]
19. Bruzzone, O.; Easdale, M.H. Rhythm of Change of Trend-Cycles of Vegetation Dynamics as an Early Warning Indicator for Land Management. *Ecol. Indic.* **2021**, *126*, 107663. [CrossRef]
20. Huete, A.R. A Soil-Adjusted Vegetation Index (SAVI). *Remote Sens. Environ.* **1988**, *25*, 295–309. [CrossRef]
21. Xue, J.; Su, B. Significant Remote Sensing Vegetation Indices: A Review of Developments and Applications. *J. Sens.* **2017**, *2017*, 1353691. [CrossRef]
22. Rouse, J.W.; Haas, R.H.; Schell, J.A.; Deering, D.W. Monitoring vegetation system in the great plains with ERTS. In Proceedings of the Third Earth Resources Technology Satellite-1 Symposium, Greenbelt, MD, USA, 1 January 1974; pp. 3010–3017, NASA SP-351.
23. Seydi, S.T.; Amani, M.; Ghorbanian, A. A Dual Attention Convolutional Neural Network for Crop Classification Using Time-Series Sentinel-2 Imagery. *Remote Sens.* **2022**, *14*, 498. [CrossRef]
24. Amani, M.; Mahdavi, S.; Kakooei, M.; Ghorbanian, A.; Brisco, B.; Delancey, E.; Toure, S.; Reyes, E.L. Wetland Change Analysis in Alberta, Canada Using Four Decades of Landsat Imagery. *IEEE J. Sel. Top. Appl. Earth Obs. Remote Sens.* **2021**, *14*, 10314–10335. [CrossRef]
25. Petersen, L.K. Real-Time Prediction of Crop Yields from MODIS Relative Vegetation Health: A Continent-Wide Analysis of Africa. *Remote Sens.* **2018**, *10*, 1726. [CrossRef]
26. Karan, S.K.; Samadder, S.R.; Maiti, S.K. Assessment of the Capability of Remote Sensing and GIS Techniques for Monitoring Reclamation Success in Coal Mine Degraded Lands. *J. Environ. Manag.* **2016**, *182*, 272–283. [CrossRef]
27. Yan, J.; Zhang, G.; Ling, H.; Han, F. Comparison of Time-Integrated NDVI and Annual Maximum NDVI for Assessing Grassland Dynamics. *Ecol. Indic.* **2022**, *136*, 108611. [CrossRef]
28. Zhang, Y.; He, Y.; Li, Y.; Jia, L. Spatiotemporal Variation and Driving Forces of NDVI from 1982 to 2015 in the Qinba Mountains, China. *Environ. Sci. Pollut. Res.* **2022**, *29*, 52277–52288. [CrossRef]
29. Gao, W.; Zheng, C.; Liu, X.; Lu, Y.; Chen, Y.; Wei, Y.; Ma, Y. NDVI-Based Vegetation Dynamics and Their Responses to Climate Change and Human Activities from 1982 to 2020: A Case Study in the Mu Us Sandy Land, China. *Ecol. Indic.* **2022**, *137*, 108745. [CrossRef]
30. Huete, A.; Didan, K.; Miura, T.; Rodriguez, E.P.; Gao, X.; Ferreira, L.G. Overview of the Radiometric and Biophysical Performance of the MODIS Vegetation Indices. *Remote Sens. Environ.* **2002**, *83*, 195–213. [CrossRef]
31. Matsushita, B.; Yang, W.; Chen, J.; Onda, Y.; Qiu, G. Sensitivity of the Enhanced Vegetation Index (EVI) and Normalized Difference Vegetation Index (NDVI) to Topographic Effects: A Case Study in High-Density Cypress Forest. *Sensors* **2007**, *7*, 2636–2651. [CrossRef]
32. Didan, K.; Munoz, A.B.; Solano, R.; Huete, A. *MODIS Vegetation Index User's Guide (MOD13 Series)*; Vegetation Index and Phenology Lab, University of Arizona: Tucson, AZ, USA, 2015.

33. Kumari, N.; Srivastava, A.; Dumka, U.C. A Long-Term Spatiotemporal Analysis of Vegetation Greenness over the Himalayan Region Using Google Earth Engine. *Climate* **2021**, *9*, 109. [CrossRef]
34. Jamali, S.; Seaquist, J.; Eklundh, L.; Ardö, J. Automated Mapping of Vegetation Trends with Polynomials Using NDVI Imagery over the Sahel. *Remote Sens. Environ.* **2014**, *141*, 79–89. [CrossRef]
35. Liu, C.; Huang, H.; Sun, F. A Pixel-Based Vegetation Greenness Trend Analysis over the Russian Tundra with All Available Landsat Data from 1984 to 2018. *Remote Sens.* **2021**, *13*, 4933. [CrossRef]
36. Sen, P.K. Estimates of the Regression Coefficient Based on Kendall's Tau. *J. Am. Stat. Assoc.* **1968**, *63*, 1379–1389. [CrossRef]
37. Emamian, A.; Rashki, A.; Kaskaoutis, D.G.; Gholami, A.; Opp, C.; Middleton, N. Assessing Vegetation Restoration Potential under Different Land Uses and Climatic Classes in Northeast Iran. *Ecol. Indic.* **2021**, *122*, 107325. [CrossRef]
38. Gholamnia, M.; Khandan, R.; Bonafoni, S.; Sadeghi, A. Spatiotemporal Analysis of MODIS NDVI in the Semi-Arid Region of Kurdistan (Iran). *Remote Sens.* **2019**, *11*, 1723. [CrossRef]
39. Verbesselt, J.; Hyndman, R.; Newnham, G.; Culvenor, D. Detecting Trend and Seasonal Changes in Satellite Image Time Series. *Remote Sens. Environ.* **2010**, *114*, 106–115. [CrossRef]
40. Sedighifar, Z.; Motlagh, M.G.; Halimi, M. Investigating Spatiotemporal Relationship between EVI of the MODIS and Climate Variables in Northern Iran. *Int. J. Environ. Sci. Technol.* **2020**, *17*, 733–744. [CrossRef]
41. Eskandari Dameneh, H.; Gholami, H.; Telfer, M.W.; Comino, J.R.; Collins, A.L.; Jansen, J.D. Desertification of Iran in the Early Twenty-First Century: Assessment Using Climate and Vegetation Indices. *Sci. Rep.* **2021**, *11*, 20548. [CrossRef] [PubMed]
42. Lenton, T.M.; Held, H.; Kriegler, E.; Hall, J.W.; Lucht, W.; Rahmstorf, S.; Schellnhuber, H.J. Tipping Elements in the Earth's Climate System. *Proc. Natl. Acad. Sci. USA* **2008**, *105*, 1786–1793. [CrossRef]
43. Lambin, E.F.; Geist, H.J.; Lepers, E. Dynamics of Land-Use and Land-Cover Change in Tropical Regions. *Annu. Rev. Environ. Resour.* **2003**, *28*, 205–241. [CrossRef]
44. Barnosky, A.D.; Hadly, E.A.; Bascompte, J.; Berlow, E.L.; Brown, J.H.; Fortelius, M.; Getz, W.M.; Harte, J.; Hastings, A.; Marquet, P.A.; et al. Approaching a State Shift in Earth's Biosphere. *Nature* **2012**, *486*, 52–58. [CrossRef] [PubMed]
45. Takaku, J.; Tadono, T.; Tsutsui, K. Generation of High Resolution Global Dsm from Alos Prism. *ISPRS Ann. Photogramm. Remote Sens. Spat. Inf. Sci.* **2014**, *2*, 243–248. [CrossRef]
46. Ghorbanian, A.; Kakooei, M.; Amani, M.; Mahdavi, S.; Mohammadzadeh, A.; Hasanlou, M. Improved Land Cover Map of Iran Using Sentinel Imagery within Google Earth Engine and a Novel Automatic Workflow for Land Cover Classification Using Migrated Training Samples. *ISPRS J. Photogramm. Remote Sens.* **2020**, *167*, 276–288. [CrossRef]
47. Mansouri Daneshvar, M.R.; Ebrahimi, M.; Nejadsoleymani, H. An Overview of Climate Change in Iran: Facts and Statistics. *Environ. Syst. Res.* **2019**, *8*, 7. [CrossRef]
48. Fallah, B.; Sodoudi, S.; Russo, E.; Kirchner, I.; Cubasch, U. Towards Modeling the Regional Rainfall Changes over Iran Due to the Climate Forcing of the Past 6000 Years. *Quat. Int.* **2017**, *429*, 119–128. [CrossRef]
49. Justice, C.O.; Townshend, J.R.G.; Vermote, E.F.; Masuoka, E.; Wolfe, R.E.; Saleous, N.; Roy, D.P.; Morisette, J.T. An Overview of MODIS Land Data Processing and Product Status. *Remote Sens. Environ.* **2002**, *83*, 3–15. [CrossRef]
50. Sobrino, J.A.; Julien, Y. Trend Analysis of Global MODIS-Terra Vegetation Indices and Land Surface Temperature between 2000 and 2011. *IEEE J. Sel. Top. Appl. Earth Obs. Remote Sens.* **2013**, *6*, 2139–2145. [CrossRef]
51. Wang, D.; Morton, D.; Masek, J.; Wu, A.; Nagol, J.; Xiong, X.; Levy, R.; Vermote, E.; Wolfe, R. Impact of Sensor Degradation on the MODIS NDVI Time Series. *Remote Sens. Environ.* **2012**, *119*, 55–61. [CrossRef]
52. MOD13Q1 V006. Available online: https://lpdaac.usgs.gov/products/mod13q1v006/ (accessed on 5 April 2022).
53. MODIS Validation Strategy. Available online: https://modis-land.gsfc.nasa.gov/MODLAND_val.html?_ga=2.3712953.804892803.1651590227-365777974.1651413819 (accessed on 5 April 2022).
54. General Accuracy Statement for Vegetation Indices. Available online: https://modis-land.gsfc.nasa.gov/ValStatus.php?ProductID=MOD13&_ga=2.70272921.804892803.1651590227-365777974.1651413819 (accessed on 5 April 2022).
55. Sabziparvar, A.A.; Ghahfarokhi, S.M.M.; Khorasani, H.T. Long-Term Changes of Surface Albedo and Vegetation Indices in North of Iran. *Arab. J. Geosci.* **2020**, *13*, 117. [CrossRef]
56. Ebrahimi-Khusfi, Z.; Mirakbari, M.; Khosroshahi, M. Vegetation Response to Changes in Temperature, Rainfall, and Dust in Arid Environments. *Environ. Monit. Assess.* **2020**, *192*, 691. [CrossRef]
57. Gorelick, N.; Hancher, M.; Dixon, M.; Ilyushchenko, S.; Thau, D.; Moore, R. Google Earth Engine: Planetary-Scale Geospatial Analysis for Everyone. *Remote Sens. Environ.* **2017**, *202*, 18–27. [CrossRef]
58. Amani, M.; Ghorbanian, A.; Ahmadi, S.A.; Kakooei, M.; Moghimi, A.; Mirmazloumi, S.M.; Moghaddam, S.H.A.; Mahdavi, S.; Ghahremanloo, M.; Parsian, S.; et al. Google Earth Engine Cloud Computing Platform for Remote Sensing Big Data Applications: A Comprehensive Review. *IEEE J. Sel. Top. Appl. Earth Obs. Remote Sens.* **2020**, *13*, 5326–5350. [CrossRef]
59. Sulla-Menashe, D.; Gray, J.M.; Abercrombie, S.P.; Friedl, M.A. Hierarchical Mapping of Annual Global Land Cover 2001 to Present: The MODIS Collection 6 Land Cover Product. *Remote Sens. Environ.* **2019**, *222*, 183–194. [CrossRef]
60. Zhao, F.; Wu, Y.; Sivakumar, B.; Long, A.; Qiu, L.; Chen, J.; Wang, L.; Liu, S.; Hu, H. Climatic and Hydrologic Controls on Net Primary Production in a Semiarid Loess Watershed. *J. Hydrol.* **2019**, *568*, 803–815. [CrossRef]
61. Jin, K.; Wang, F.; Zong, Q.; Qin, P.; Liu, C.; Wang, S. Spatiotemporal Differences in Climate Change Impacts on Vegetation Cover in China from 1982 to 2015. *Environ. Sci. Pollut. Res.* **2022**, *29*, 10263–10276. [CrossRef]

62. Jiang, S.; Chen, X.; Smettem, K.; Wang, T. Climate and Land Use Influences on Changing Spatiotemporal Patterns of Mountain Vegetation Cover in Southwest China. *Ecol. Indic.* **2021**, *121*, 107193. [CrossRef]
63. Zheng, K.; Tan, L.; Sun, Y.; Wu, Y.; Duan, Z.; Xu, Y.; Gao, C. Impacts of Climate Change and Anthropogenic Activities on Vegetation Change: Evidence from Typical Areas in China. *Ecol. Indic.* **2021**, *126*, 107648. [CrossRef]
64. Zheng, K.; Wei, J.-Z.; Pei, J.-Y.; Cheng, H.; Zhang, X.-L.; Huang, F.-Q.; Li, F.-M.; Ye, J.-S. Impacts of Climate Change and Human Activities on Grassland Vegetation Variation in the Chinese Loess Plateau. *Sci. Total Environ.* **2019**, *660*, 236–244. [CrossRef] [PubMed]
65. Notaro, M.; Liu, Z.; Gallimore, R.G.; Williams, J.W.; Gutzler, D.S.; Collins, S. Complex Seasonal Cycle of Ecohydrology in the Southwest United States. *J. Geophys. Res. Biogeosci.* **2010**, *115*. [CrossRef]
66. Chen, Z.; Wang, W.; Fu, J. Vegetation Response to Precipitation Anomalies under Different Climatic and Biogeographical Conditions in China. *Sci. Rep.* **2020**, *10*, 830. [CrossRef] [PubMed]
67. Xu, X.; Liu, H.; Jiao, F.; Gong, H.; Lin, Z. Nonlinear Relationship of Greening and Shifts from Greening to Browning in Vegetation with Nature and Human Factors along the Silk Road Economic Belt. *Sci. Total Environ.* **2021**, *766*, 142553. [CrossRef] [PubMed]
68. Holtvoeth, J.; Vogel, H.; Valsecchi, V.; Lindhorst, K.; Schouten, S.; Wagner, B.; Wolff, G.A. Linear and Non-Linear Responses of Vegetation and Soils to Glacial-Interglacial Climate Change in a Mediterranean Refuge. *Sci. Rep.* **2017**, *7*, 8121. [CrossRef] [PubMed]
69. Jeong, S.-J.; Ho, C.-H.; Brown, M.E.; Kug, J.-S.; Piao, S. Browning in Desert Boundaries in Asia in Recent Decades. *J. Geophys. Res. Atmos.* **2011**, *116*. [CrossRef]
70. Yang, Q.; Qin, Z.; Li, W.; Xu, B. Temporal and Spatial Variations of Vegetation Cover in Hulun Buir Grassland of Inner Mongolia, China. *Arid Land Res. Manag.* **2012**, *26*, 328–343. [CrossRef]
71. Eastman, J.R.; Sangermano, F.; Machado, E.A.; Rogan, J.; Anyamba, A. Global Trends in Seasonality of Normalized Difference Vegetation Index (NDVI), 1982–2011. *Remote Sens.* **2013**, *5*, 4799–4818. [CrossRef]
72. Kazemzadeh, M.; Noori, Z.; Jamali, S.; Abdi, A.M. Four Decades of Air Temperature Data over Iran Reveal Linear and Nonlinear Warming. *J. Meteorol. Res.* **2022**, *36*. [CrossRef]
73. Kazemzadeh, M.; Hashemi, H.; Jamali, S.; Uvo, C.B.; Berndtsson, R.; Huffman, G.J. Linear and Nonlinear Trend Analyzes in Global Satellite-Based Precipitation, 1998–2017. *Earth's Future* **2021**, *9*, e2020EF001835. [CrossRef]
74. Jamali, S.; Klingmyr, D.; Tagesson, T. Global-Scale Patterns and Trends in Tropospheric NO2 Concentrations, 2005–2018. *Remote Sens.* **2020**, *12*, 3526. [CrossRef]
75. Sahadevan, A.S.; Joseph, C.; Gopinath, G.; Ramakrishnan, R.; Gupta, P. Monitoring the Rapid Changes in Mangrove Vegetation of Coastal Urban Environment Using Polynomial Trend Analysis of Temporal Satellite Data. *Reg. Stud. Mar. Sci.* **2021**, *46*, 101871. [CrossRef]
76. Zhang, Y.; Liang, S.; Xiao, Z. Observed Vegetation Greening and Its Relationships with Cropland Changes and Climate in China. *Land* **2020**, *9*, 274. [CrossRef]
77. Ghebrezgabher, M.G.; Yang, T.; Yang, X.; Sereke, T.E. Assessment of NDVI Variations in Responses to Climate Change in the Horn of Africa. *Egypt. J. Remote Sens. Sp. Sci.* **2020**, *23*, 249–261. [CrossRef]
78. Wingate, V.R.; Phinn, S.R.; Kuhn, N. Mapping Precipitation-Corrected NDVI Trends across Namibia. *Sci. Total Environ.* **2019**, *684*, 96–112. [CrossRef]
79. Braswell, B.H.; Schimel, D.S.; Linder, E.; Moore Iii, B. The Response of Global Terrestrial Ecosystems to Interannual Temperature Variability. *Science* **1997**, *278*, 870–873. [CrossRef]
80. Kong, D.; Zhang, Q.; Singh, V.P.; Shi, P. Seasonal Vegetation Response to Climate Change in the Northern Hemisphere (1982–2013). *Glob. Planet. Chang.* **2017**, *148*, 1–8. [CrossRef]
81. Easdale, M.H.; Bruzzone, O.; Mapfumo, P.; Tittonell, P. Phases or Regimes? R Evisiting NDVI Trends as Proxies for Land Degradation. *Land Degrad. Dev.* **2018**, *29*, 433–445. [CrossRef]
82. Zhang, X.; Xiao, X.; Wang, X.; Xu, X.; Chen, B.; Wang, J.; Ma, J.; Zhao, B.; Li, B. Quantifying Expansion and Removal of Spartina Alterniflora on Chongming Island, China, Using Time Series Landsat Images during 1995–2018. *Remote Sens. Environ.* **2020**, *247*, 111916. [CrossRef]
83. Jiang, L.; Liu, Y.; Wu, S.; Yang, C. Analyzing Ecological Environment Change and Associated Driving Factors in China Based on NDVI Time Series Data. *Ecol. Indic.* **2021**, *129*, 107933. [CrossRef]
84. Yuan, Z.; Wang, Y.; Xu, J.; Wu, Z. Effects of Climatic Factors on the Net Primary Productivity in the Source Region of Yangtze River, China. *Sci. Rep.* **2021**, *11*, 1376. [CrossRef] [PubMed]
85. Chen, Y.; Cao, R.; Chen, J.; Liu, L.; Matsushita, B. A Practical Approach to Reconstruct High-Quality Landsat NDVI Time-Series Data by Gap Filling and the Savitzky–Golay Filter. *ISPRS J. Photogramm. Remote Sens.* **2021**, *180*, 174–190. [CrossRef]
86. Dhillon, M.S.; Dahms, T.; Kuebert-Flock, C.; Borg, E.; Conrad, C.; Ullmann, T. Modelling Crop Biomass from Synthetic Remote Sensing Time Series: Example for the DEMMIN Test Site, Germany. *Remote Sens.* **2020**, *12*, 1819. [CrossRef]
87. Khare, S.; Latifi, H.; Khare, S. Vegetation Growth Analysis of Unesco World Heritage Hyrcanian Forests Using Multi-Sensor Optical Remote Sensing Data. *Remote Sens.* **2021**, *13*, 3965. [CrossRef]
88. Dhillon, M.S.; Dahms, T.; Kübert-Flock, C.; Steffan-Dewenter, I.; Zhang, J.; Ullmann, T. Spatiotemporal Fusion Modelling Using STARFM: Examples of Landsat 8 and Sentinel-2 NDVI in Bavaria. *Remote Sens.* **2022**, *14*, 677. [CrossRef]

89. Zhou, Q.; Rover, J.; Brown, J.; Worstell, B.; Howard, D.; Wu, Z.; Gallant, A.L.; Rundquist, B.; Burke, M. Monitoring Landscape Dynamics in Central Us Grasslands with Harmonized Landsat-8 and Sentinel-2 Time Series Data. *Remote Sens.* **2019**, *11*, 328. [CrossRef]
90. Wang, Y.; Zhang, Z.; Chen, X. Quantifying Influences of Natural and Anthropogenic Factors on Vegetation Changes Based on Geodetector: A Case Study in the Poyang Lake Basin, China. *Remote Sens.* **2021**, *13*, 5081. [CrossRef]
91. Xu, X.; Liu, H.; Lin, Z.; Jiao, F.; Gong, H. Relationship of Abrupt Vegetation Change to Climate Change and Ecological Engineering with Multi-Timescale Analysis in the Karst Region, Southwest China. *Remote Sens.* **2019**, *11*, 1564. [CrossRef]
92. Guo, E.; Wang, Y.; Wang, C.; Sun, Z.; Bao, Y.; Mandula, N.; Jirigala, B.; Bao, Y.; Li, H. NDVI Indicates Long-Term Dynamics of Vegetation and Its Driving Forces from Climatic and Anthropogenic Factors in Mongolian Plateau. *Remote Sens.* **2021**, *13*, 688. [CrossRef]

remote sensing

MDPI

Article

Spatial-Temporal Characteristics and Driving Forces of Aboveground Biomass in Desert Steppes of Inner Mongolia, China in the Past 20 Years

Nitu Wu [1], Guixiang Liu [2], Deji Wuyun [3], Bole Yi [4], Wala Du [2] and Guodong Han [1,*]

1 Key Laboratory of Grassland Resources of the Ministry of Education, College of Grassland, Resources and Environment, Inner Mongolia Agricultural University, Hohhot 010011, China
2 Institute of Grassland Research, Chinese Academy of Agricultural Sciences, Hohhot 010010, China
3 Institute of Agricultural Resources and Regional Planning, Chinese Academy of Agricultural Sciences, Beijing 100081, China
4 College of Forestry, Inner Mongolia Agricultural University, Hohhot 010011, China
* Correspondence: hanguodong@imau.edu.cn

Abstract: The desert steppe serves as a transitional zone between grasslands and deserts, and long-term monitoring of aboveground biomass (AGB) in the desert steppe is essential for understanding grassland changes. While AGB observation techniques based on multisource remote-sensing data and machine-learning algorithms have been widely applied, research on monitoring methods specifically for the desert steppe remains limited. In this study, we focused on the desert steppe of Inner Mongolia, China, as the study area and used field sampling data, MODIS data, MODIS-based vegetation indices (VI), and environmental factors (topography, climate, and soil) to compare the performance of four commonly used machine-learning algorithms: multiple linear regression (MLR), partial least-squares regression (PLS), random forest (RF), and support vector machine (SVM) in AGB estimation. Based on the optimal model, the spatial–temporal characteristics of AGB from 2000 to 2020 were calculated, and the driving forces of climate change and human activities on AGB changes were quantitatively analyzed using the random forest algorithm. The results are as follows: (1) RF demonstrated outstanding performance in terms of prediction accuracy and model robustness, making it suitable for AGB estimation in the desert steppe of Inner Mongolia; (2) VI contributed the most to the model, and no significant difference was found between soil-adjusted VIs and traditional VIs. Elevation, slope, precipitation, and temperature all had positive effects on the model; (3) from 2000 to 2020, the multiyear average AGB in the study area was 58.34 g/m^2, exhibiting a gradually increasing distribution pattern from the inner region to the outer region (from north to south); (4) from 2000 to 2020, the proportions of grassland with AGB slightly and significantly increasing trend in the study area were 87.08% and 5.13%, respectively, while the proportions of grassland with AGB slightly and significantly decreasing trend were 7.76% and 0.05%, respectively; and (5) over the past 20 years, climate change, particularly precipitation, has been the primary driving force behind AGB changes of the study area. This research holds reference value for improving desert steppe monitoring capabilities and the rational planning of grassland resources.

Keywords: desert steppe; aboveground biomass; remote sensing; machine learning; random forest

1. Introduction

The Desert steppe in Inner Mongolia is an important component of the temperate grasslands in Central Asia and plays a crucial role in regulating the regional climate, preserving soil and water resources, and providing habitats for numerous species [1–3]. However, climate change and inappropriate land use have put these areas under severe pressure [4]. Aboveground biomass (AGB) is a valuable indicator for assessing terrestrial ecosystem health and understanding its spatial–temporal characteristics and underlying factors is critical for effectively conserving and managing the desert steppe ecosystem [5,6]. Field observation and remote sensing are the two main methods for grassland AGB estimation. Compared with traditional field-based methods, remote-sensing-based techniques have emerged as a valuable tool for efficiently and accurately tracking AGB changes over large areas [5,7]. In recent decades, a multitude of remote-sensing-based AGB observation technologies has been developed and found to be suitable for vegetation observation problems at various scales [7–9]. However, constructing a fitting model that accurately captures changes in AGB depends heavily on the selection of both feature variables and appropriate modeling algorithms.

Feature selection is a critical step in remote-sensing-based AGB monitoring [10]. Currently, the main variables used to retrieve vegetation AGB are derived from spectral reflectance and vegetation indices (VI) [11,12]. As VIs are calculated based on the reflectance of multiple spectral bands, they exhibit high sensitivity to surface features, including vegetation, making them the preferred variables for biomass observation [13]. In addition to traditional VIs, such as the normalized difference vegetation index (NDVI) and the enhanced vegetation index (EVI), modified indices, such as the soil-adjusted vegetation index (SAVI), the modified soil-adjusted vegetation index (MSAVI), the transformed soil-adjusted vegetation index (TSAVI), the optimized soil-adjusted vegetation index (OSAVI), and the adjusted transformed soil-adjusted vegetation index (ATSAVI), have been proposed and widely used in vegetation observation [14–20]. However, studies showed that some soil-adjusted VIs may not perform better than traditional indices in sparsely vegetated dry grasslands [21,22]. In addition, relying solely on VI in biomass monitoring models may result in regional differences in form and accuracy. To address this issue, some researchers integrated environmental variables such as geography, topography, and climate into single-factor models and developed multivariate regression models that significantly improve model-fitting accuracy [23–26]. Nevertheless, further research is necessary to determine which variables are suitable for estimating AGB in the desert steppe.

Modeling algorithm selection is a crucial aspect of the biomass estimation process [27]. Currently, there are numerous algorithms available for retrieving AGB, which can be broadly categorized into parametric and nonparametric algorithms based on whether they require assumptions about the model structure. Studies used VIs as the observed variable and constructed single-variable models, such as linear, exponential, and power to fit vegetation biomass [28–30]. However, the accuracy of these simple models is prone to be affected by soil and spectral saturation effects, particularly in heterogeneous regions [31,32]. To improve the generalization ability and accuracy, studies emphasized the establishment of multiple linear regression (MLR) models by increasing the feature variables [33–35]. In recent years, MLR models incorporating dimensionality reduction techniques such as the partial least squares regression (PLS) have also gained popularity [36,37]. However, as vegetation growth is a complex process, parametric models often lack the ability to fit this complex relationship, resulting in a lack of ecological significance in the entire operation process. In contrast, nonparametric models do not require assumptions about the form of the objective function and can thus fit complex relationships, making them the most popular machine-learning algorithms for estimating AGB [5,13]. Some studies showed that these algorithms, such as the random forest (RF), and support vector machine (SVM) perform better than traditional parametric models in estimating grassland AGB [38,39]. However, no indepth comparative study of AGB in Inner Mongolia's desert steppe has been conducted. The desert steppe in Inner Mongolia, China, is characterized by sparse vegetation and complex environmental conditions, leading to uncertainty in the performance of these algorithms in fitting vegetation AGB. Therefore, further research is needed to evaluate the performance of these models in estimating AGB in this region.

Human activities and climate change are widely considered to be the main factors affecting grassland health, but quantifying their dominance is challenging [40–43]. Some studies used linear regression coefficients and variance ratios to assess the impact of external drivers on grasslands [43–45]. However, these methods are prone to overlooking ecological processes and cannot reflect complex ecological relationships [46]. Other studies attempted to establish quantitative analysis conditions from an ecological perspective by using ecological process models to explain driving factors [47–49]. However, this approach makes it difficult to provide data references for some specific driving forces, resulting in a lack of indepth analysis. Furthermore, some recent studies attempted to use nonparametric models such as RF [50] and boosted regression trees [51] to analyze the impacts of external driving forces on land changes. Similarly, we believe that this approach is also applicable to the topic of our research. The relationship between vegetation and external factors is complex and diverse, making nonparametric algorithms more advantageous in simulating this complex relationship. In addition, selecting appropriate algorithms and building grassland AGB inversion models that are suitable for the characteristics of the study area is key to accurately reflecting the spatial–temporal characteristics and driving forces of desert steppe in Inner Mongolia.

To fill the knowledge gaps mentioned above, this study focused on the desert steppe of Inner Mongolia, China, and carried out the following studies: (1) compared the performance of four commonly used machine-learning algorithms (MLR, PLS, RF, and SVM) on grassland AGB estimation, and generated a suitable modeling method; (2) extracted the spatial–temporal feature of grassland AGB in past 20 years (2000–2020); and (3) analyzed the driving forces of grassland AGB change in past 20 years.

2. Materials and Methods

2.1. Study Area

This research focuses on five regions (counties) in central Inner Mongolia, namely Sunite East Banner (SE), Sunite West Banner (SW), Erenhot City (ER), Siziwang Banner (SZW), and Damao Banner (DM) (Figure 1). The region is predominantly characterized by desert grasslands, covering an approximate area of 56,200 km^2, and represents the main body of the desert steppe in Inner Mongolia. The study area spans from 37°N to 46°N and 105°E to 115°E, with an elevation range of 800–2100 m and an average elevation of 1100 m. The climate in this area is characterized by a continental arid climate with an annual precipitation of 100–200 mm and an average annual temperature of 2–5 °C [2]. The grassland vegetation in this region is typified by a low diversity of species and a simple community structure. The drought-tolerant bunchgrasses are the dominant grass species, often accompanied by numerous drought-tolerant shrubs. The common herbaceous plants include *Stipa klemenzii*, *Stipa breviflora*, *Artemisia frigida*, and *Cleistogenes songorica* [2,52].

2.2. Data Used

2.2.1. Field Measurement Data

During the peak growing season (July and August) from 2018 to 2020, 97 field plots from various terrains and vegetation coverage were collected (Figure 1). The plot size was 500 m × 500 m, with each corner aligned as closely as possible with the corresponding MODIS pixel. Within each plot, 3–5 samples of 1 m × 1 m were selected along the diagonal, and the average value of these sample points was used as the AGB for that plot (Figure 1). The distance between adjacent plots was no less than 1 km, and the recorded information included fresh biomass, latitude, longitude, and elevation.

2.2.2. Remote-Sensing Data

- Multispectral Bands and Vegetation Indices

The MOD09GA product, which provides surface reflectance data, was acquired from the Google Earth Engine (GEE) platform. The time period covers from July 2000 to September 2020. The MOD09GA product records the surface spectral reflectance of MODIS bands 1 through 7 at a spatial resolution of 500 m, corrected for atmospheric conditions such as gases, aerosols, and Rayleigh scattering [53]. Specifically, the bands included in the MOD09GA are red (band 1), near infrared (band 2), blue (band 3), green (band 4), shortwave infrared 1 (band 6), and shortwave infrared 2 (band 7). In this study, in addition to selecting the above-mentioned six bands, seven MODIS-based VIs were considered. These VIs include two traditional VIs (NDVI, EVI) and five soil-adjusted vegetation indices (SAVI, MSAVI, TSAVI, OSAVI, and ATSAVI). The specific calculation formulas for these indices are given in Table 1.

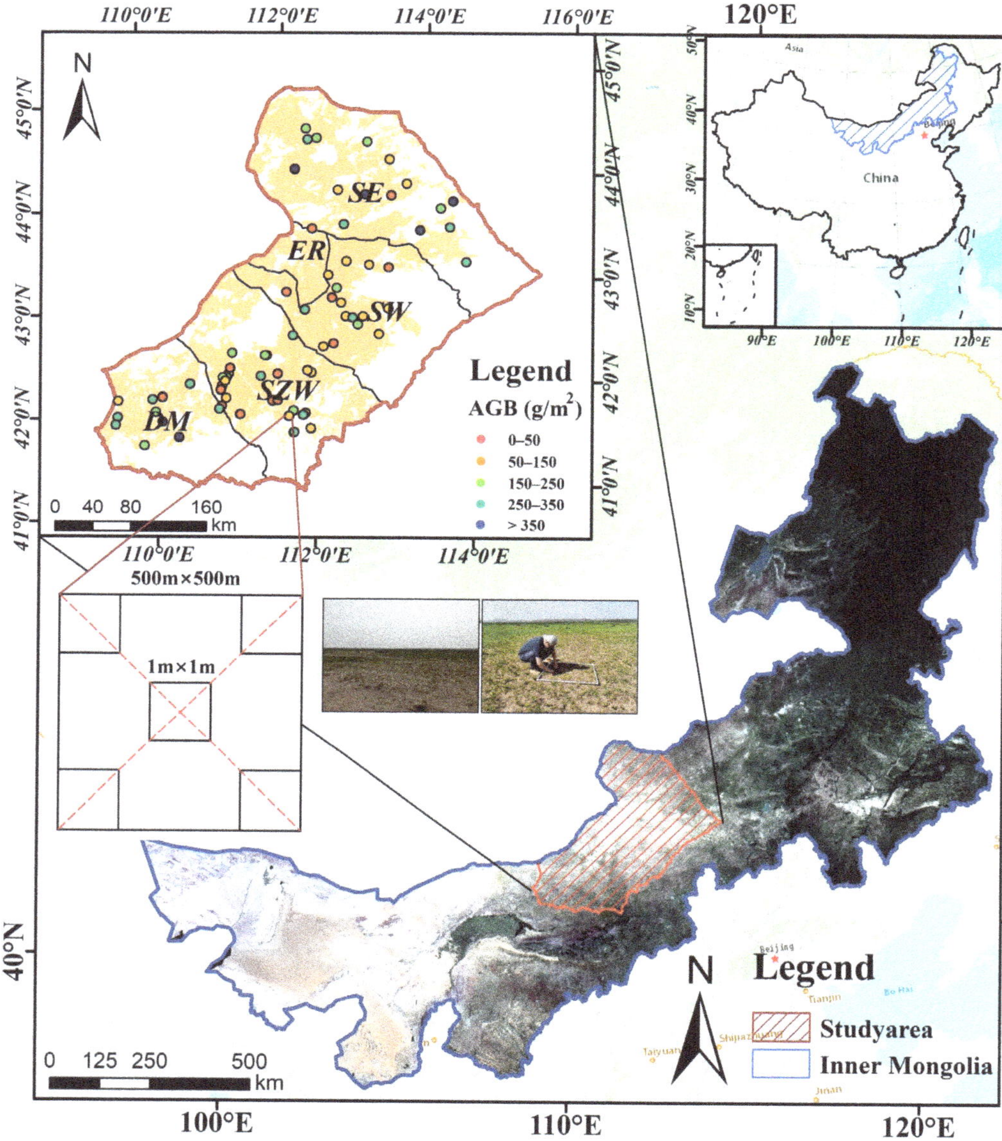

Figure 1. Study area and field-samples distribution.

As some of these indices require correction parameters, 30 barren sample sites were selected in the study area to establish a soil line using the reflectance of soil in the near-infrared and red spectral bands, and obtain the correction parameters a and b.

Table 1. Selected vegetation indices.

Vegetation Indices Type	Description
Traditional vegetation indices	$NDVI = \frac{\rho_{NIR} - \rho_{RED}}{\rho_{NIR} + \rho_{RED}}$ $EVI = 2.5 \times \frac{\rho_{NIR} - \rho_{RED}}{\rho_{NIR} + 6 \times \rho_{RED} - 7.5 \times \rho_{BLUE} + 1}$
Soil-adjusted vegetation indices	$SAVI = (1 + L) \times \frac{\rho_{NIR} - \rho_{RED}}{\rho_{NIR} + \rho_{RED} + L}$ $MSAVI = 0.5 \times [(2 \times \rho_{NIR} + 1) - \sqrt{(2 \times \rho_{NIR} + 1)^2 - 8 \times (\rho_{NIR} - \rho_{RED})}]$ $TSAVI = \frac{a \times (\rho_{NIR} - a \times \rho_{RED} - b)}{a \times \rho_{NIR} + \rho_{RED} + a \times b}$ $OSAVI = 1.16 \times \frac{\rho_{NIR} - \rho_{RED}}{\rho_{NIR} + \rho_{RED} + 0.16}$ $ATSAVI = \frac{a \times (\rho_{NIR} - a \times \rho_{RED} - b)}{a \times \rho_{NIR} + \rho_{RED} - a \times b + 0.08 \times (1 + a^2)}$

- Terrain data

The terrain variables, elevation, slope, and aspect were from ASTER GDEM version 2, which is a global DEM generated through a collaboration between NASA and Japan's Ministry of Economy, Trade, and Industry (METI). Despite variations in accuracy depending on terrain and other factors, the ASTER GDEM v2 is generally considered a high-quality DEM with a root mean square error (RMSE) of less than 10 m [54]. In this study, we applied the GEE platform to complete preprocessing operations such as projection transformation, resampling, and clipping.

- Meteorological data

Annual total precipitation data were obtained from the Climate Hazards Group InfraRed Precipitation with Station data (CHIRPS), which is a quasi-global rainfall dataset spanning over 30 years. CHIRPS combines 0.05° resolution satellite imagery with in situ station data to generate gridded rainfall time series suitable for trend analysis and seasonal drought monitoring [55].

Annual mean temperature and annual mean wind data were sourced from the ERA5-Land dataset, the latest global meteorological reanalysis dataset released by the European Centre for Medium-Range Weather Forecasts (ECMWF). The dataset provides continuous meteorological records since 1950 at a pixel resolution of up to 9 km and encompasses all conventional meteorological variables [56].

Annual total evapotranspiration data were extracted from the MOD16A2 Version 6 product, which is an 8-day composite dataset produced at a 500 m pixel resolution. The algorithm utilized for the MOD16 data product collection is based on the Penman-Monteith equation and incorporates inputs of daily meteorological reanalysis data, along with MODIS remotely sensed data products such as vegetation property dynamics, albedo, and land cover [57]. In this study, we employed the GEE platform to preprocess the aforementioned factors, including projection conversion, resampling, cropping, and unit conversion. The datasets cover the time period from 2000 to 2020.

- Soil data

The soil factors, including soil bulk density, soil pH, clay content, and sand content at 10 cm depth, were obtained from the OpenLandMap soil dataset, which is an environmental data portal that provides global-scale data on land cover, vegetation, soil, climate, terrain, and other related information to the public [58]. The soil dataset in OpenLandMap is established through the integration of global soil profile data points and machine-learning algorithms and provides multiple soil properties data at six standard depths (0, 10, 30, 60, 100, and 200 cm) with a resolution of 250 m [59]. In this study, we utilized the GEE platform to extract each factor and performed preprocessing operations such as projection transformation, resampling, and cropping.

- Population data

The population data was sourced from the WorldPop project, which is an interdisciplinary research group dedicated to improving the evidence base for spatial demographics and utilizing the data for health and development applications [60,61]. The project takes recent population counts from censuses and breaks them down into grid cells of approximately 100 m × 100 m using machine-learning techniques. These techniques exploit the relationships between population densities and different geospatial covariate layers. In this study, we employed the GEE platform to carry out preprocessing tasks. The datasets used encompass the period from 2000 to 2020.

2.2.3. Statistical Data

The statistical data involved in this study include annual livestock quantities, agricultural gross domestic product (GDP), industrial GDP, and consumer GDP for each county from 2000 to 2020. These data were obtained from local statistical yearbooks. To capture the multiyear trends of these statistical variables, we employed the least squares method and assigned the resulting trends to the corresponding administrative boundary polygons.

2.3. Analysis Methods

The technical route of this study can be divided into four parts: data preprocessing, optimal model construction, extraction of AGB spatial–temporal features, and analysis of driving forces (Figure 2). As mentioned in the previous section, the data used in this study mainly consist of field-sample data, remote-sensing data, and statistical data. First, extract the raster values corresponding to the coordinates of the sample plots for the year 2020 to build a sample dataset. Then, split the dataset into training and testing datasets using a 70:30 ratio. Next, the training dataset is used in combination with leave-one-out cross-validation (LOOCV) to train four different models: MLR, PLSR, RF, and SVM. The performance of these models is compared on the testing dataset, and the best model and variable combination are selected. Then, using the selected model, a spatial–temporal dataset of AGB in the desert steppe of the study area is generated for the period from 2000 to 2020. Following this, change detection and trend analysis methods are employed to characterize the changes in desert steppe AGB over the past 20 years. Lastly, the RF algorithm is utilized to quantitatively analyze the potential driving forces behind the AGB changes.

2.3.1. Modeling Algorithms

MLR is the most used parametric statistical algorithm for relating linear relationships between various predictors and independent variables [62]. Stepwise regression is a statistical approach used to build an MLR model by selecting the most significant predictors from a larger set. The algorithm adds or removes variables from the model in a stepwise manner based on predetermined criteria until the optimal subset of variables is achieved. The final model consists of the most significant predictors, and the coefficients represent the strength and direction of the relationship between predictors and the outcome variable [63].

PLS is a statistical method that is similar to principal components regression in that it finds a linear regression model by projecting the predicted variables and the observable variables to a new space rather than finding hyperplanes of maximum variance between the response and independent variables [64]. This algorithm is especially useful when the predictor matrix contains more variables than observations and there is multicollinearity among the variables.

RF is a powerful machine-learning algorithm used for regression analysis that falls under the category of ensemble learning algorithms. The algorithm works by creating multiple decision trees on randomly sampled subsets of the training data and averaging their predictions to obtain the final output [65]. Generally, RF involves two important hyperparameters: the number of decision trees (n_estimators) and the maximum number of features considered when splitting a node in each decision tree (max_features) [65]. The n_estimators refer to the number of decision trees used to construct the random forest, and the max_features refer to the maximum number of features considered when splitting a node in each decision tree.

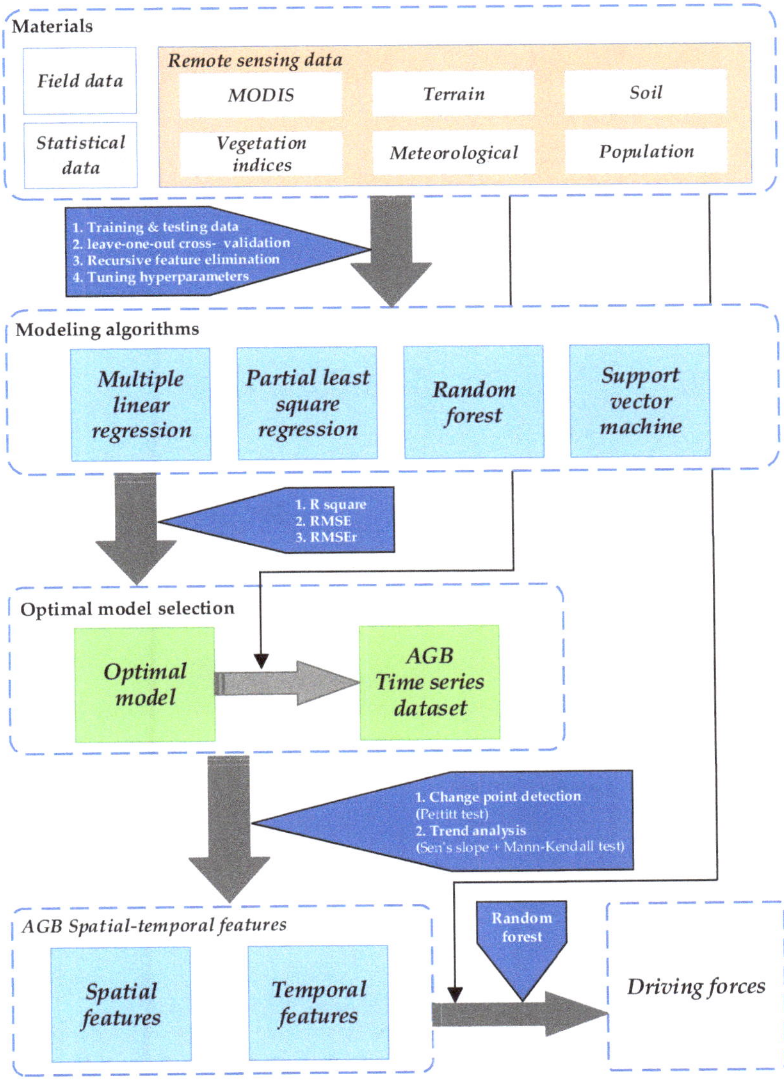

Figure 2. Technical route.

SVM is a popular machine-learning technique for predicting continuous output values. The algorithm aims to find the best hyperplane that minimizes the error between the predicted and true output values, with a penalty for the weight magnitude [66]. It can handle nonlinear data using kernel functions such as the radial basis function (RBF), which maps data into higher dimensional feature spaces. The SVM-RBF algorithm is capable of modeling nonlinear relationships between input and output variables and is robust to noise and outliers. Two hyperparameters need to be adjusted: penalty factor (c) and kernel parameter (gamma). The C parameter controls the trade-off between achieving a low error on the training data and minimizing the weights' mean, while the gamma parameter controls the distance of influence of a single training point.

Feature selection has always been an important step in model training. In this study, the recursive feature elimination (RFE) method was employed to iteratively select feature variables for both RF and SVM. RFE is a feature selection technique that aims to iteratively eliminate features with lower predictive contributions to the model. In this method, the model is repeatedly trained, and less informative features are removed until a subset of the most predictive features is obtained [67].

2.3.2. Model Assessment

In this study, three validation standards, namely R^2, RMSE, and RMSEr, were selected to evaluate model accuracy, with the following specific formulas:

$$R^2 = 1 - \frac{\sum_{i=1}^{n}(y_i - \hat{y}_i)^2}{\sum_{i=1}^{n}(y_i - \bar{y})^2} \tag{1}$$

$$RMSE = \sqrt{\sum_{i=1}^{n}\frac{(\hat{y}_i - y_i)^2}{n}} \tag{2}$$

$$RMSEr = \frac{RMSE}{\bar{y}} \times 100 \tag{3}$$

Here, $\hat{y}_i$ represents the measured AGB value of the i-th sample, y_i represents the predicted AGB value of the i-th sample, $\bar{y}$ represents the average AGB value of the samples, and n represents the total number of samples.

2.3.3. Change Point Detection

The Pettitt test was used in this study to extract the abrupt change point of grassland AGB time-series data. The Pettitt test is a nonparametric test based on Mann–Whitney two independent samples and its hypothesis is that H_0 states that if two samples have no difference, then their central positions will be the same, while H_1 states that if there is a difference, then their central positions will not be the same [68].

The nonparametric statistic is defined as follows:

$$\hat{U} = \max\left|U_k\right| \tag{4}$$

$$U_k = 2\sum_{i=1}^{k} r_i - k(n+1), k = 1, 2, \ldots, n \tag{5}$$

Here, $\hat{U}$ is the test statistic, and its value is equal to the maximum absolute value of the vector U_k; r_i is the order of the series data, and k is the possible change year.

The formula for the p-value of Pettitt's test:

$$p = 2\exp\left(\frac{-6\hat{U}^2}{n^3 + n^2}\right) \tag{6}$$

Here, if p is less than a given significance level β, reject the null hypothesis, namely series data has a change point; On the contrary, if p is greater than a given significance level β, accept the null hypothesis, namely the series data without change point.

2.3.4. Trend Analysis

In this study, the grassland AGB changing trend was analyzed using Sen's slope method. The Mann–Kendall test was employed to test the significance of the sequence trend. The Mann–Kendall test was preferred as it does not make any assumptions about the distribution of the sequence and is robust to outliers [69].

Sen's slope formula is:

$$\beta = Median(\frac{AGB_j - AGB_i}{j - i}) \tag{7}$$

Here, Median(x) represents the median function, and β is the Sen's slope value. A positive value of β indicates an increasing trend of AGB, while a negative value of β indicates a decreasing trend.

The Mann–Kendall test involves testing two hypotheses: H_0 assumes that the sequence data are randomly arranged and there is no significant change trend, while H_1 assumes that the sequence data exhibit an upward or downward trend. The formula for the test statistic S is:

$$S = \sum_{i=1}^{n-1} \sum_{j=i+1}^{n} sgn(AGB_j - AGB_i) \tag{8}$$

$$sgn(AGB_j - AGB_i) = \begin{cases} 1 \; AGB_j - AGB_i > 0 \\ 0 \; AGB_j - AGB_i = 0 \\ -1 \, AGB_j - AGB_i < 0 \end{cases} \tag{9}$$

Here, sgn(x) is the sign function that returns 1, −1, or 0, depending on whether x is positive, negative, or zero, respectively.

The values of Sen's slope (β) and the Mann–Kendall test statistic (p) were used to determine the levels of the AGB change trend, which included significantly increasing, slightly increasing, significantly decreasing, and slightly decreasing (Table 2).

Table 2. Levels of AGB change the trend of grassland.

AGB Change Trend and Test Results	Levels of AGB Change Trend
$\beta > 0, p < 0.05$	significantly increasing
$\beta > 0, p > 0.05$	slightly increasing
$\beta < 0, p < 0.05$	significantly decreasing
$\beta < 0, p > 0.05$	slightly decreasing

2.3.5. Driving Factors Analysis

The RF algorithm was employed in this study to analyze the driving factors behind changes in grassland AGB quantitatively. The following steps were taken in the analysis: (1) nine potential driving forces, from climate and human activity categories, were selected as independent variables to investigate their impact on grassland AGB changes in the study area (Table 3); (2) 300 sample points were randomly selected, and the corresponding independent variable and the AGB change trend (dependent variable) values were extracted for each point; (3) the sample points were divided into training and testing datasets in a 70:30 ratio, and an RF regression model was optimized using the training dataset and a 10-fold cross-validation method; (4) the feature importance was extracted to quantitatively evaluate the driving factors.

RF provides two metrics for measuring the importance of features: mean decrease impurity (MDI) and mean decrease accuracy (MDA). The MDI metric measures the number of times a feature is selected in the decision tree and its contribution to the impurity reduction, which can evaluate the contribution of a feature to classification. On the other hand, the MDA metric calculates the change in model accuracy on a test set by randomly shuffling the values of a feature, where a greater decrease in accuracy indicates a higher feature importance. It should be noted that although both metrics can be used to measure feature importance, they are not always consistent.

Table 3. Selected potential driving forces of AGB changes.

Factor Type	Factor Name	Description	No.	Unit
Climate change	Precipitation trends	Annual precipitation trends from 2000 to 2020	D1	mm
	Temperature trends	Annual temperature trends from 2000 to 2020	D2	°C
	Evapotranspiration trends	Annual evapotranspiration trends from 2000 to 2020	D3	mm
	Wind trends	Annual wind trends from 2000 to 2020	D4	m
Human activity	Population density trends	Annual population density trends from 2000 to 2020	D5	%
	Livestock growth rate	Annual average livestock growth rate	D6	%
	Primary industry GDP growth rate	Averaged annual growth rate of primary industry GDP from 2000 to 2020	D7	%
	Secondary industry GDP growth rate	Averaged annual growth rate of secondary industry GDP from 2000 to 2020	D8	%
	Tertiary industry GDP growth rate	Averaged annual growth rate of tertiary industry GDP from 2000 to 2020	D9	%

3. Results

3.1. The Performance of the Selected Algorithms

Figure 3 presents the relationship between algorithm performances and the number of variables (components). The cross-validation accuracy of these algorithms on the training dataset was unstable, indicating that an excessive number of variables does not improve the performance of the model. MLR retained eight variables for optimal performance after stepwise feature testing, indicating the presence of multicollinearity among the selected features. The optimal performance of PLS was achieved by retaining only one principal component. After evaluating each feature using the RFE method, RF and SVM chose 17 and 22 effective variables, respectively, to achieve the best performance.

Table 4 lists the detailed features selected by the modeling algorithms. MLR selected the feature variables MIR, SWIR, MSAVI, ATSAVI, soil bulk density, clay content, precipitation, and temperature to construct the AGB regression model, expressed as AGB = 35.14 × MIR − 33.69 × SWIR − 200.18 × MSAVI + 229.79 × ATSAVI + 18.44 × soil bulk density − 21.27 × clay content + 22.94 × precipitation − 12.82 × temperature + 73.41. PLS retained all variables since it integrates dimensionality reduction techniques. RF identified the effective variables as blue, green, red, NIR, MIR, SWIR, NDVI, EVI, SAVI, TSAVI, OSAVI, MSAVI, ATSAVI, elevation, slope, precipitation, and temperature, while SVM selected all variables.

Figure 4 displays the model performance of RF and SVM under different hyperparameter combinations at the optimal variable conditions. For the RF, the model achieved high accuracy when the n_estimators was set between 500 and 700, and the best model performance was obtained when the n_estimators and max_features were set to 700 and 4, respectively. As for the SVM, the model achieved high prediction accuracy when gamma was set between 0.02 and 0.05, and the best model performance was achieved when gamma and cost were set to 0.03 and 8, respectively.

Figure 5 compares the prediction accuracy of selected algorithms on the training and testing datasets under effective feature and hyperparameter conditions. The results indicate that nonparametric models outperformed linear models significantly. On the testing dataset, RF achieved the highest prediction accuracy (R^2 = 0.66, RMSE = 39.85, RMSEr = 39.12), followed by SVM (R^2 = 0.60, RMSE = 52.16, RMSEr = 51.21), PLS (R^2 = 0.59, RMSE = 44.47, RMSEr = 43.67), and MLR (R^2 = 0.52, RMSE = 48.85, RMSEr = 47.96). These findings suggest that RF has the best predictive ability for unseen data, which is consistent with the inversion requirements of AGB in Inner Mongolia's desert steppe.

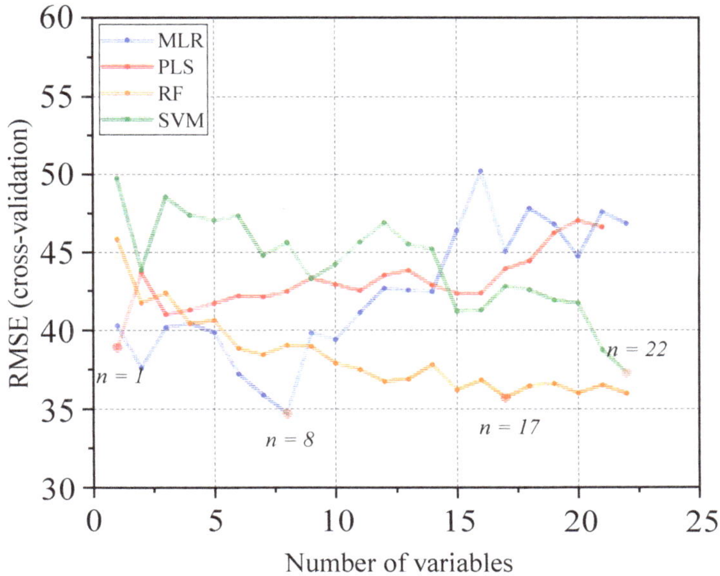

Figure 3. Relationship between the algorithm performances and the number of variables.

Table 4. Selected effective variables for models.

Algorithms	Selected Variables	No. of Variables
MLR	MIR, SWIR, MSAVI, ATSAVI, soil bulk density, clay content, precipitation, temperature	8
PLS	All	22
RF	Blue, green, red, NIR, MIR, SWIR, NDVI, EVI, SAVI, TSAVI, OSAVI, MSAVI, ATSAVI, elevation, slope, precipitation, temperature	17
SVM	All	22

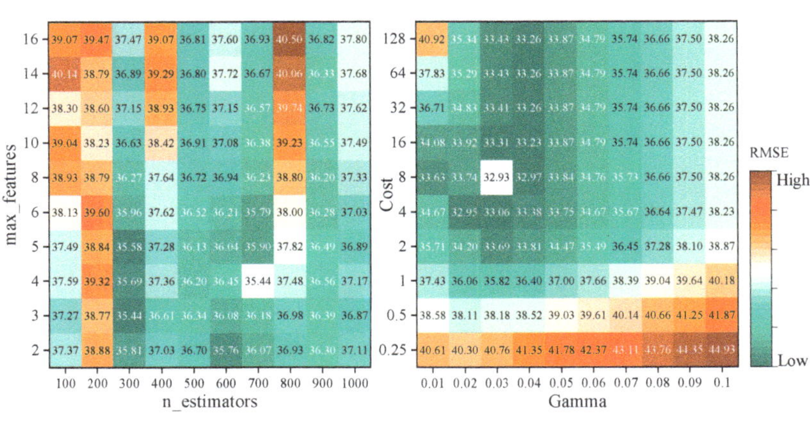

Figure 4. Hyperparameter tuning with grid search for RF and SVM models.

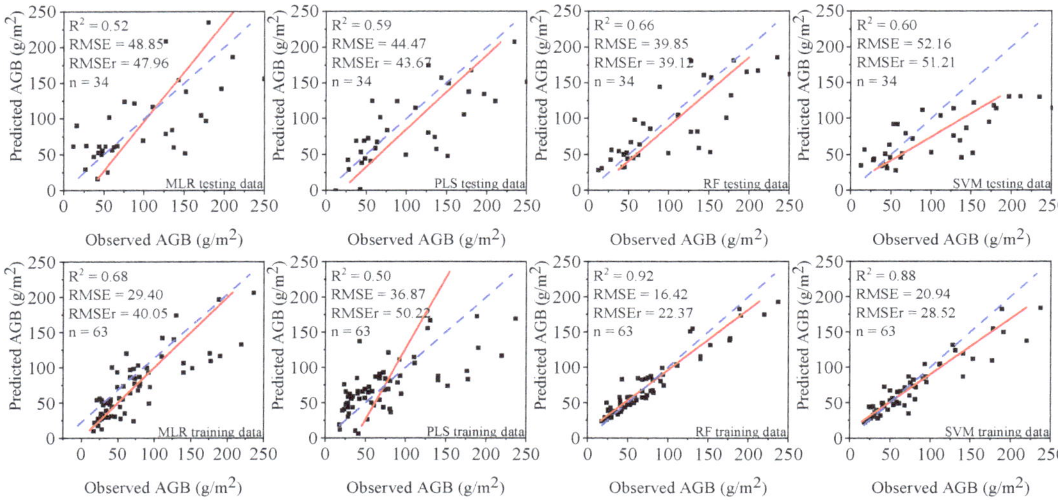

Figure 5. Prediction accuracy of optimal models on training and testing datasets.

3.2. *Spatial–Temporal Feature of Desert Steppe AGB in Inner Mongolia during the Past 20 Years*

As seen in Figure 6, from 2000 to 2020, the vegetation cover of the desert steppe in the study area exhibited a gradually increasing distribution pattern from north to south, with an average AGB of 58.34 g/m^2. Among them, the SZW had the highest average AGB of 68.91 g/m^2, followed by SZ (60.90 g/m^2), DM (56.52 g/m^2), SY (49.82 g/m^2), and ER (35.32 g/m^2). Within the entire study area, the AGB was highest in the southern parts of SZW and DM, exceeding 120 g/m^2, while the vegetation in most parts of ER, the northern part of SZW, and the western part of SE was sparse, with an AGB ranging from 20–50 g/m^2.

Figure 7a,b demonstrates that while AGB changed in most of the study area, only a small number of regions experienced significant changes ($p < 0.05$). This suggests that over the past 20 years, the AGB changes in the study area have been comparatively stable. Figure 7c shows that the areas with significant changes over the past 20 years were mainly distributed in the southern SZW and central DM, and the years with significant changes mainly occurred in 2010, 2011, and 2012.

Figure 8 displays the spatiotemporal variation of AGB in the study area from 2000 to 2020. Overall, AGB in desert steppe increased with an average annual growth rate of 1.33 g/m^2 during the past two decades (Figure 8a,d). According to the significance testing results (Figure 8b), grassland with a significant increase in AGB accounted for 5.13%, primarily found in the central and southern SZW and central DM, while grassland with a slight increase in AGB accounted for 87.08%, found throughout the study area. The proportion of grassland with a significant decrease in AGB was less than 0.5% and the proportion of grassland with a slight decrease was only 7.76%, with the majority concentrated in the eastern SW, central SE, and northeastern ER (Figure 8).

At the county level, SZW had the most significant restoration, with an annual increase in grassland AGB of 1.84 g/m^2, and the proportion of grassland with a significant increase in AGB was 9.95%, while the proportion of grassland with a slight increase in AGB was 87.17%. DM followed, with an annual increase in grassland AGB of 1.44 g/m^2, with 94.21% of grassland showing an increase in AGB. In SE, the annual increase in grassland AGB was 1.30 g/m^2 and the proportion of grassland with an increase in AGB was 92.38%. In ER, the annual increase in grassland AGB was 0.89 g/m^2, with 90.09% of grassland showing an increase in AGB. In SW, the annual increase in grassland AGB was 0.84 g/m^2 and the proportion of grassland with an increase in AGB was 85.57% (Figure 8c,d).

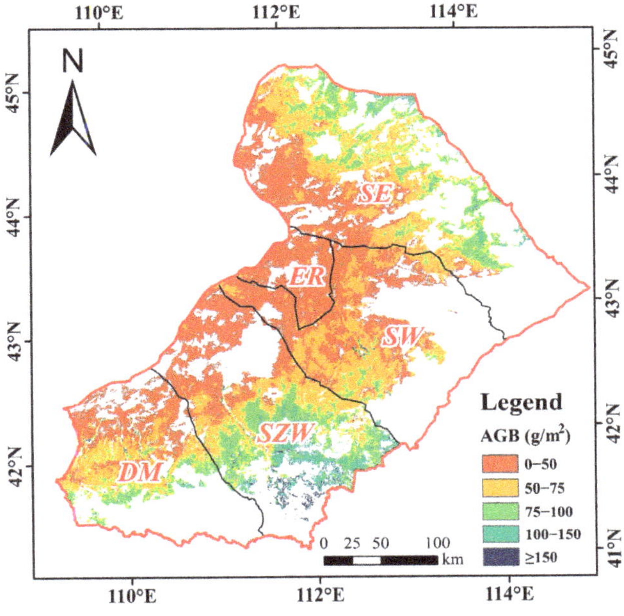

Figure 6. Distribution of AGB in the study area from 2000 to 2020.

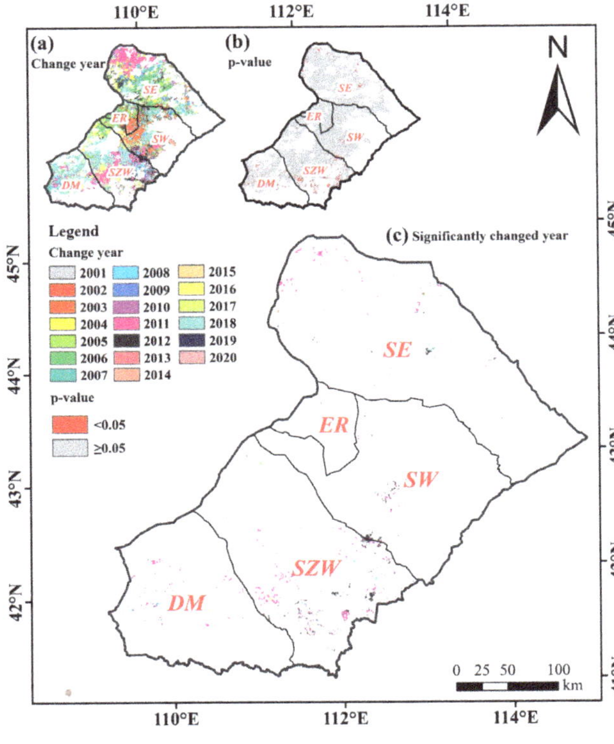

Figure 7. The years with significant changes in AGB in the study area from 2000 to 2020.

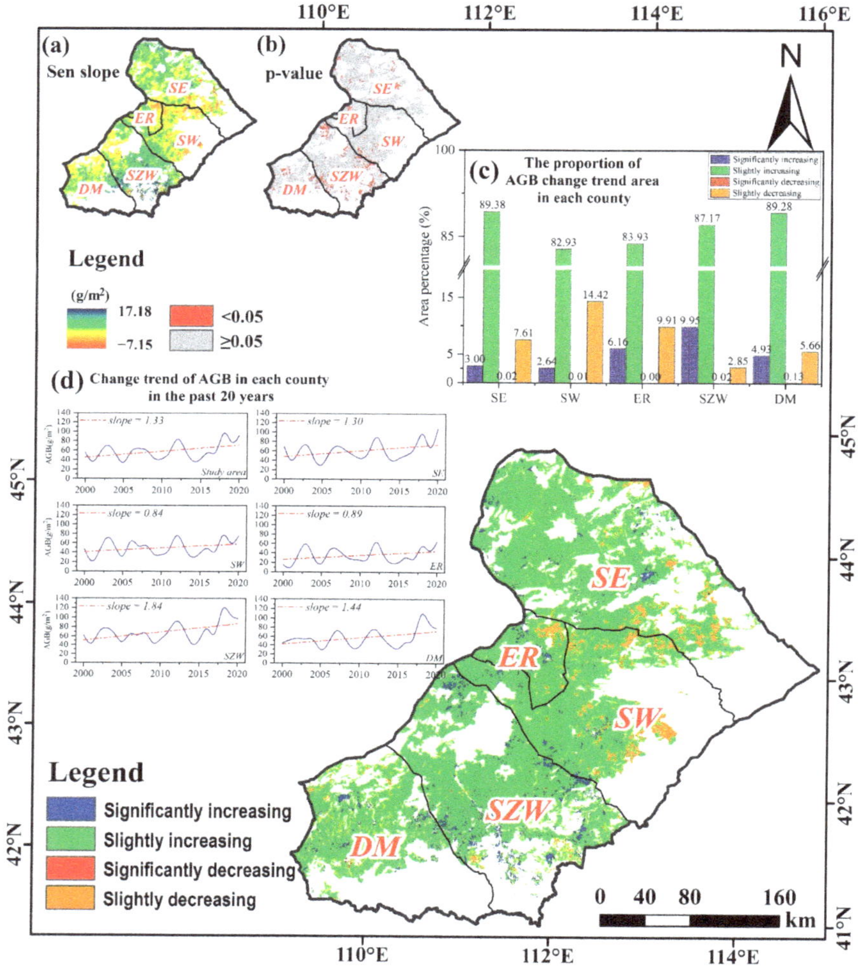

Figure 8. Spatial–temporal variation of desert steppe AGB in the study area from 2000 to 2020.

3.3. Driving Factors for Desert Steppe AGB in Inner Mongolia during the Past 20 Years

Figure 9 displays the drivers behind the desert steppe AGB changes in the study area from 2000 to 2020. Both indicators provide similar rankings. In the MDA ranking, precipitation trends were the most important, followed by evapotranspiration trends > wind trends > temperature trends > livestock growth rate > secondary industry GDP growth rate > population growth rate > tertiary industry GDP growth rate > primary industry GDP growth rate. Similarly, in the MDI, precipitation trends were the most important, followed by temperature trends > wind trends > population growth rate > livestock growth rate > secondary industry GDP growth rate > tertiary industry GDP growth rate > primary industry GDP growth rate. Combining the two indicators, it can be concluded that climate change is the primary factor influencing the changes in AGB in the study area's desert steppe and human activities have a much smaller impact compared to climate change.

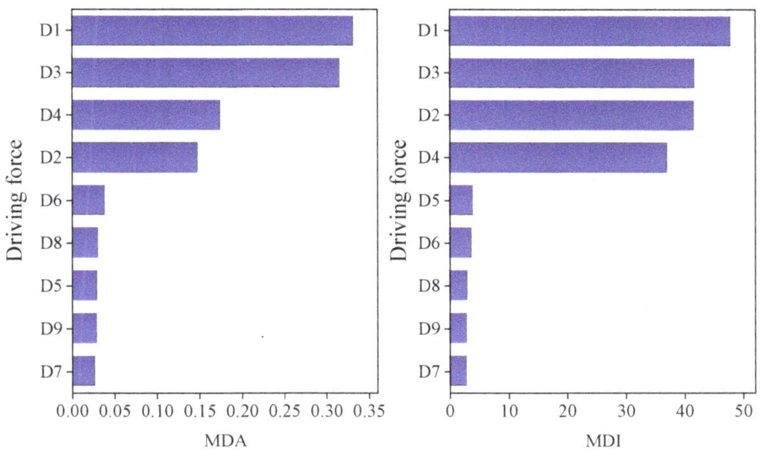

Figure 9. Drivers for the desert steppe AGB changes in the study area from 2000 to 2020.

4. Discussion

4.1. Modeling Algorithm and Feature Selection for Estimating AGB in the Desert Steppe of Inner Mongolia

The desert steppe of Inner Mongolian is characterized by sparse vegetation and simple structure, which poses challenges for reflecting the spatial and temporal characteristics of AGB in the region. In this study, the performance of four popular machine-learning algorithms, MLR, PLS, RF, and SVM, were compared for AGB estimation. From the perspective of prediction accuracy, RF performed the best. Moreover, RF also outperformed other algorithms in terms of noise resistance. This fully demonstrates that RF is suitable for AGB inversion tasks in the Inner Mongolian desert steppe. Similarly, some other studies have observed the same results [24,70]. RF predicts by combining multiple decision-tree models, and due to the ensemble effect of the model, RF has high accuracy and good handling capabilities for high-dimensional and missing data, as well as advantages in interpretability and noise resistance [65]. This makes it one of the commonly used algorithms in machine learning. However, these nonparametric models also have disadvantages, such as low interpretability and reduced model portability due to the influence of spatial heterogeneity [27]. This uncertainty raises the issue of whether the trained RF model is applicable to other desert steppe areas outside the study area, which requires further research for generalization. In comparison, parametric models such as MLR and PLS have advantages in model interpretability, flexibility, and ease of operation. However, based on the results of this study, these models are easily affected by multicollinearity, leading to the exclusion of some key variables from the modeling process. In conclusion, RF is the optimal model for obtaining AGB in the Inner Mongolian desert steppe.

Among the tested variables, VIs contributed the most to AGB inversion in the desert steppe, particularly in nonparametric models, where these indices were all selected as effective variables. However, in the MLR model, only MSAVI and ATSAVI were chosen, indicating the presence of multicollinearity among these VIs. A Spearman correlation test revealed significant correlations among the chosen vegetation indices (Figure A1). This suggests that soil-adjusted VIs do not significantly enhance the spatial heterogeneity representation of AGB in the desert steppe at MODIS spatial resolution scale. Similarly, Baghi and Oldeland found comparable results when comparing soil-adjusted VIs and standard VIs for AGB inversion in Northern Iran's semiarid grasslands using MODIS data [22]. Ren et al. emphasized that the performance of soil-adjusted VIs based on MODIS data was inferior to traditional VIs [21]. This could be attributed to the image resolution, as most of these indices were developed using Landsat imagery, which may affect their performance on coarser spatial-resolution images. On the other hand, soil–vegetation indices might lose their ability to describe the green vegetation information in areas with low green-vegetation cover [21]. In terms of environmental variables, elevation, slope, precipitation, and temperature all contributed positively to AGB inversion in the desert steppe. This implies that the selected samples exhibit spatial differences in these features. The selection of soil factors should be further determined based on the modeling algorithm.

4.2. Spatial–Temporal Feature and Driving Forces of AGB Changes in the Desert Steppe of Inner Mongolia

Over the past 20 years, the multiyear average AGB of the desert steppe in the study area was 58.34 g/m^2 and it exhibited a gradually decreasing trend from the outer region towards the inner region. This pattern is consistent with the distribution characteristics of annual precipitation in this area. Similarly, some previous studies have also observed the same distribution patterns [24,71]. This indicates that precipitation is the primary factor determining the vegetation distribution characteristics in the region. In terms of the change trend, the AGB exhibited an increasing tendency, with an average annual growth of 1.33 g/m^2, and most regions have not experienced abrupt shifts. That is, the desert steppe in Inner Mongolia has undergone a stable recovery during these 20 years. Human activities and climate change have long been considered the primary factors driving changes in grasslands [44]. However, quantitatively analyzing their impacts on grasslands presents challenges. This study considered that the interaction mechanisms between external factors and grassland vegetation are complex and diverse. Therefore, employing nonparametric algorithms to address these intricate relationships is more consistent with the actual situation. Based on the feature importance ranking from the RF model, climate change emerges as the dominant factor influencing AGB changes in the grasslands of the study area, particularly the contributions of precipitation and evapotranspiration. During the past 20 years, most of the regions in the study area have experienced an upward trend in precipitation, with annual precipitation surpassing annual evapotranspiration. This implies that the water required for vegetation growth has been adequately replenished, promoting vegetation productivity. The desert steppe is primarily dominated by xerophytic, perennial bunchgrasses, but during wet years, annual hydrophilic vegetation can easily form a dominant layer, positively affecting community biomass [46]. Moreover, results show that temperature and wind speed are also major factors causing AGB changes. Nevertheless, the variations in these two factors over the past 20 years have been relatively minor, implying that their associations with AGB are more prominently manifested in spatial distribution patterns. Compared with climatic factors, human activities exert a less pronounced impact on AGB within the desert steppe of the study area. However, it is worth noting that quantifying social statistical indicators such as GDP and livestock numbers for spatial use, particularly on smaller scales, may be challenging and could lead to uncertainties in the results. Therefore, the study of spatial transformation methods for socioeconomic indicators is a key issue for future quantitative analysis of driving forces.

5. Conclusions

This study employed field-survey data and multisource remote-sensing data to compare the performance of four commonly used machine-learning algorithms, including linear and nonparametric methods (MLR, PLS, RF, and SVM), in estimating desert steppe AGB in Inner Mongolia, China. The optimal model was used to analyze the spatiotemporal characteristics of AGB from 2000 to 2020, and the driving forces, including climate change and human activities, on AGB change were quantitatively assessed based on the RF algorithm. The main conclusions are as follows: (1) RF outperformed other algorithms regarding prediction accuracy and model robustness, rendering it suitable for AGB estimation in the desert steppe of Inner Mongolia; (2) among the tested variables, VI contributed the most to the model and no significant difference was found between soil-adjusted VIs and traditional VIs. Elevation, slope, precipitation, and the temperature had positive effects on model construction; (3) from 2000 to 2020, the multiyear average AGB in the study area was 58.34 g/m^2 and it exhibited a gradually decreasing trend from the outer region towards the inner region; (4) from 2000 to 2020, grassland with a slight increase in AGB accounted for 87.08%, significantly increased accounted for 5.13%, significantly decreased accounted for only 0.05%, and slightly decreased accounted for 7.76% in the study area; (5) over the past two decades, climate change, particularly precipitation, is the dominant driving force behind AGB change in the desert steppe of the study area.

Author Contributions: Conceptualization, N.W. and G.H.; methodology, N.W.; software, N.W.; validation, G.L., D.W. and W.D.; formal analysis, G.L.; writing—original draft preparation, N.W.; writing—review and editing, B.Y. and D.W.; project administration, G.H.; funding acquisition, G.H. All authors have read and agreed to the published version of the manuscript.

Funding: This research was funded by the Innovative Research Team of the Ministry of Education of China (No. IRT_17R59) and the Inner Mongolia Key Project (No. ZDZX2018020).

Data Availability Statement: Not applicable.

Conflicts of Interest: The authors declare no conflict of interest.

Appendix A

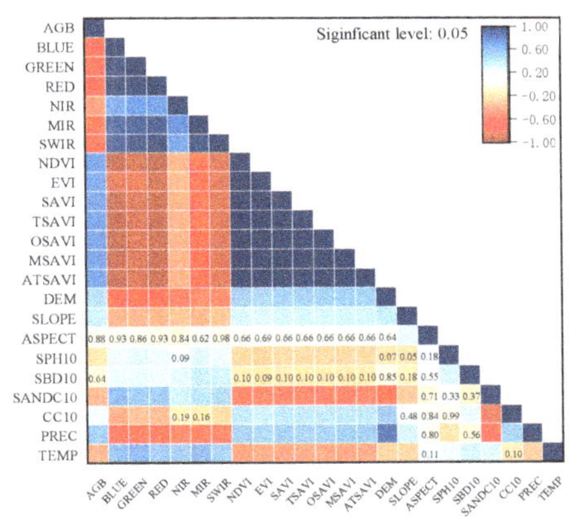

Figure A1. Spearman correlation between variables.

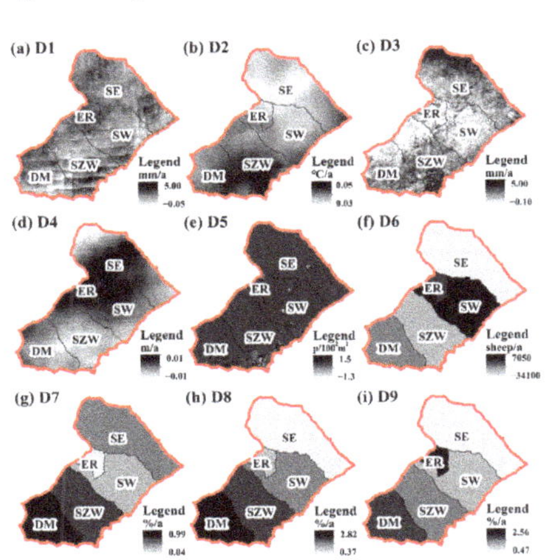

Figure A2. Spatial distribution of driving forces.

References

1. Zhang, R.; Wang, Z.; Han, G.; Schellenberg, M.P.; Wu, Q.; Gu, C. Grazing induced changes in plant diversity is a critical factor controlling grassland productivity in the Desert Steppe, Northern China. *Agric. Ecosyst. Environ.* **2018**, *265*, 73–83. [CrossRef]
2. Yu, D.; Li, Y.; Yin, B.; Wu, N.; Ye, R.; Liu, G. Spatiotemporal variation of net primary productivity and its response to drought in Inner Mongolian desert steppe. *Glob. Ecol. Conserv.* **2022**, *33*, e01991. [CrossRef]
3. Li, L.; Chen, J.; Han, X.-G.; Zhang, W.; Shao, C. Desert Steppe Ecosystem. In *Grassland Ecosystems of China*; Springer Berlin/Heidelberg, Germany, 2020; pp. 249–283.
4. Akiyama, T.; Kawamura, K. Grassland degradation in China: Methods of monitoring, management and restoration. *Grassl. Sci* **2007**, *53*, 1–17. [CrossRef]
5. Ali, I.; Greifeneder, F.; Stamenkovic, J.; Neumann, M.; Notarnicola, C. Review of Machine Learning Approaches for Biomass and Soil Moisture Retrievals from Remote Sensing Data. *Remote Sens.* **2015**, *7*, 16398–16421. [CrossRef]
6. Xu, K.; Su, Y.; Liu, J.; Hu, T.; Jin, S.; Ma, Q.; Zhai, Q.; Wang, R.; Zhang, J.; Li, Y.; et al. Estimation of degraded grassland aboveground biomass using machine learning methods from terrestrial laser scanning data. *Ecol. Indic.* **2020**, *108*, 105747 [CrossRef]
7. Wang, Z.; Ma, Y.; Zhang, Y.; Shang, J. Review of Remote Sensing Applications in Grassland Monitoring. *Remote Sens.* **2022**, *14*, 2903. [CrossRef]
8. Kuang, Q.; Yuan, Q.-Z.; Han, J.-C.; Leng, R.; Wang, Y.-S.; Zhu, K.-H.; Lin, S.; Ren, P. A remote sensing monitoring method for alpine grasslands desertification in the eastern Qinghai-Tibetan Plateau. *J. Mt. Sci.* **2020**, *17*, 1423–1437. [CrossRef]
9. Ehlers, D.; Wang, C.; Coulston, J.; Zhang, Y.; Pavelsky, T.; Frankenberg, E.; Woodcock, C.; Song, C. Mapping Forest Aboveground Biomass Using Multisource Remotely Sensed Data. *Remote Sens.* **2022**, *14*, 1115. [CrossRef]
10. Luo, M.; Wang, Y.; Xie, Y.; Zhou, L.; Qiao, J.; Qiu, S.; Sun, Y. Combination of Feature Selection and CatBoost for Prediction: The First Application to the Estimation of Aboveground Biomass. *Forests* **2021**, *12*, 216. [CrossRef]
11. Villacrés, J.; Fuentes, A.; Reszka, P.; Cheein, F.A. Retrieval of Vegetation Indices Related to Leaf Water Content from a Single Index: A Case Study of *Eucalyptus globulus* (Labill.) and *Pinus radiata* (D. Don.). *Plants* **2021**, *10*, 697. [CrossRef]
12. Maimaitijiang, M.; Sagan, V.; Sidike, P.; Maimaitiyiming, M.; Hartling, S.; Peterson, K.T.; Maw, M.J.W.; Shakoor, N.; Mockler, T.; Fritschi, F.B. Vegetation Index Weighted Canopy Volume Model (CVMVI) for soybean biomass estimation from Unmanned Aerial System-based RGB imagery. *ISPRS J. Photogramm. Remote Sens.* **2019**, *151*, 27–41. [CrossRef]
13. Lourenço, P. Biomass Estimation Using Satellite-Based Data. In *Forest Biomass*; Gonçalves, A.C., Sousa, A., Malico, I., Eds.; IntechOpen: London, UK, 2021.
14. Tucker, C.J. Red and photographic infrared linear combinations for monitoring vegetation. *Remote Sens. Environ.* **1979**, *8*, 127–150. [CrossRef]
15. Huete, A.R.; Liu, H.Q.; Batchily, K.; Leeuwen, W.V. A comparison of vegetation indices over a global set of TM images for EOS-MODIS. *Remote Sens. Environ.* **1997**, *59*, 440–451. [CrossRef]
16. Huete, A.R. A soil-adjusted vegetation index (SAVI). *Remote Sens. Environ.* **1988**, *25*, 295–309. [CrossRef]
17. Qi, J.G.; Chehbouni, A.R.; Huete, A.R.; Kerr, Y.H.; Sorooshian, S. A Modified Soil Adjusted Vegetation Index. *Remote Sens. Environ.* **1994**, *48*, 119–126. [CrossRef]
18. Baret, F.; Guyot, G.; Major, D.J. TSAVI: A vegetation index which minimizes soil brightness effects on LAI and APAR estimation. In Proceedings of the 12th Canadian Symposium on Remote Sensing Geoscience and Remote Sensing Symposium, Vancouver, BC, Canada, 10–14 July 1989.
19. Baret, S.F. Optimization of soil-adjusted vegetation indices. *Remote Sens. Environ.* **1996**, *55*, 95–107. [CrossRef]
20. Baret, F.; Guyot, G. Potentials and limits of vegetation indices for LAI and APAR assessment. *Remote Sens. Environ.* **1991**, *35*, 161–173. [CrossRef]
21. Ren, H.; Feng, G. Are soil-adjusted vegetation indices better than soil-unadjusted vegetation indices for above-ground green biomass estimation in arid and semi-arid grasslands? *Grass Forage Sci.* **2015**, *70*, 611–619. [CrossRef]
22. Baghi, N.G.; Oldeland, J. Do soil-adjusted or standard vegetation indices better predict above ground biomass of semi-arid, saline rangelands in North-East Iran? *Int. J. Remote Sens.* **2019**, *40*, 8223–8235. [CrossRef]
23. Ali, I.; Cawkwell, F.; Dwyer, E.; Green, S. Modeling Managed Grassland Biomass Estimation by Using Multitemporal Remote Sensing Data—A Machine Learning Approach. *IEEE J. Sel. Top. Appl. Earth Obs. Remote Sens.* **2017**, *10*, 3254–3264. [CrossRef]
24. Ge, J.; Hou, M.; Liang, T.; Feng, Q.; Meng, X.; Liu, J.; Bao, X.; Gao, H. Spatiotemporal dynamics of grassland aboveground biomass and its driving factors in North China over the past 20 years. *Sci. Total Environ.* **2022**, *826*, 154226. [CrossRef] [PubMed]
25. Yang, S.; Feng, Q.; Liang, T.; Liu, B.; Zhang, W.; Xie, H. Modeling grassland above-ground biomass based on artificial neural network and remote sensing in the Three-River Headwaters Region. *Remote Sens. Environ.* **2018**, *204*, 448–455. [CrossRef]
26. Wu, C.; Shen, H.; Shen, A.; Deng, J.; Gan, M.; Zhu, J.; Xu, H.; Wang, K. Comparison of machine-learning methods for above-ground biomass estimation based on Landsat imagery. *J. Appl. Remote Sens.* **2016**, *10*, 35010. [CrossRef]
27. Morais, T.G.; Teixeira, R.F.M.; Figueiredo, M.; Domingos, T. The use of machine learning methods to estimate aboveground biomass of grasslands: A review. *Ecol. Indic.* **2021**, *130*, 108081. [CrossRef]
28. Xu, B.; Yang, X.; Tao, W.; Qin, Z.; Liu, H.; Miao, J. Remote sensing monitoring upon the grass production in China. *Acta Ecol. Sin.* **2007**, *27*, 405–413. [CrossRef]

29. Wylie, B.K.; Meyer, D.J.; Tieszen, L.L.; Mannel, S. Satellite mapping of surface biophysical parameters at the biome scale over the North American grasslands: A case study. *Remote Sens. Environ.* **2002**, *79*, 266–278. [CrossRef]
30. Xu, B.; Yang, X.C.; Tao, W.G.; Qin, Z.H.; Liu, H.Q.; Miao, J.M.; Bi, Y.Y. MODIS-based remote sensing monitoring of grass production in China. *Int. J. Remote Sens.* **2010**, *29*, 5313–5327. [CrossRef]
31. Wang, G.; Liu, S.; Liu, T.; Fu, Z.; Yu, J.; Xue, B. Modelling above-ground biomass based on vegetation indexes: A modified approach for biomass estimation in semi-arid grasslands. *Int. J. Remote Sens.* **2018**, *40*, 3835–3854. [CrossRef]
32. Mutanga, O.; Skidmore, A. Narrow band vegetation indices overcome the saturation problem in biomass estimation. *Int. J. Remote Sens.* **2004**, *25*, 3999–4014. [CrossRef]
33. Lu, D. The potential and challenge of remote sensing-based biomass estimation. *Int. J. Remote Sens.* **2007**, *27*, 1297–1328. [CrossRef]
34. Magiera, A.; Feilhauer, H.; Waldhardt, R.; Wiesmair, M.; Otte, A. Modelling biomass of mountainous grasslands by including a species composition map. *Ecol. Indic.* **2017**, *78*, 8–18. [CrossRef]
35. Liang, T.; Yang, S.; Feng, Q.; Liu, B.; Zhang, R.; Huang, X.; Xie, H. Multi-factor modeling of above-ground biomass in alpine grassland: A case study in the Three-River Headwaters Region, China. *Remote Sens. Environ.* **2016**, *186*, 164–172. [CrossRef]
36. Ohsowski, B.; Dunfield, K.; Klironomos, J.; Hart, M. Improving Plant Biomass Estimation in the Field Using Partial Least Squares Regression and Ridge Regression. *Botany* **2016**, *94*, 501–508. [CrossRef]
37. Otgonbayar, M.; Atzberger, C.; Chambers, J.; Damdinsuren, A. Mapping pasture biomass in Mongolia using Partial Least Squares, Random Forest regression and Landsat 8 imagery. *Int. J. Remote Sens.* **2018**, *40*, 3204–3226. [CrossRef]
38. Zhang, B.; Zhang, L.; Xie, D.; Yin, X.; Liu, C.; Liu, G. Application of Synthetic NDVI Time Series Blended from Landsat and MODIS Data for Grassland Biomass Estimation. *Remote Sens.* **2015**, *8*, 10. [CrossRef]
39. Wang, J.; Xiao, X.; Bajgain, R.; Starks, P.; Steiner, J.; Doughty, R.B.; Chang, Q. Estimating leaf area index and aboveground biomass of grazing pastures using Sentinel-1, Sentinel-2 and Landsat images. *ISPRS J. Photogramm. Remote Sens.* **2019**, *154*, 189–201. [CrossRef]
40. Wu, X.; Li, P.; Jiang, C.; Liu, P.; He, J.; Hou, X. Climate changes during the past 31 years and their contribution to the changes in the productivity of rangeland vegetation in the Inner Mongolian typical steppe. *Rangel. J.* **2014**, *36*, 519–526. [CrossRef]
41. Zhou, W.; Gang, C.; Zhou, L.; Chen, Y.; Li, J.; Ju, W.; Odeh, I. Dynamic of grassland vegetation degradation and its quantitative assessment in the northwest China. *Acta Oecologica* **2014**, *55*, 86–96. [CrossRef]
42. Zhou, W.; Yang, H.; Huang, L.; Chen, C.; Lin, X.; Hu, Z.; Li, J. Grassland degradation remote sensing monitoring and driving factors quantitative assessment in China from 1982 to 2010. *Ecol. Indic.* **2017**, *83*, 303–313. [CrossRef]
43. Yang, Y.; Wang, Z.; Li, J.; Gang, C.; Zhang, Y.; Zhang, Y.; Odeh, I.; Qi, J. Comparative assessment of grassland degradation dynamics in response to climate variation and human activities in China, Mongolia, Pakistan and Uzbekistan from 2000 to 2013. *J. Arid Environ.* **2016**, *135*, 164–172. [CrossRef]
44. Li, D.; Xu, D.; Wang, Z.; You, X.; Zhang, X.; Song, A. The dynamics of sand-stabilization services in Inner Mongolia, China from 1981 to 2010 and its relationship with climate change and human activities. *Ecol. Indic.* **2018**, *88*, 351–360. [CrossRef]
45. Ma, Y.; Fan, S.; Zhou, L.; Dong, Z.; Zhang, K.; Feng, J. The temporal change of driving factors during the course of land desertification in arid region of North China: The case of Minqin County. *Environ. Geol.* **2007**, *51*, 999–1008. [CrossRef]
46. Wu, N.; Liu, A.; Ye, R.; Yu, D.; Du, W.; Chaolumeng, Q.; Liu, G.; Yu, S. Quantitative analysis of relative impacts of climate change and human activities on Xilingol grassland in recent 40 years. *Glob. Ecol. Conserv.* **2021**, *32*, e01884. [CrossRef]
47. Zhang, Y.; Wang, Q.; Wang, Z.; Yang, Y.; Li, J. Impact of human activities and climate change on the grassland dynamics under different regime policies in the Mongolian Plateau. *Sci. Total Environ.* **2020**, *698*, 134304. [CrossRef]
48. Liu, Y.; Wang, Q.; Zhang, Z.; Tong, L.; Wang, Z.; Li, J. Grassland dynamics in responses to climate variation and human activities in China from 2000 to 2013. *Sci. Total Environ.* **2019**, *690*, 27–39. [CrossRef]
49. Xu, D.Y.; Kang, X.W.; Zhuang, D.F.; Pan, J.J. Multi-scale quantitative assessment of the relative roles of climate change and human activities in desertification—A case study of the Ordos Plateau, China. *J. Arid Environ.* **2010**, *74*, 498–507. [CrossRef]
50. Wu, H.; Lin, A.; Xing, X.; Song, D.; Li, Y. Identifying core driving factors of urban land use change from global land cover products and POI data using the random forest method. *Int. J. Appl. Earth Obs. Geoinf.* **2021**, *103*, 102475. [CrossRef]
51. Li, Z.; Liu, M.; Hu, Y.; Xue, Z.; Sui, J. The spatiotemporal changes of marshland and the driving forces in the Sanjiang Plain, Northeast China from 1980 to 2016. *Ecol. Process.* **2020**, *9*, 24. [CrossRef]
52. Ye, R.; Liu, G.; Chang, H.; Shan, Y.; Mu, L.; Wen, C.; Te, R.; Wu, N.; Shi, L.; Liu, Y.; et al. Response of plant traits of Stipa breviflora to grazing intensity and fluctuation in annual precipitation in a desert steppe, northern China. *Glob. Ecol. Conserv.* **2020**, *24*, e01237. [CrossRef]
53. Vermote, E.; Wolfe, R. MOD09GA MODIS/Terra Surface Reflectance Daily L2G Global 1km and 500m SIN Grid V006. Available online: https://ladsweb.modaps.eosdis.nasa.gov/api/v1/productPage/product=MOD09GA (accessed on 19 March 2023).
54. Tachikawa, T.; Kaku, M.; Iwasaki, A.; Gesch, D.B.; Oimoen, M.J.; Zhang, Z.; Danielson, J.J.; Krieger, T.; Curtis, B.; Haase, J.; et al. *ASTER Global Digital Elevation Model Version 2—Summary of Validation Results*; NASA: Washington, DC, USA, 2011; p. 27.
55. Funk, C.; Peterson, P.; Landsfeld, M.; Pedreros, D.; Verdin, J.; Shukla, S.; Husak, G.; Rowland, J.; Harrison, L.; Hoell, A.; et al. The climate hazards infrared precipitation with stations—A new environmental record for monitoring extremes. *Sci. Data* **2015**, *2*, 150066. [CrossRef] [PubMed]

56. Muñoz-Sabater, J.; Dutra, E.; Agustí-Panareda, A.; Albergel, C.; Arduini, G.; Balsamo, G.; Boussetta, S.; Choulga, M.; Harrigan, S Hersbach, H.; et al. ERA5-Land: A state-of-the-art global reanalysis dataset for land applications. *Earth Syst. Sci. Data* **2021**, *13* 4349–4383. [CrossRef]
57. Mu, Q.; Heinsch, F.A.; Zhao, M.; Running, S.W. Development of a global evapotranspiration algorithm based on MODIS anc global meteorology data. *Remote Sens. Environ.* **2007**, *111*, 519–536. [CrossRef]
58. OpenLandMap. 2020. Available online: https://opengeohub.org/about-openlandmap (accessed on 20 June 2022).
59. Hengl, T. OpenLandMap: Using Machine Learning for Global Good. Available online: https://opengeohub.org/article/openlandmap-using-machine-learning-global-good (accessed on 20 June 2022).
60. Gaughan, A.E.; Stevens, F.R.; Linard, C.; Jia, P.; Tatem, A.J. High Resolution Population Distribution Maps for Southeast Asia in 2010 and 2015. *PLoS ONE* **2013**, *8*, e55882. [CrossRef]
61. WorldPop. What Is WorldPop? Available online: https://www.worldpop.org/ (accessed on 22 May 2022).
62. Aiken, L.S.; West, S.G.; Pitts, S.C. Multiple Linear Regression. In *Handbook of Psychology*; Wiley: Hoboken, NJ, USA, 2012; pp. 481–507.
63. Hair, J.F. Multivariate Data Analysis: An Overview. In *International Encyclopedia of Statistical Science*; Lovric, M., Ed.; Springer: Berlin/Heidelberg, Germany, 2011; pp. 904–907.
64. Wold, S.; Sjöström, M.; Eriksson, L. PLS-regression: A basic tool of chemometrics. *Chemom. Intell. Lab. Syst.* **2001**, *58*, 109–130. [CrossRef]
65. Breiman, L. Random Forests. *Mach. Learn.* **2001**, *45*, 5–32. [CrossRef]
66. Cortes, C.; Vapnik, V. Support-vector networks. *Mach. Learn.* **1995**, *20*, 273–297. [CrossRef]
67. Guyon, I.; Elisseeff, A. An Introduction to Variable and Feature Selection. *J. Mach. Learn. Res.* **2003**, *3*, 1157–1182.
68. Pettitt, A.N. A Non-Parametric Approach to the Change-Point Problem. *J. R. Stat. Society. Ser. C (Appl. Stat.)* **1979**, *28*, 126–135. [CrossRef]
69. Akritas, M.G.; Murphy, S.A.; LaValley, M.P. The Theil-Sen Estimator With Doubly Censored Data and Applications to Astronomy. *J. Am. Stat. Assoc.* **1995**, *90*, 170–177. [CrossRef]
70. Wang, Y.; Wu, G.; Deng, L.; Tang, Z.; Wang, K.; Sun, W.; Shangguan, Z. Prediction of aboveground grassland biomass on the Loess Plateau, China, using a random forest algorithm. *Sci. Rep.* **2017**, *7*, 6940. [CrossRef] [PubMed]
71. Gao, T.; Xu, B.; Yang, X.; Jin, Y.; Ma, H.; Li, J.; Yu, H. Using MODIS time series data to estimate aboveground biomass and its spatio-temporal variation in Inner Mongolia's grassland between 2001 and 2011. *Int. J. Remote Sens.* **2013**, *34*, 7796–7810. [CrossRef]

MDPI AG
Grosspeteranlage 5
4052 Basel
Switzerland
Tel.: +41 61 683 77 34

Remote Sensing Editorial Office
E-mail: remotesensing@mdpi.com
www.mdpi.com/journal/remotesensing

www.ingramcontent.com/pod-product-compliance
Lightning Source LLC
Chambersburg PA
CBHW040801250926
51850CB00007BA/1049
* 9 7 8 3 7 2 5 8 5 9 6 9 6 *